Living with Energy Poverty

Living with Energy Poverty: Perspectives from the Global North and South expands our collective understanding of energy poverty and deepens our recognition of the phenomenon by engaging with the lived experiences of energy-poor households across different contexts.

Understanding the lived experience of energy poverty is an essential component in the design of any effort to alleviate what is fundamentally a deep-rooted, multi-faceted, wickedly complex problem. This requires a nuanced understanding of the causal factors and the research methods that can respond to the flexible spatial and temporal nature of the condition, as well as its well-being and justice implications. Drawing together the expertise and connectedness of authors from the Global South and North, this book presents novel approaches to understanding the often hidden forms of domestic energy deprivation. Case studies from 20 countries provide critical perspectives on this phenomenon while analysing the policy practices, government strategy, and sustainability implications of divergent manifestations. The book takes a multidimensional perspective, challenging the bias towards energy production and service provision, which often do not align with the aspirations and realities of energy households across global contexts, thus facilitating a useful dialogue on the nature of energy poverty.

The book is a timely source for policymakers, practitioners, and scholars seeking fresh, diverse insights into the everyday reality of energy poverty and wanting to better understand the challenges a people-centred, just energy transition can present.

Paola Velasco-Herrejón is a Marie Skłodowska-Curie actions (MSCA) Research Fellow in the Department of Technology Systems, University of Oslo, Norway. Her research adopts an interdisciplinary perspective to examine socio-technical aspects of the global energy transition and decarbonisation, focusing on well-being, energy poverty, participation, power, and social justice. She was a co-investigator of the EnergyMeasures H2020 project, which aims to better understand the lived experience of energy poverty and support energy-poor households across seven European countries.

Breffní Lennon is a Research Fellow at the Cleaner Production Promotion Unit, Environmental Research Institute at University College Cork, Ireland. He is a geographer researching the energy geographies of climate change and the energy transition. His research focuses on emerging cultural ecologies of place, human-environment dimensions of energy, and the intersectional experiences of citizens in relation to environmental and energy justice. He is a co-Principal Investigator of the EnergyMeasures H2020 project and a co-investigator on the ENCLUDE H2020 project.

Niall P. Dunphy is a Senior Research Fellow in the School of Engineering and Architecture and the Environmental Research Institute at University College Cork, Ireland. Niall has over 20 years' experience in multidisciplinary research; he leads the Cleaner Production Promotion Unit, a research group working at the intersection of science and engineering with the social sciences and humanities. His scholarship has a focus on the social dimensions of energy, with the unifying thread being the place of the energy citizen. Niall is the project Coordinator of EnergyMeasures (H2020 2020–2024), which is working to provide tailored advice and bespoke small measures to energy-vulnerable households across seven European countries.

Routledge Explorations in Energy Studies

Assembling Petroleum Production and Climate Change in Ecuador and Norway
Elisabeth Marta Tómmerbakk

International Law and Renewable Energy Investment in the Global South
Avidan Kent

Local Energy Governance
Opportunities and Challenges for Renewable and Decentralised Energy in France and Japan
Edited by Magali Dreyfus and Aki Suwa

Local Energy Communities
Emergence, Places, Organizations, Decision Tools
Edited by Gilles Debizet, Marta Pappalardo & Frédéric Wurtz

Building Resilient Energy Systems
Lessons from Japan
Jennifer F. Sklarew

Regulatory Support for Off-Grid Renewable Electricity
Edited by Ngozi Chinwa Ole, Eduardo G. Pereira, Peter Kayode Oniemola and Gustavo Kaercher Loureiro

Northern Indigenous Community-Led Disaster Management and Sustainable Energy
Ranjan Datta, Margot Hurlbert and William Marion

Energy Policy Design in the South-Eastern Mediterranean Basin
A Roadmap to Energy Efficiency
Bertug Ozarisoy and Hasim Altan

Living with Energy Poverty
Perspectives from the Global North and South
Edited by Paola Velasco-Herrejón, Breffní Lennon and Niall P. Dunphy

For more information about this series, please visit: www.routledge.com/Routledge-Explorations-in-Energy-Studies/book-series/REENS

Living with Energy Poverty

Perspectives from the Global North and South

Edited by Paola Velasco-Herrejón, Breffní Lennon, and Niall P. Dunphy

Routledge
Taylor & Francis Group
LONDON AND NEW YORK

earthscan
from Routledge

Designed cover image: © Erik Mclean

First published 2024
by Routledge
4 Park Square, Milton Park, Abingdon, Oxon OX14 4RN

and by Routledge
605 Third Avenue, New York, NY 10158

Routledge is an imprint of the Taylor & Francis Group, an informa business

British Library Cataloguing-in-Publication Data
A catalogue record for this book is available from the British Library

ISBN: 978-1-032-52818-2 (hbk)
ISBN: 978-1-032-52819-9 (pbk)
ISBN: 978-1-003-40853-6 (ebk)

DOI: 10.4324/9781003408536

The Open Access version of chapters 1 and 22 was funded by EU's Horizon 2020 Research and Innovation Programme.

Contents

Figures and Tables

Figures

Tables

Preface

The origins of this book can be found in the EnergyMeasures Horizon 2020 project[1] (2020–2024) in which the editors collaborated. This project operated across seven European countries to better understand the lived experience of energy poverty, and to support energy-poor households by providing tailored small measures and behaviour change advice. The work in the project prompted reflection on the challenges of identifying and, most importantly, engaging energy-poor households. We noted with interest the varying experiences of energy poverty across participating countries and differences in the effectiveness of approaches to finding and supporting the energy poor.

Cognisant that knowledge developed within EnergyMeasures was Euro-centric, we were interested to explore how observations of energy poverty in Europe compared with the situation in other places in the world. We, therefore, proposed a panel for the Development Studies Association Conference (DSA 2022) entitled "*Understanding the lived experiences of energy poverty in the Global North and South*". There was a great deal of interest in the call, enough to hold two conference panels. The participants came from five continents and the topics presented challenged many perceptions of energy poverty – including for example, the dichotomous view that in the global south it is solely about access while in the global north it is an affordability issue, and linked the condition with wider sustainability issues.

The fascinating discussions at the DSA 2022 panels prompted us to draw together the various voices found in this volume. We purposefully took an inclusive approach, focusing not on English proficiency or writing style[2] but rather on seeking diverse arguments and experiences that fit with the central book concept.

The book is then the result of the hard work and insightful experiences of researchers from the Global South and North. We hope that this edited collection provides deep contemporary insights into our everyday relationship with energy poverty, as well as its well-being and justice implications while considering the environmental consequences of these actions.

We would like to thank the European Commission for supporting projects such as Energy Measures, the Development Studies Association for being an engaging and inclusive platform that provided the seed idea for this book, Routledge for their interest in expanding literature on energy from diverse contexts, and most importantly, the authors and their research participants whose effort, passion, and stories made this book possible.

Notes

1 More information on the EnergyMeasures Project can be found at https://www.doi.org/10.3030/894759
2 Which can very often act as barriers to voices from the Global South, and any perceived shortcomings can of course be resolved during the editing process.

Foreword

This year marks the mid-point of the United Nations global 2030 Agenda with its 17 Sustainable Development Goals (SDGs). It serves as an important inflection point for assessing progress and renewing commitment to achieving the goals. The Agenda recognises the importance of energy poverty through Sustainable Development Goal 7, which aims to ensure access to affordable, reliable, sustainable, and modern energy for all by 2030. While energy poverty is a global challenge, the way that it manifests varies by location, population, and context. In the Global North, energy poverty has largely been a matter of affordability and efficiency rather than access. Whereas, in much of the Global South, considerations of availability, reliability and affordability are just as relevant. In all contexts, however, energy poverty severely impedes the ability of people to live a decent life by depriving them of essential energy services needed for human well-being.

Despite progress having been made in addressing energy poverty globally over the last couple of decades, the recent COVID-19 pandemic and Russian invasion of Ukraine have resulted in disruptions to global supply chains and tighter financial conditions that have led to significant setbacks. In all regions and among all people, the most vulnerable are generally the ones that are being left behind. Appreciating the varied lived experiences of energy poverty, as well as the opportunities and constraints the energy-vulnerable face is critically important to developing effective policies and a better way forward to help such populations.

This book provides a timely set of contributions that provide new understanding of the lived experiences of energy poverty across different contexts in the Global North and South. With its focus on identifying, appreciating, and addressing energy poverty, the book includes learnings from 20 case studies. It highlights the multifaceted nature of the phenomenon and yields new insights for how policy can guide solutions that take a holistic view of energy vulnerabilities and injustices to help eradicate energy poverty everywhere.

Dr. Shonali Pachauri
Transformative Institutional and Social Solutions (TISS) Research Group
Energy, Climate, and Environment (ECE) Program
International Institute for Applied Systems Analysis
1972-2022 Schlossplatz 1, A-2361 Laxenburg, Austria

Contributors

James Adams is a PhD candidate in the Department of Anthropology and a member of EcoGovLab at the University of California, Irvine (UCI). His research and teaching focus on current debates in energy and environmental ethics, experimental ethnography, and the politics of knowledge production. Adams's work examines the multi-scalar dynamics of socio-technical change and considers their implications for developing more just and effective modes of environmental and energy governance. His current book project, The Transition is Out of Joint, shows how sedimented histories of racial exclusion and petro-capitalist development continue to haunt Austin's renewable energy transition.

Nino Antadze is an Associate Professor of Environmental Studies at the University of Prince Edward Island, Canada. Dr. Antadze studies environmental planning processes with an emphasis on environmental justice and large-scale environmental change with a focus on climate justice, sustainability transitions, and social innovations. Dr. Antadze earned a PhD in urban and regional planning from the University of Waterloo, Canada. She also holds an MSc in Environmental Management and Policy from Lund University, Sweden and an MSc in Environmental Sciences and Policy from Central European University, Hungary.

Paz Araya is a doctoral researcher at the IRI THESys of the Humboldt University, Berlin. Member of the Energy Poverty Network in Chile, also of the Núcleo de Estudios Sistémicos Transdisciplinarios (NEST)in Chile and of the Red Ecofeminista por la Transición Energética also in Chile. She has a Master's degree in Energy Economics and a bachelor's degree in Mechanical Engineering. For 12 years, She worked on studies on energy efficiency policies, energy poverty, energy education strategies and prospective support for decision-makers for sustainable energy policies. She continues to collaborate with background and inputs for the public debate on issues of energy poverty, just energy transition and the right to energy. She is currently working on my PhD project, "Living in cold houses, ¿fitting or retrofitting? Practices of thermal comfort and improvement of housing in southern Chile", which aims to understand thermal comfort-related social practices for homes under energy poverty in southern Chile.

Roberto Barrella is currently working as a research assistant professor at the Institute for Research in Technology (IIT). His research is carried out in collaboration with the Chair of Energy and Poverty. His teaching activities are mainly developed in the Department of Mechanical Engineering of the ICAI School of Engineering (Comillas Pontifical University).

Susana Batel is an assistant researcher and invited Lecturer at the University Institute of Lisbon (ISCTE-IUL). Her research adopts a critical and interdisciplinary perspective to examine the relationship between people, territories, and the climate crisis, specifically around energy transformations towards carbon neutrality, and associated socio-environmental justice and political participation issues.

He obtained a Bachelor's Degree in Energy Engineering in 2015 and a master's degree in Energy Engineering in 2018, both from the Università degli Studi di Roma La Sapienza. He developed his master's thesis at the Universidad Carlos III de Madrid. In July 2022 he defended his PhD thesis proposing an integrated approach to energy poverty developed at Comillas Pontifical University. During his PhD, he carried out an international exchange at the Center for Environmental and Sustainability Research (CENSE), Universidade NOVA de Lisboa (Portugal).

Tania Berger is a trained architect and planner with a strong focus on interlinkages between societies and the built environment; she heads the centre for Social sPACe-based research in built Environment (SPACE) in the Department of Building and Environment at Danube University Krems, which researches global urbanisation processes, precarious housing and environmental justice. She coordinates international cooperation projects with a focus on informal settlements in India and Ethiopia. www.donau-uni.ac.at/dbu/space_en

Julian Bloomer is an assistant lecturer in Geography at Mary Immaculate College (MIC). His research focuses on human-environment relationships, livelihoods, and development using a political ecology lens – with a particular interest in sub-Saharan Africa. Julian has carried out research in Asia (Laos), southern Africa (Lesotho) and Ireland.

Dickson Boateng is a PhD student in the Department of Geography at Mary Immaculate College. He holds a master's degree in Sustainable Environmental Management from the University of Greenwich-UK, and a BA in Geography and Rural Development from the KNUST-Ghana. He has a keen research interest in energy poverty, political ecology, just transitions, rural development and environmental management.

Betina Cardoso is a researcher at the National Council for Scientific and Technical Research (CONICET), Argentina. Patagonian Andean Institute of Biological and Geoenvironmental Technologies (IPATEC). National University of Comahue (UNCo). PhD in Biology in the Ethnoecology discipline. She has worked on the relationship between traditional knowledge as a tool for analysis and the generation of public policies in rural communities in Patagonia. She is a collaborator at the IIES-UNAM in Mexico, where she works on the adoption of biomass ecotechnologies. She currently studies topics about energy efficiency, energy poverty, socio-environmental injustice and fuel adoption.

Marine Cornelis is the founder of Next Energy Consumer, an international policy consultancy that focuses on the social aspects of energy and climate transitions. Combining her expertise in consumer rights advocacy with a commitment to decentralisation, democratisation, digitalisation, and decarbonisation of energy systems, Marine aims to empower people and communities and create a more sustainable future. Her work is particularly focused on underrepresented groups, such as women and ethnic minorities, and she acts as a bridge between civil society, scientific communities, businesses, and policymakers. Marine hosts Energ'Ethic, a podcast that profiles inspiring individuals working to improve energy and the environment. She has been recognised as one of the first ambassadors of the European Climate Pact, and her work takes her across Europe, Africa, Latin America, and the Middle East.

Dominik David is a PhD student at the Department of International Relations and European Studies, FSS, MU, where he completed Master's degrees in Security and Strategic Studies and International Relations and Energy Security. His research focuses on the study of energy poverty and energy vulnerability from the "bottom-up" perspective; his research is mainly based on in-depth interviews. Recently, he published research on the contrasting views between official narratives and lived experiences of energy poverty in the Czech Republic.

Ina Dervishi is a lecturer and Scientific Researcher in the field of sustainable Architecture and urban design. He graduated from Middle East Technical University (BSc), Epoka University (MSc) and La Sapienza University (PhD).

Sokol Dervishi is a building physicist and international consultant in energy efficiency with considerable academic and professional experience. He graduated from Istanbul Technical University (BSc) and Vienna University of Technology (MSc and PhD). In 1014, he received the title Associate professorship (habilitation) at the Vienna University of Technology. Mr. Dervishi conducts advanced scientific research, teaching, consulting services and project design in different European countries. Areas of expertise include Building Physics and Urban Physics, Energy and Buildings, Building Performance and Diagnostics, Human Ecology, Thermal Performance of Traditional Architecture, Industrial Buildings and Underground cities, Energy Poverty.

Federico Dubois is an energy engineer from the Universidad Nacional del General San Martín (UNSAM). His thesis was on energy poverty and how energy efficiency could be used as a tool to alleviate the problem in households in Buenos Aires Province, Argentina. He is currently an Assistant Professor at the School of Science and Technology of UNSAM and works as an energy efficiency policy analyst at the Secretary of Energy in the National Ministry of Economy of Argentina.

Niall P. Dunphy is a Senior Research Fellow in the School of Engineering and Architecture and the Environmental Research Institute at University College Cork, Ireland. Niall has over 20 years' experience in multidisciplinary research; he leads the Cleaner Production Promotion Unit, a research group working at the intersection of science and engineering with the social sciences and humanities. His group conducts engaged research on the themes of Society, Sustainability and Energy, with a particular emphasis on people's relationship with energy. Niall has been the principal investigator on numerous large collaborative research projects. Examples of significant projects relevant to this topic include: ENTRUST (H2020 2015–2018, as project coordinator), exploring the human dimension of the energy system; ENCLUDE (H2020 2021–2024), examining and conceptualising ideas of energy citizenship; and EnergyMeasures (H2020 2020–2024, as project coordinator), which is working to provide tailored advice and bespoke small measures to energy vulnerable households across seven European countries.

Edmond VII Mballa Elanga holds a PhD in sociology, is a lecturer, teacher-researcher in the Human and Social Sciences Research Laboratory, Faculty of Letters and Human Sciences of the University of Douala, and is the coordinator of the Center for Applied Research in Human and Social Sciences (CRASSH). He is also a member of the International Society for the Sciences of Religions (SISR), the International Association of French-speaking Sociologists (AISLF), the Council for Development Social Science Research in Africa (CODESRIA) and the American Political Science Association (APSA), among others. He is the author of several articles and books and a member of the LEAP-RE team (Long-Term Joint Research and Innovation Partnership on Renewable Energy between the European Union and the African Union).

Mariëlle Feenstra is a senior researcher on gender-responsive innovation at TU Delft and an experienced policy advisor. Trained in public policy and public administration, she worked for 15 years as a policy advisor for municipalities in the Netherlands specialising in European Affairs and funding. Her PhD thesis, "Gender Just Energy Policy: engendering the energy transition in Europe" highlighted her academic work on gender-aware energy policy, focusing on policy design and formulation in the policy process cycle. Mariëlle is the chair of the ENGAGER network of energy poverty researchers in Europe. Since 2021, she has been the scientific director and co-founder of the Dutch foundation 75InQ, whose expertise centres on gender and energy and promotes more visibility for diversity in the energy transition. Her contribution as a gender-energy pioneer has been acknowledged both in academia and in practice by invitations to participate in policy formulation activities.

Daniel Torrego Gómez is a postdoctoral Researcher at Universidad Politécnica de Madrid, via Margarita Salas fellowships, contracted by Universidad de Alicante. His research focuses on adaptation to urban heat, assessing technical and socio-cultural passive cooling strategies. He is also co-manager of UMBRAL, an architecture studio oriented towards bioclimatic design and low-carbon materials. For his PhD, he studied material agencies within architectural practice. He has published six articles as main author and participated in four international conferences as a speaker. He has been awarded in public competitive calls for the implementation of participatory projects focused on the cultural dimension of urban transformations (Imagina Madrid, Veranos De la Villa, Ayudas a la Creación) and urban heat (FPRN-EPEC). He works on UIA-EPIU and H2020 Cooltorise projects.

Marta Gayoso Heredia is a PhD candidate at the Universidad Politécnica de Madrid. MSc in gender studies. Her work linked to energy poverty started in 2019, as a researcher within the project FEMENMAD (Feminization of energy poverty in the city of Madrid). Currently, she continues her work as a researcher collaborating with two European projects, UIA-EPIU (Energy Poverty Intelligence Unit) and H-2020 COOLTORISE (Raising summer energy poverty awareness to reduce cooling needs). She is developing her PhD research focused on energy poverty and qualitative research methods. Author of 1 JCR Q2, she has participated in 4 international conferences. She has collaborated as an expert for EPAH Technical Assistance. She develops social-climatic cartographies, granted by FPRN-EPEC 2022. Since 2021, she has been studying Group Facilitation to promote eco-social transformations and innovative processes.

Pável Soriano Hernández is a PhD student and research assistant in the Department of Geography at the University of Victoria. He completed a Master's degree in Energy, Society and Sustainability at the University of Edinburgh and a Bachelor's degree in Physics at the Universidad Nacional Autónoma de México. His research focuses on the study of the justice aspects of renewable energy transitions, with renewable energy clusters as a particular case. Pavel's research mostly uses mixed approaches to determine the impact of the on-going renewable energy transition on society and the study of energy poverty. Furthermore, he has studied inequality of income, wealth, and energy distributions among households and companies using statistical physics tools, heterodox economics, and quantitative methods. Previously, Pavel worked for different companies in diverse sectors as a data scientist, including sustainable finance and market research.

Laurent Jodoin is a doctor in philosophy of sciences (Sorbonne) and an engineer with more than 15 years of experience in the area of sustainability as a researcher as well as a project manager. His research focuses on the explanatory strategies of complex systems in physics and biology, especially those involving the concept of entropy, and on the ethical issues of sustainable development, especially those bearing on energy justice. He has delivered many conferences in Europe, Africa and North America; he teaches at the University of Montreal and Collège Lionel-Groulx, and he is an active member of the CIRST and the GRÉEA.

Alison Kenner is an associate professor of Politics and a faculty member in the Center for Science, Technology, and Society at Drexel University. Her first book, Breathtaking: Asthma Care in a Time of Climate Change (University of Minnesota Press, 2018), documents how care is materialised at different scales to address the U.S. asthma epidemic. From 2014 to 2020, Kenner led the Philadelphia Health and Environment Ethnography Lab (PHEEL), which facilitated collaborative projects between Drexel students, governmental and nongovernmental partners, and community organisations, including Climate Ready Philly, a climate education workshop series hosted in neighbourhoods across the city. Her latest research, The Energy Rights Project, is a National Science Foundation funded study that looks at how organisations address energy vulnerability in the U.S. Mid-Atlantic region. She is an editorial collective member of Engaging Science Technology, and Society, the open access journal of the Society for Social Studies of Science.

Hedvika Koďousková is an assistant professor at the Department of International Relations and European Studies, FSS, MU, and a researcher at the International Institute of Political Science. She is a member of the Centre for Energy Studies. From 2019 to 2022, she took part in the WG1 "ENGAGER COST Action" research network on energy poverty (CA16232). Her majors cover energy and society, energy policies, and energy poverty. Recently, she has published several articles and a book chapter on energy poverty and energy vulnerability in Czechia and Slovakia. She is currently mapping the socio-spatial distribution of energy vulnerability in Czechia as part of an applied research project funded by the Technology Agency of the Czech Republic (TK03010086). Next to her majors, Hedvika is also interested in energy policies in Asia.

Luca Lamonaca has a multidisciplinary background. He holds an MSc in City Planning and Real Estate Development and an MSc in Cognitive Science and Artificial Intelligence and Humanities. Apart from research, he has work experience in various fields, such as education, consultancy and blue-collar jobs. He has a keen interest in carrying out research and policy-making in energy transition, justice and new technologies and AI.

Breffní Lennon is a Research Fellow at the Cleaner Production Promotion Unit, Environmental Research Institute at University College Cork, Ireland. He is an environmental geographer researching the energy geographies of climate change and the energy transition. His research focuses on emerging cultural ecologies of place, human-environment dimensions of energy, and the intersectional experiences of citizens in relation to environmental and energy justice. He is a co-Principal Investigator of the EnergyMeasures H2020 project and a co-investigator on the ENCLUDE H2020 project.

Briana Leone (she/her) holds a BS in Criminology and Sociology from Drexel University (2021) and a M.S. in Criminology from St. Joseph's University (2023). She has served as the Policy Lead for the Energy Rights Project. Her research has focused on utility moratoria policies and social inequities. She also coordinated the Energy Rights Project's pedagogy project in 2020–2021.

Ingmar Lippert engages in research and teaching in the field of Science and Technology Studies (STS), building on a PhD in Sociology and a background in environmental management studies. He currently works at Goethe University Frankfurt; earlier he worked as an associate professor at the IT University of Copenhagen, at Brandenburg University of Technology as well as at the National University of Singapore. His research addresses the digital infrastructures mobilised for sustainability governance. He has conducted research on the phenomenon of corporate carbon accounting and on metrics for energy poverty. He published widely on ethnographic methods of STS, data practices and politics, on social studies of numbers, and on infrastructures and discourses of ecological modernisation, specifically on corporate environmental management – in journals like Geoforum, Big Data & Society, Science as Culture, Science & Technology Studies, Environmental Politics, Global Transitions, Leviathan or the Journal for Discourse Studies.

Carelle Mang-Benza holds a PhD in geography and environment from the University of Western Ontario, where her research explored the energy transition in the context of relationships between Indigenous and non-Indigenous Canadians. Carelle also holds a MSc in chemical engineering from Universite Laval in Quebec City. She is currently the policy lead at Co-operation Canada, focusing on international development and humanitarian work. Prior to that, Carelle accumulated 20 years of experience working in climate change mitigation and adaptation at the United Nations in Germany and environmental policy with the US State Department in West and Central Africa.

María María Ibáñez Martín holds a PhD in Economics from the Universidad Nacional del Sur (UNS), is a researcher at CONICET, Secretary for Graduate Studies and Research and a professor in the Department of Economics at UNS. Ibañez specialises in the areas of Economic Development and Energy Economics, with an emphasis on the development of indicators and measurements of social phenomena and impacts. She has actively participated in national and international research groups. Together with Dr. Carina Guzowski and Dr. Florencia Zabaloy, she is the coordinator of the Higher Diploma in Renewable Energy Economics at UNS. Additionally, she served as a consultant to international organisations in work related to clean energy, renewable energy and the energy transition.

Gliti Mazniku is a Researcher and Architect in the field of sustainable design. He graduated from Epoka University, Tirana (MSc).

Nthabiseng Mohlakoana is currently a researcher at the Delft University of Technology in the Netherlands. She has over 20 years of experience in research focusing on energy policy implementation, energy access and use as well as gender mainstreaming in the energy sector. Her research at TU Delft focuses on European Energy Communities and how community-driven interventions can facilitate affordable and clean energy for all residents. Her in-depth work experience in South Africa includes her time at Stellenbosch University's Centre for Sustainability Transitions (CST) where she focused on research concerning the energy transition with a focus on socio-economic issues such as energy access, energy poverty, jobs, skills development, gender as well as impact on communities and their livelihoods.

Adolfo Mejía Montero is a Lecturer in Energy, Society and Sustainability at the School of Social and Political Sciences in the Department of Science, Technology and Innovation Studies. He holds a bachelor's in physics from the Autonomous National University of Mexico (UNAM) and an MSc in sustainable energy systems from the University of Edinburgh School of Engineering. He completed his PhD at the University of Edinburgh by studying utility-scale wind power development in southern Mexico through a social science approach and the conceptual lenses of energy justice. Previous to his current role, Adolfo worked as a research fellow at the University of Birmingham, supporting the Energy Solidarity in Latin America (ESLatinA) project and as a teaching fellow at the University of Edinburgh. Adolfo also has experience leading and participating in different interdisciplinary research projects and as a consultant on renewable energy projects in Mexico, Chile and the UK.

John Morrissey is a Lecturer in Geography at Mary Immaculate College (MIC). His research is focused on issues of sustainable development, particularly sustainability transitions, low-carbon development and the challenges of a low carbon economy for urban and coastal communities. John's work is informed by transdisciplinary approaches with a focus on the spatial and social differentiation of transition processes between communities. John has experience researching sustainability issues in Ireland, Australia, New Zealand and the UK.

Ahmed Noaman holds a master's degree in environmental and resource management, and a bachelor's degree in mechanical engineering. Ahmed Noaman has two and a half years of working experience in energy-oriented research with Energypedia UG, with a keen interest in social research and qualitative analysis.

Tamara Oyarzún is a psychologist from the University of Chile and has a Master's degree in Community Psychology from the same university. Her work focuses on the link between energy poverty, ageing processes, and well-being, with an emphasis on experiences and practices related to energy. She is currently the Executive Coordinator of the Energy Poverty Network in Chile (RedPE), and is also a member of the NGO Network for Energy Equity (NGO RedPE) and the Nucleus of Transdisciplinary Systemic Studies (NEST-r3).

Miguel Núñez Peiró is a postdoctoral Researcher at the Universidad Politécnica de Madrid. His research focuses on the fields of urban climate, energy vulnerability and buildings' energy performance. In the past few years, he has led the technical development of several monitoring campaigns of the urban heat island phenomenon in Madrid. He has participated in several national and international research projects and networks. Author of 12 high impact articles (8 JCR Q1, 1 JCR Q2), 8 conference communications, 4 indexed book chapters, 2 reports for public institutions and 1 patent. Over 340 citations, an h-index of ten. Since 2019, he has been an associate lecturer in the Degree in Architecture at Universidad CEU San Pablo, and he currently supervises a doctoral thesis.

Luc Richaud is an electrical engineer specialising in electrical networks, energy transition and strategic marketing. He is the Project Director of Odit-e, a company using data to modernise electricity networks and put them at the service of the energy transition. He oversees Odit-e's digital solution deployment in its projects in Europe, Africa and Asia, such as EURICA, funded by the LEAP-RE program which has received funding from the European Union's Horizon 2020 Research and Innovation Program under Grant Agreement 963530.

Coralie Robert holds a PhD in sociology and is an urban policy specialist who works in the Centre de Recherche sur l'Habitat in the research unit LAVUE at the University of Paris Nanterre. Her research focuses on urban sociology and the ecological transition, particularly energy issues. Her work mainly deals with the public policies designed to combat energy poverty in France, raise awareness about energy savings and encourage energy renovation. It examines their development, their reception and their appropriation by their targeted audiences.

José Carlos Romero is an Assistant Professor in the Department of Industrial Organisation of ICAI School of Engineering at Comillas University in Madrid and a researcher at the Institute of Technological Research. He is a graduated Industrial Engineer from the University of Malaga and defended his doctoral thesis on sustainable energy system models at Comillas. Throughout his career, he has published and participated in research and consultancy projects related to indicators of energy sustainability at an urban, regional and national level; the adaptation of the energy system to climate change; the technical-economic analysis of energy poverty and the development of future energy scenarios. Currently, he is coordinating the Chair of Energy and Poverty, and I am a collaborator of the BP Chair of Energy and Sustainability.

Andrew Rosenthal (he/him) holds a B.A. in Political Science from Drexel University (2021). He joined the Energy Rights Project in January 2020 and serves as the project's Data Lead. His work has focused on the qualitative and quantitative analysis of interview data as well as the development of data visualisations. He is currently pursuing a graduate degree in Science and Technology Studies at Virginia Tech.

Carmen Sánchez-Guevara Sánchez Associate Professor at the Construction Department in the School of Architecture at the Technical University of Madrid (UPM). MSc Architecture and International PhD (2016). MSc in Bioclimatic Architecture in a Sustainable Environment (2008). Her research focuses on energy poverty from the right to habitability and thermal comfort perspective. She is currently the Principal Investigator of three European projects: EPIU, which is an Urban Innovative Action, COOLTORISE (H2020) and BUILDSPACE (HORIZON). She is also collaborating with the Energy Poverty HUB from the EC advising municipalities on energy poverty. She has worked on a total of 17 competitive projects. She has published 28 indexed scientific papers (14 Q1 – JCR), participated in six books and 21 communications to national or international conferences. She has 667 citations and H index of 13 and an i10 of 17 in Google Scholar.

Morgan Sarao (she/her) holds a B.A. in Political Science from Drexel University (2019). She has worked for the Energy Rights Project since December 2019 and is the team's Fieldwork Lead. In this position, she coordinated the team's fieldwork, recruited research participants for data collection, and supported community partners throughout these efforts. She currently works at Energy Outreach Colorado as the Community Solar Program Coordinator, where she helps manage limited-income community solar gardens across the state.

Dinita Setyawati is an advisor at Just Transition Network and an energy specialist with more than 10 years of experience in the areas of policy and sustainability. She was the lead representative for the Toyota Foundation Research Grant Program 2019 and the author of papers on energy policy and energy justice.

Ana Stojilovska is a research fellow at the Centre for Social Sciences, Budapest. She is a social scientist and holds a PhD in Environmental Sciences and Policy from the Central European University (2021). Additionally, she is an OSUN Teaching Fellow. Her work has primarily focused on expanding knowledge about energy poverty by exploring the lived experience of energy-vulnerable households in the context of a socially just energy transition, as well as investigating questions of (in)justice and equity related to infrastructure, heating, and fuel use. She also investigates new governance models to uncover the systematic drivers of energy vulnerability. Dr. Stojilovska has over 13 years of experience working on energy policies in Central and Eastern Europe. She was part of the COST-funded project ENGAGER. Her previous work experience includes a position as a think tank researcher at Analytica in Skopje and as a teaching fellow at ELTE in Budapest.

Alessia Tafani is a researcher and architect in the field of sustainable design. She graduated from Epoka University, Tirana (MSc).

Abby Tzinberg (she/her) is a second-year student at Drexel University. She is a Psychology major with a minor in Science, Technology, and Society. She joined the Energy Rights Project team in the summer of 2022. In her time as a research assistant, she has taken steps to make the project's findings and activities more accessible using social media. Most recently, her research has focused on the City of Philadelphia's efforts to transition away from natural gas infrastructure.

Paola Velasco-Herrejón is a Marie Skłodowska-Curie actions (MSCA) Research Fellow in the Department of Technology Systems, University of Oslo, Norway. Her research adopts an interdisciplinary perspective to examine socio-technical aspects of the global energy transition and decarbonisation, focusing on well-being, energy poverty, participation, power, and social justice. She was a co-investigator of the EnergyMeasures H2020 project, which aims to better understand the lived experience of energy poverty and support energy poor households across seven European countries.

Peta Wolpe has more than 15 years' experience working in the urban energy and climate space in South Africa and sub-Saharan Africa. She has worked on a variety of projects, including energy poverty, multi-level governance, political economy, policy, training and capacity building. Her current focus is on a Just Transition and energy poverty. She works at the interface with the government and other key stakeholders, including communities and workers. Prior to returning to South Africa in 1996, Peta worked in London as a senior psychiatric social worker for 13 years, and her experience in this field has assisted in her facilitation skills and ability to hold complex processes. She has a degree in Sociology from the University of Essex and a Master of Science in Social Administration and Social Work Studies from the London School of Economics and Political Science.

María Florencia Zabaloy is an Assistant Professor in the Department of Economics at Universidad Nacional del Sur (UNS) and a postdoctoral fellow at the Institute for Economic and Social Research of the South (IIESS) (CONICET-UNS). She has a PhD in Economics from the Universidad Nacional del Sur and she specialised in the areas of Energy Economics and Public Finance. Her latest publications and research projects address the topics of energy transition, energy efficiency, clean and renewable energy and the carbon tax, among others. Her research is focused on the construction of indicators for the evaluation of public policies based on quantitative and qualitative techniques, both for Argentina and other Latin American countries.

Introduction

1 The Global Face of Energy Poverty

Paola Velasco-Herrejón, Breffní Lennon and Niall P. Dunphy

1.1 Introduction

For the first time in decades, the price pressures and economic upheaval primarily caused by the global energy crisis – sparked by Russia's invasion of Ukraine and post-COVID pandemic supply chain stress – have led to a rise in the number of people without access to energy (IEA, 2022).

Although there has been a steady increase in the global availability of electricity over recent years, with access rates increasing from 83% in 2010 to 91% in 2020, it is estimated that the number of people without access to what is increasingly an essential resource is likely to reach 774 million in 2023 (UN, 2022). Of those without access, 80% live in sub-Saharan Africa, where the number of people excluded from the electricity grid is almost back to its record high from 2013. At the same time, the number of households that spend at least 10% of their disposable income on energy has increased by 160 million from 2019 to 2022 (IEA, 2022).

In 2020, 4 billion people still relied on inefficient and polluting cooking systems, with 40% of these households in sub-Saharan Africa and 55% in developing Asia. Surging fuel prices have caused an estimated 100 million people to switch back from LPG to cooking with dirtier fuels such as traditional biomass, coal, and kerosene in 2022 (UN, 2022).

Whilst this situation markedly affects emerging markets and developing economies, energy poverty is also to be found in more developed countries (albeit in somewhat different manifestations), particularly in marginalised communities. For instance, the problem of energy poverty is widespread in Europe, with an estimated 50–125 million people (depending on the source) who are energy poor (Economidou et al. 2019). We can expect this estimation to worsen when data for 2022 is published.

The human consequences of energy poverty include a significant deterioration in physical health and mental well-being, along with premature death due to severe winter and summer conditions, unhealthy and/or restricted lifestyles, and social exclusion (Thomson et al., 2016). There have been efforts to reduce energy poverty in both developing and developed countries over the past few decades. While there have been substantial efforts to analyse energy poverty at the macro and meso levels, there remains a gap in our collective knowledge of how best to understand and engage with energy-poor households at the local level. This is true across both the Global South and the Global North.

This book aims to engender a conversation that builds an understanding of the lived experiences of energy poverty primarily at the household level across developed and developing contexts. This includes exploring a range of methods to identify energy-poor households, measure energy poverty, and characterise those most at risk or vulnerable to energy poverty, including instances of so-called "hidden energy poverty". Finally, the book will also discuss

DOI: 10.4324/9781003408536-2

the most appropriate practices for supporting energy-poor households across various regions, explore the links between energy poverty and sustainability, and outline actions to achieve SDG 7 – *Ensure access to affordable, reliable, sustainable, and modern energy for all* (UN, 2022). Addressing the worsening energy poverty crisis will be essential to achieve people-centred, just energy transitions.

1.2 Energy poverty in the Global North and South: is the divide still relevant?

The need for electricity grid expansion has long been articulated in developing and developed contexts since the first half of the 20th century (Harrison, 2013; Hughes, 1993). More recently, global energy markets and their effects on energy availability and prices, as well as pressing environmental concerns, have induced a surge in research on energy deprivation, a condition that previously had little public recognition. The relationship between energy use at the household level and well-being is now a prevalent topic that has been widely explored in different ways within academic analysis, policy, and advocacy work. These views have tended to differ in Global North and Global South contexts.

1.2.1 Expressions of energy poverty in the Global North

Even with increased access to energy, concerns about the under-consumption of energy gained momentum in the Global North in the 1980s. This phenomenon was first described through the concept of "fuel poverty", coined in the United Kingdom by Brenda Boardman in 1991, who sought to raise awareness about the affordability of heating services and its relationship with its negative effects on health (Boardman, 1991). Fuel-poor households were initially defined as needing to spend more than 10% of their income on heating and other energy services (Boardman, 2013) and moved on to also refer to the inadequacy of the energy efficiency of homes, heating systems, and appliances (Boardman, 2013; Bouzarovski & Petrova, 2015). The "fuel poverty" term was influential at the time of its origin, and it is still used today in research and policy on energy and well-being in other Global North contexts (Day et al., 2016).

Bouzarovski and Petrova (2015) argue that all forms of household-scale energy deprivation, regardless of it being referred to as fuel or energy poverty, share the same consequence: a lack of adequate energy services in the dwelling. Therefore, they claim that fuel poverty and energy poverty can be considered under the same conceptual umbrella: "*as a set of domestic energy circumstances that do not allow for participating in the lifestyles, customs and activities that define membership of society*" (Bouzarovski & Petrova, p. 32). This points to the obsolescence of the notion of "fuel poverty" as placing emphasis on the supply of energy carriers to the home is not aligned with new conceptual debates about the phenomenon. Therefore, there is reason to advocate for the worldwide adoption of the term energy poverty.

More recently, notions of energy poverty have driven research in the Global North to investigate a wider configuration of material, socio-economic, and political conditions. For instance, the introduction of the term "energy vulnerability" (Bouzarovski & Petrova, 2015) has raised awareness that energy needs can be flexible and temporal and vary from one household to another. Also, fuel poverty has varying impacts on different people within a household such as the elderly, the disabled, and the very young (Day et al., 2016).

Support for energy vulnerability has been exemplified by the development of the ENGAGER Cost Action[1] in 2017 and the establishment of the EU Energy Poverty Observatory (EPOV) in 2018,[2] which are seen as pivotal moments for the advancement of a policy agenda addressing this societal challenge within the European Union (Middlemiss et al., 2018).

Moreover, there has been significant funding for Horizon 2020 projects[3] that address energy vulnerability, including EnergyMeasures. However, these new approaches and support measures are still rare and need further development (Bouzarovski & Petrova, 2015; Day et al., 2016; Day & Walker, 2013).

Overall, the main discourse around energy poverty in the Global North tends to converge on thermal comfort, primarily warmth through household heating as a critical energy service. Concerns largely focus on poor health and excess deaths during the winter season. However, the focus on fuel poverty might weaken the relationship between energy poverty and wider understandings of well-being deprivation (Day et al., 2016). This link could be strengthened by exploring perspectives and approaches to energy poverty in the Global South, which will be briefly explored in the following section.

1.2.2 Manifestations of energy poverty in the Global South

Energy poverty manifestations in the Global South have mainly been focused on supply-side issues related to inadequate access to clean and reliable energy, particularly for lighting, cooling, and cooking. Despite high political attention, 2.4 billion people in African and Asian developing countries are still using traditional biomass for cooking and heating (IEA, 2022). This situation is referred to in research, policy and advocacy spaces as energy poverty and has been associated with other development dimensions including economic productivity, infrastructure, health and education, and social equity (Bazilian et al., 2014; Modi et al., 2005; Pachauri et al., 2004). This reflects how energy poverty in the Global South signifies a more comprehensive understanding of how energy and energy services relate to well-being and quality of life (Day et al., 2016).

Outcomes of energy poverty in the Global South are often associated with the burning of biomass and the use of low-grade fuels (wood, coal, dung, and waste material) for cooking and heating. These fuels are usually burned directly in the home with highly inefficient devices (Pachaurai & Spreng, 2011). Lighting needs are also met using candles and kerosene lamps. These practices have substantial effects on health as they are associated with high levels of indoor air pollution, which is responsible for numerous respiratory and cardiovascular diseases, as well as cases of lung cancer. Women, children, and the elderly, who spend more time in dwellings and are more exposed to harmful concentrations of pollutants resulting from the incomplete combustion of biomass fuels, are thereby more likely to suffer obstructive pulmonary diseases and acute infections of the respiratory system (Bruce et al., 2000; Rehfuess, 2006; Pachaurain & Spreng, 2011). Research shows that more people die from indoor pollution than from malaria and tuberculosis (IEA, 2010). Furthermore, women and children are also exposed to health-related risks linked to the time-intensive process of gathering fuel. Not only does this activity involve gathering wood and toxic manure, but it can also often carry additional risks including back and foot injuries, wounds, sexual assault, and exposure to extreme weather (Masud et al., 2007).

Energy poverty can also create environmental impacts related to deforestation and changes in land use, as well as the related greenhouse gas emissions from burning biomass. Even when entire trees are not felled, the collection of branches, roots, and bark can deplete forest ecosystems and soils of their nutrients and overall biodiversity (Alam et al., 1991). Conversely, deforestation can result in severe price increases for fuelwood as shown in a study in Bangladesh, that revealed how the increase in the price of wood meant that a typical household was spending about 50% of their annual income on fuel (Biswas et al., 2001).

Increasing access to clean energy in the Global South has been associated with positive developments to well-being. The provision of modern energy services can improve health by facilitating access to potable water, cleaner cooking facilities, and refrigeration (Larson & Kartha, 2000). Pachauri (2007) provides evidence of a significant correlation between an increase in food calorie intake and access to modern cooking energy sources in India such as kerosene or LPG. In education, statistics show that populations with higher levels of electricity have higher literacy rates, lower student dropouts, and increased study time (González-Eguino, 2015; Porcaro & Takanda, 2005). Electricity can also improve gender equality by saving time on gendered household work and improving women's health (Sovacool, 2012). Modern energy services also expand the opportunities to set up income-generating activities including lighting for factories and shops, power for processing crops, and refrigeration for preserving products (Larson & Kartha, 2000). Providing universal access to modern, cleaner energy sources can be reached by investing 35 billion USD a year for 20 years, a figure that is not high in relation to the 1 trillion USD given annually in subsidies to fossil fuels (IEA, 2022).

1.2.3 The global access-affordability divide

Traditionally, research on the divide between the Global North and South concludes that the driving forces of energy and fuel poverty, and the policies to address them, focus on the lack of heating vs. the lack of access to clean energy, respectively. At the same time, in both cases, though the consequences of domestic energy deprivation are primarily focused on health impacts, fuel poverty is focused on cold air exposure and energy poverty is preoccupied with indoor air pollution (Bouzarovski & Petrova, 2015).

Current research, however, is challenging the access-affordability divide, and the term energy poverty has become the overarching term used to encompass questions of access, infrastructure, affordability, efficiency and equity, and well-being and the environment. For instance, middle-income families can now face both infrastructural access and affordability issues while households in higher-income countries are still facing access problems to clean energy sources. At the same time, questions of justice, security, well-being, and clean energy transitions are becoming prominent in the energy poverty literature as explored below.

1.3 Definitions and perspectives on energy poverty

Experiences of energy poverty in the Global North and South have driven a number of authors to propose frameworks to analyse the concept and its relationship with affordability and efficiency (e.g., Chester & Morris, 2011; Katsoulakos, 2011; Petrova et al., 2013) as well as well-being and development (e.g., Bazilian et al., 2014; Sovacool, 2014; Zulu & Richardson, 2013).

Early definitions of energy poverty referred to levels of energy consumption that are sufficient to meet basic needs, as well as the lack of physical access to modern energy (IEA, 2010; UNDP and WHO 2009 cited in Pachauri & Spreng, 2011). González-Eguino (2015) puts forward a typology that summarises these approaches. The *technological threshold* approach is based on the idea that energy poverty is mainly a problem of access to "modern" fuels. Modern fuels may include kerosene, LPG, biofuels, and electricity, which are seen as more reliable, efficient, and cleaner than other "traditional" fuels such as wood or dung (UNDP, 2005). The idea of the "energy ladder" is commonly used to define a spectrum of fuel cleanliness, where moving "up" the ladder means increasing the level of prosperity in the household (Hosier & Dowd, 1987; UNDP, 2005; Sovacool, 2012). A *physical threshold* approach refers to estimates of minimum energy consumption that are associated with basic needs. Researchers have aimed to define minimum acceptable levels of energy consumption based on a set of goods and services (e.g., Foster et al., 2000; Modi et al., 2005; Mirza & Szirmai, 2010). Finally, an *economic threshold* approach attempts to establish a maximum percentage of income that is reasonable to spend on energy. Examples of this threshold include the fuel poverty indicator that sets a percentage of 10% of available income; or Barnes et al. (2010) who propose the minimum end-use energy (MEE) that presupposes there is a level of income below which energy consumption does not meet the needs for cooking and lighting. Linked to these thresholds, the concept of hidden energy poverty has also been introduced by Meyer et al. (2018), which refers to circumstances when households self-restrain their energy consumption.

While these approaches have been widely used in the literature of energy poverty, they have also been recently criticised since their highly standardised approach to people's needs can be characterised as simplistic (Day et al., 2016) and, therefore, do not fully capture the intricacies of how households engage with energy to attain well-being. Addressing the technological, physical, and economic barriers to increasing overall levels of consumption can be a necessary measure, but not a sufficient one, to improve people's quality of life. Energy poverty can hamper many forms of well-being and therefore needs to be seen from a more holistic perspective to avoid (re)creating vicious circles of deprivation.

More recently, a number of authors and organisations have moved away from the diagnosis of energy poverty that emphasises sufficiency, efficiency and affordability, to focus instead on securing adequate "energy services" (e.g., Bouzarovski & Petrova, 2015; Modi et al., 2005; Sovacool et al., 2014; UNDP, 2005; Wilhite et al., 2000). These can be understood as "the benefits that energy carriers produce for human wellbeing" (Modi et al., 2005, p. 9), to emphasise that "people do not demand energy per se, but instead, they would like to have access to mobility, washing, heating, cooking, cooling and lighting" (Haas et al., 2008, p. 4013). Focusing on energy services opens the possibility of addressing the demand-side of energy insecurity within an integrated approach to meet the needs of the population (Jansen & Seebregts, 2010). The energy services approach recognises that energy is needed for achieving a range of well-being outcomes. Varying priorities among households, communities, and regions implies that "the driving forces of domestic energy deprivation are multidimensional and hybrid" and therefore extend beyond Global North and South distinctions (Bouzarovski & Petrova, 2015). This multi-faceted, hybrid approach can be associated with the multidimensional perspective of energy poverty and development, which provides "a more nuanced approach than 'access' or 'total energy' approaches" (Day et al., 2016, p. 258).

The energy access-consumption matrix introduced by Pachaurai et al. (2004) provides a means to visualise the level of energy services a population can enjoy. Similarly, Nussbaumer et al. (2012) proposed an indicator of Multidimensional Energy Poverty (MEP) using indicators such as modern cooking fuel, access to electricity for lighting, having a TV/radio (for entertainment and education), having a fridge, and having a phone (for communication). In this same line, the NGO Practical Action proposed an indicator of Total Energy Access that measures households against a minimum range of services that people need such as lighting, cooking, water heating, space heating, food cooling, and ICT[4] (Practical Action, 2010).

Reddy et al. (2000) also refer to the access to energy services, while at the same time introducing the idea of the absence of sufficient choice "in accessing adequate, affordable, reliable, high-quality, safe and environmentally benign energy services to support economic and human development" (p. 44).

Bouzarovski and Petrova (2015) introduce the idea of energy vulnerability to the concept of energy services, which integrates the temporal dynamics of the phenomenon of energy poverty. This term also encapsulates the risk factors that add to the precariousness of particular spaces and groups of people. To this concept, Middlemiss and Gillard (2015) again suggest integrating the notion of multidimensionality to encapsulate how people experience energy vulnerability in different ways.

Several authors have proposed the capability approach to transpose ideas of energy needs across different contexts, which moves away from necessary energy sources, amounts, and even services, promoting instead the idea of focusing on sets of outcomes, or capabilities (e.g., Day & Walker, 2011; Jodoin, 2021; Nussbaumer et al., 2012; Sovacool et al., 2014; Walker & Day, 2012). The capabilities approach introduced by Amartya Sen and Martha Nussbaum (Nussbaum, 2001, 2003; Sen, 2001; Sen & Nussbaum, 1993) provides a theoretical justification to pay attention to wider human flourishing, and on what people can do and achieve.[5] Particularly, Day et al. (2016) propose a systematic way of looking at the link between energy needs and capabilities, defining energy poverty as the "*inability to realise essential capabilities as a direct or indirect result of insufficient access to affordable, reliable and safe energy services, and taking into account available reasonable alternative means of realising these capabilities*" (p. 260). This novel definition can be adaptable and sensitive to local circumstances as it highlights the diverse ways in which energy can be instrumentally important for acquiring well-being, recognising thus the multiple adverse impacts of being in a situation of energy poverty in both Global North and South contexts.

Similarly, the application of justice theories to the understanding of energy poverty has gained increasing traction in policy and research circles. In their seminal contribution, Walker and Day (2012) argue that at its core, energy poverty is "*fundamentally a complex problem of distributive injustice*" (p. 69), and suggested that this injustice is underpinned by other injustices related to fairness in decision-making procedures (procedural justice) and the degree of respect given to different socio-cultural identities (recognition justice). Further studies have unpacked the philosophical and moral foundations of the emerging concept and framing of "energy justice" that considers poverty as a form of injustice within the broader energy justice paradigm (e.g., Heffron & McCauley, 2017; Jenkins et al., 2016; McCauley et al., 2019) focusing particularly on inequalities between social groups as well as social forms of disadvantage (Bouzarovski, 2018; Heffron & de Fontenelle, 2023).

The diverse understandings of energy poverty reviewed above point to different perspectives on the issue. Analysing energy poverty through the supply or demand side, as well as through outcomes such as energy services, capabilities, and (in)justices, can provide a wide analysis of drivers of energy deprivation at the household level and the risks and vulnerabilities people face as a result.

However, given the increasing incidence of the condition and its relationship to improving other well-being dimensions and tackling environmental degradation, we need to continue expanding our understandings of the phenomenon while innovating with methods and more holistic views that remain under researched.

1.4 Coverage of the book

The overarching purpose of this book is to expand our collective understanding of energy poverty and deepen our recognition of the phenomenon by engaging with the lived experiences of energy-poor households across different contexts.

Understanding how energy poverty is experienced is central to any efforts to alleviate it. This requires a definition that can respond to the flexible spatial and temporal nature of the condition, measurements that align with that definition, and monitoring that speaks to diverse audiences. The book, therefore, gathers the expertise and connectedness of authors from the Global North and South, which provide deep contemporary insights into our everyday relationship with energy poverty, as well as its well-being and justice implications. Their diverse contributions help move the discussion away from perspectives born in the Global North while also further challenging the Global North and South research divide.

Part I – Methodological approaches – aims to explore new ways of identifying forms of energy poverty which underpin well-known and novel approaches towards the study of domestic energy deprivation. The chapters challenge the existing quantitative methods such as the way energy underconsumption is conceived while also expanding qualitative methodologies such as ethnography and action research, which remain a gap in the energy poverty literature. This allows data to be collected in unstructured ways, which can allow people experiencing energy poverty to introduce their own categories, needs, and ideas into the research. Furthermore, qualitative research is needed to understand the changing experience of energy poverty, with a particular emphasis on how external shocks such as the pandemic or the energy crisis affect people's experiences of the condition.

Chapter 2 opens this part of the book with Dunphy et al. presenting findings from the EnergyMeasures H2020 project in identifying and recruiting energy poor households in seven participating countries across Europe. While the geographic scope may be limited, the range and depth of practical knowledge of participants allowed for the uncovering of a range of nuanced and considered approaches to identifying and recruiting energy-vulnerable households that are cognisant of the historical, cultural, and environmental factors specific to individual locales. In keeping with the overall theme of the book, the chapter considers how to appropriately identify energy poor households and draws from both the literature and experiences in the field.

In Chapter 3, Ibañez Martín et al. highlight the limiting effect the focus on energy expenditure as an indicator has had on developing appropriate policy responses. Using data from Argentina's National Household Expenditure Survey, the authors offer a more nuanced approach for capturing the real numbers for those dealing with energy poverty there.

In Chapter 4, Barrella and Romero critique the quantitative methods used to examine energy poverty in Europe. Using Spain as a test case, they propose using indicators that more accurately capture the complexities of energy poverty at the household level (in particular, how to account for underspending or hidden energy poverty) that are adequately addressed in official statistics.

In Chapter 5, Mejía-Montero and Soriano-Hernández build on earlier work, presented in a recent research paper, and call for greater methodological diversity when studying the topic of energy poverty. They take us through the results of a quantitative characterisation of Energy Poverty (EP) in Mexico.

In Chapter 6, Gayoso Heredia et al. outline the qualitative methods they applied as part of their work for the COOLTORISE H2020 project. Comprising focus group workshops and participatory action research, the authors uncover the lived experiences of Summer Energy Poverty in vulnerable communities in Madrid, Spain.

In Chapter 7, Antadze describes how she coordinated an investigation into the lived experience of energy-vulnerable households in the Republic of Georgia, using video ethnography as a key research method. Her approach, partly in response to travel restrictions arising from the COVID-19 pandemic, generated novel insights from her team's work with energy-vulnerable householders in Georgia.

Part II – Lived experiences – moves the focus to the material and social aspects of lived experiences of energy poverty as a critical way of identifying the different ways in which the phenomenon is lived and experienced in different contexts, and in a particular place. This experience has implications for improving our awareness of vulnerability, which enables a nuanced understanding of complex issues that contribute to and reinforce a situation of energy poverty.

Chapters in this section use a capabilities approach to analyse energy poverty or engage with the framework to reflect on situations in which energy poverty influences people's capabilities in varying contexts. These works provide novel contributions on how to address the methodological challenges that introducing the idea of capability supposes. This section also offers more grounded work on the links between energy justice and the lived experiences of energy poverty, making references not only to distributive justice but also to procedural and recognition justice. These contributions strengthen notions that injustices experienced at the household level can be influenced by mechanisms operated in the wider energy system (Jenkins et al., 2016; McCauley et al., 2013).

This section begins with Chapter 8, where Jodoin and Mang-Benza adopt a normative framework based on the concept of energy justice to analyse gendered experiences of energy poverty in rural areas of three sub-Saharan African countries (Benin, Senegal, and Togo). Their findings point to the importance of operationalising the capability approach to interpret justice principles and identifying how certain groups of people may be more vulnerable to energy system impacts.

In Chapter 9, David and Koďousková analyse the reaction of vulnerable households to the bust of energy suppliers in the Czech Republic and the transfer of these households to a "supplier of last instance" in 2021, resulting in higher energy prices. By doing so, they look at the interplay among households' material culture, practices, and norms (energy cultures framework), highlighting the need to focus on energy poverty prevention by considering external and internal factors that make households more susceptible to falling into the condition.

In Chapter 10, Setyawati draws on the lived experiences of street vendors in Indonesia who must negotiate access to electricity both through formal and informal channels. The author shows that even among the most vulnerable energy users, there is a hierarchy of needs impacting their levels of resilience and empowerment.

In Chapter 11, Mohlakoana and Wolpe outline the social and historical contexts of energy poverty in South Africa. They illustrate through a sample of electrified households how marginalised communities continue to bear the brunt of on-going market ideologies of the state some 30 years after the end of apartheid and the foundation of democratic institutions in the country. The authors also explain how despite policies in place and good intentions government has continued to fail to meet their energy poverty targets and obligations over this period.

In Chapter 12, Robert examines the coping strategies of energy-vulnerable people in France dealing with the three overarching impacts of limited financial resources, rising energy prices, and the poor energy performance of the buildings they occupy. The author applies a capability theory perspective to examine how the instincts of volunteerism and resilience frame the coping strategies of energy-vulnerable communities there.

In Chapter 13, Berger provides a comparative study of the lived experience of energy-vulnerable households in Austria and India and the differing challenges faced by households in each country. Coping with thermal (dis)comfort in energy poor households is a common feature in both countries despite differences in climate, the built environment, and the social configurations they must contend with.

In Chapter 14, Araya et al. document research conducted with indigenous peoples in Argentine Patagonia, a small town in southern Chile, and elderly people from an old neighbourhood in the Chilean capital, Santiago. Looking at the practices and coping strategies of these disparate groups, the authors propose a suite of targeted measures for tackling the specific energy poverty challenges faced in each locale.

This section ends with Chapter 15, where Stojilovska and Feenstra provide a study of women's experience of energy poverty in North Macedonia. Using an energy justice lens, the authors incorporate a strong gendered perspective focusing on the coping mechanisms of women as they pertain to health, age, and ethnicity/migrant status. The authors rightly acknowledge that climate and energy transition policies in the Global North continue to reflect broader societal inequalities with women's experiences still not adequately addressed or often ignored.

Part III – Practices, policy and sustainability – pushes an analysis of policy practices, government strategy, and sustainability implications of holistic views of energy poverty. Chapters in this section challenge the idea of programmes being focused on service provision that is not in line with the aspirations and realities of the targeted participants and how they exercise choice about fuels, stoves, and retrofit approaches. Provision alone will not imply sustained use and the improvement of comfort levels and well-being. They also look at the nexus between energy poverty and the urban energy transition, which is a nexus that plays a prominent role given the urbanisation processes, particularly in the Global South.

Kenner et al. begin the section (Chapter 16) by describing a relevant practice, the Neighborhood Energy Centers (NECs) to address energy vulnerability in Philadelphia, the United States. Through budget counselling and referrals, energy conservation workshops and supporting applications for energy assistance programmes, NECs are making energy poverty more visible in the United States and while helping those that find it more difficult to ask for assistance such as senior citizens. Drawing from interviews, authors show how NECs can provide an important layer of flexibility amid an existing well-stratified and ossified approach to government energy assistance, offering a network of embedded and responsive care actors.

In Chapter 17, Cornelis et al. investigate two sub-Saharan African countries' (Burkina Faso and Madagascar) experiences of service-based electricity and flexibility/approaches to electrification, particularly energy as a service and demand flexibility, to know to what extent these innovative tools can address energy poverty in the African continent. These experiences shed light on promoting more sustainable consumption practices in Europe, by moving away from a traditional consumption logic to a consumption logic passed on through resource sharing.

In Chapter 18, Dervishi et al. present research on a number of potential retrofit scenarios and building measures available to energy-vulnerable households in residential housing in Albania.

In Chapter 19, Noaman El-Sherbini and Lippert use discourse analysis to examine the European Union member states' National Energy and Climate Plans (NECPs) to discuss whether these indicate ambiguity, complexity, and multidimensionality in the way energy poverty is expressed in their language and frame. They suggest changing the nature of the diagnosis of energy poverty from measuring and defining to one that aims to identify the "vulnerable consumer", reasons for their vulnerability, and measures to tackle specific types of vulnerability.

In Chapter 20, Lamonaca and Batel adopt a lived experience lens to empirically examine the role of recognition justice in smart city projects in social housing in the Global North (Bolzano, Italy). The authors show the role of symbolic and psychosocial aspects in promoting inclusive and democratic accountability when promoting clean energy, as well as highlight the importance of addressing misrecognition to empower citizens to enable a more active and autonomous energy citizenship.

Boatend et al., in Chapter 21, explore factors affecting the use of wood fuels in Ghana to then examine possible uses of cleaner alternative for cooking and how these would affect rural livelihoods. By doing a qualitative survey, they propose having a holistic approach that ensures enhanced access to cleaner cooking fuels in accordance with availability, cost, and culture. In the case of the Ghanaian context, they suggest the use of improved cookstoves as a first step in the pathway to introducing LPG.

Reflecting on the preceding contributions and informed by their experiences in realising the EnergyMeasures project, Dunphy et al. conclude the volume (Chapter 22) with perspectives on the phenomenon of energy poverty and forward ideas on what these might mean for attempts to support the energy poor.

Acknowledgement

This work has been conducted in the context of the EnergyMeasures Project funded by the European Union's Horizon 2020 Programme for Research and Innovation under Grant Agreement No. 894759. Further information about the project can be found at https://doi.org/10.3030/894759.

Notes

1 Energy Poverty Action: Agenda Co-Creation and Knowledge Innovation (ENGAGER 2017–2021) is a research network funded via the European Co-operation in Science and Technology (COST) scheme. https://www.engager-energy.net/

2 EPOV was a project established by the European Commission to improve the measuring, monitoring, and sharing of knowledge and best practices on energy poverty. The observatory has since been integrated into the Energy Poverty Advisory Hub (EPAH), a successor project. https://www.energypoverty.eu

3 Horizon 2020 was a European Union research and innovation funding programme from 2014 to 2020. https://research-and-innovation.ec.europa.eu/funding/funding-opportunities/funding-programmes-and-open-calls/horizon-2020_en

4 Information Communications Technology

5 Central to this approach are the concepts of "functionings" and "capabilities". Functionings are "beings and doings" that people value and actually achieve – activities (e.g., reading or dancing) or states of existence (e.g., being in good health or not being ashamed). Capabilities are the actual or real opportunities to realise given functionings, whether one chooses to at any particular time or not. Because the capabilities approach puts a high value on freedom to choose, capabilities, rather than functionings, are the object of concern.

References

Alam, M. S., Bala, B. K., Huq, A. M. Z., & Matin, M. A. (1991). A model for the quality of life as a function of electrical energy consumption. *Energy*, 16, 739–745.

Barnes, D. F., Khandker, S. R., & Samad, H. A. (2011). Energy poverty in rural Bangladesh. *Energy Policy*, 39, 894–904. https://doi.org/10.1016/j.enpol.2010.11.014

Bazilian, M., Nakhooda, S., & Van De Graaf, T. (2014). Energy Governance and Poverty. *Energy Res. Soc. Sci.*, 1, 217–225. https://doi.org/10.1016/j.erss.2014.03.006

Biswas, W. K., Bryce, P., & Diesendorf, M. (2001). Model for empowering rural poor through renewable energy technologies in Bangladesh. *Environ. Sci. Policy*, 4, 333–344. https://doi.org/10.1016/S1462-9011(01)00031-4

Boardman, B. (2013). *Fixing Fuel Poverty: Challenges and Solutions*. London & New York: Routledge.

Boardman, B. (1991). *Fuel Poverty: from Cold Homes to Affordable Warmth*. London: Belhaven Press.

Bouzarovski, S. (2018). *Energy Poverty*. Cham: Springer International Publishing. https://doi.org/10.1007/978-3-319-69299-9

Bouzarovski, S., & Petrova, S. (2015). A Global Perspective on Domestic Energy Deprivation: Overcoming the Energy Poverty-Fuel Poverty Binary. *Energy Res. Soc. Sci.*, 10(July), 31–40. https://doi.org/10.1016/j.erss.2015.06.007

Bruce, N., Perez-Padilla, R., & Albalak, R. (2000). Indoor air pollution in developing countries: a major environmental and public health challenge. *Bull. World Health Organ*., 78, 1078–1092.

Chester, L., & Morris, A. (2011). A New Form of Energy Poverty Is the Hallmark of Liberalised Electricity Sectors. *Aust. J. Soc.*, 46(4), 435–459. https://doi.org/10.1002/j.1839-4655.2011.tb00228.x

Day, R., & Walker, G. (2013). Household energy vulnerability as 'assemblage.' In: *Energy Justice in a Changing Climate: Social Equity Implications of the Energy and Low-Carbon Relationship*. London and New York: Zed Books, 14–29.

Day, R., & Walker, G. (2011). Underconsuming energy? Conceptualising household energy vulnerability in everyday life, In: *Association of American Geographers Annual Meeting*, Seattle, WA, USA. pp. 12–16.

Day, R., Walker, G., & Simcock, N. (2016). Conceptualising energy use and energy poverty using a capabilities framework. *Energy Policy*, 93, 255–264. https://doi.org/10.1016/j.enpol.2016.03.019

Economidou, M., Todeschi, V., & Bertoldi, P. (2019). *Accelerating Energy Renovation Investments in Buildings*. Luxembourg: Publications of the European Union.

Foster, V. (2000). *Measuring the Impact of Energy Reform: Practical Options*. Washington DC: World Bank.

González-Eguino, M. (2015). Energy poverty: An overview. *Renew. Sustain. Energy Rev.*, 47, 377–385. https://doi.org/10.1016/j.rser.2015.03.013

Haas, R., Nakicenovic, N., Ajanovic, A., Faber, T., Kranzl, L., Müller, A., & Resch, G. (2008). Towards Sustainability of Energy Systems: A Primer on How to Apply the Concept of Energy Services to Identify Necessary Trends and Policies. *Energy Policy*, 36(11), 4012–4021. https://doi.org/10.1016/j.enpol.2008.06.028

Harrison, C. (2013). The historical–geographical construction of power: Electricity in Eastern North Carolina. *Local Environ.*, 18, 469–486. https://doi.org/10.1080/13549839.2012.748728

Heffron, R. J., & de Fontenelle, L. (2023). Implementing Energy Justice Through a New Social Contract. *Journal of Energy & Natural Resources Law*, 41(2), 141–155. https://doi.org/10.1080/02646811.2023.2186626

Heffron, R. J., & McCauley, D. (2017). The Concept of Energy Justice Across the Disciplines. *Energy Policy*, 105(June), 658–667. https://doi.org/10.1016/j.enpol.2017.03.018

Hosier, R. H., & Dowd, J. (1987). Household fuel choice in Zimbabwe: An empirical test of the energy ladder hypothesis. *Resour. Energy*, 9, 347–361. https://doi.org/10.1016/0165-0572(87)90003-X

Hughes, T. P. (1993). *Networks of Power: Electrification in Western Society,* 1880-1930. Baltimore: John Hopkins University Press.

IEA (2022). *World Energy Outlook 2022*. Paris: International Energy Agency.

IEA (2022). *Fossil Fuels Consumption Subsidies 2022 – Analysis*. Paris: International Energy Agency. https://www.iea.org/reports/fossil-fuels-consumption-subsidies-2022

IEA (2010). *Energy Poverty. How to Make Modern Energy Access Universal?* Paris: International Energy Agency. https://www.iea.org/reports/energy-poverty-how-to-make-modern-energy-access-universal

Jansen, J. C., & Seebregts, A. J. (2010). Long-Term Energy Services Security: What Is It and How Can It Be Measured and Valued? *Energy Policy,* 38(4), 1654–1664. https://doi.org/10.1016/j.enpol.2009.02.047

Jenkins, K., McCauley, D., Heffron, R., Stephan, H., & Rehner, R. (2016). Energy Justice: A Conceptual Review. *Energy Res. Soc. Sci.*, 11, 174–182. https://doi.org/10.1016/j.erss.2015.10.004

Jodoin, L. 2021. Let Capabilities Ring: Operationalizing Energy Justice in Guinea. *Energy Res. Soc. Sci.*, 72, 101894. https://doi.org/10.1016/j.erss.2020.101894

Katsoulakos, N. (2011). Combating Energy Poverty in Mountainous Areas Through Energy-Saving Interventions. *Mt Res. Dev.*, 31(4), 284–292. https://doi.org/10.1659/MRD-JOURNAL-D-11-00049.1

Larson, E. D., & Kartha, S., (2000). Expanding roles for modernized biomass energy. *Energy Sustain. Dev.*, 4, 15–25. https://doi.org/10.1016/S0973-0826(08)60250-1

Masud, J., Sharan, D., & Lohani, B.N., (2007). *Energy for All: Addressing the Energy, Environment, and Poverty Nexus in Asia*. Manila: Asian Development Bank.

McCauley, D., Ramasar, V., Heffron, R. J., Sovacool, B. K., Mebratu, D., & Mundaca, L. (2019). Energy Justice in the Transition to Low Carbon Energy Systems: Exploring Key Themes in Interdisciplinary Research. *Appl. Energy*, 233, 916–921. https://doi.org/10.1016/j.apenergy.2018.10.005

McCauley, D., Heffron, R. J., Stephan, H., & Jenkins, K. (2013). Advancing energy justice: the triumvirate of tenets and systems thinking. *Int. Energy Law Rev.*, 32, 107–110.

Meyer, S., Laurence, H., Bart, D., Middlemiss, L., & Maréchal, K. (2018). Capturing the Multifaceted Nature of Energy Poverty: Lessons from Belgium. *Energy Research & Social Science*, 40(June), 273–283. https://doi.org/10.1016/j.erss.2018.01.017

Middlemiss, L., & Gillard, R. (2015). Fuel poverty from the bottom-up: Characterising household energy vulnerability through the lived experience of the fuel poor. *Energy Res. Soc. Sci.*, 6, 146–154. https://doi.org/10.1016/j.erss.2015.02.001

Middlemiss, L., Gillard, R., Pellicer, V., & Straver, K. (2018). Plugging the Gap Between Energy Policy and the Lived Experience of Energy Poverty: Five Principles for a Multidisciplinary Approach. In: *Advancing Energy Policy*. Cham: Springer International. pp. 15–29. https://doi.org/10.1007/978-3-319-99097-2_2

Modi, V., McDade, S., Lallement, D. & Saghir, J. (2005). *Energy Services for the Millennium Development Goals*. New York: United Nations Development Programme.

Nussbaum, M. C. (2003). Capabilities as fundamental entitlements: Sen and social justice. *Fem. Econ.*, 9, 33–59. https://doi.org/10.1080/1354570022000077926

Nussbaum, M. C. (2001). *Women and Human Development: The Capabilities Approach*. Cambridge: Cambridge University Press.

Nussbaumer, P., & Bazilian, M., Modi, V. (2012). Measuring energy poverty: Focusing on what matters. *Renew. Sustain. Energy Rev.*, 16, 231–243. https://doi.org/10.1016/j.rser.2011.07.150

Pachauri, S. (2007). *An Energy Analysis of Household Consumption: Changing Patterns of Direct and Indirect Use in India*. Dordrecht: Springer.

Pachauri, S., Mueller, A., Kemmler, A., & Spreng, D. (2004). On Measuring Energy Poverty in Indian Households. *World Dev.*, 32, 2083–2104. https://doi.org/10.1016/j.worlddev.2004.08.005

Pachauri, S., & Spreng, D. (2011). Measuring and monitoring energy poverty. *Energy Policy*, 39, 7497–7504. https://doi.org/10.1016/j.enpol.2011.07.008

Petrova, S., Gentile, M., Mäkinen, I. H., & Bouzarovski, S. (2013). *Perceptions of Thermal Comfort and Housing Quality: Exploring the Microgeographies of Energy Poverty in Stakhanov, Ukraine*. *Environ. Plan A* 45(5), 1240–1257. https://doi.org/10.1068/a45132

Porcaro, J., & Tadaka, M. (2005). *Achieving the Millennium Development Goals: The Role of Energy Services: Case Studies from Brazil, Mali and the Philippines*. New York: United Nations Development Programme.

Practical Action, (2010). *Poor People's Energy Outlook 2010*. Rugby: Practical Action Publishing.

Reddy, A. K. N., Annecke, W., Blok, K., Bloom, D., Boardman, B., Eberhard, A., & Ramakrishna, J. (2000). Energy and Social Issues. *In: World Energy Assessment: Energy and the Challenge of Sustainability*, pp. 39–60. New York: United Nations Development Programme.

Rehfuess, E, (2006). *Fuel for Life: Household Energy and Health*. Geneva: World Health Organization. https://www.who.int/publications-detail-redirect/9789241563161

Sen, A., 2001. *Development as Freedom*. Oxford: Oxford Paperbacks.

Sovacool, B. K. (2014). What Are We Doing Here? Analyzing Fifteen Years of Energy Scholarship and Proposing a Social Science Research Agenda. *Energy Res. Soc. Sci.*, 1, 1–29. https://doi.org/10.1016/j.erss.2014.02.003

Sovacool, B. K., (2012). The political economy of energy poverty: A review of key challenges. *Energy Sustain. Dev.*, 16, 272–282. https://doi.org/10.1016/j.esd.2012.05.006

Thomson, H., Snell, C., & Liddell, C., (2016). Fuel poverty in the European Union: a concept in need of definition? *People Place Policy*, 5–24. https://doi.org/10.3351/ppp.0010.0001.0002

UN. (2022). *Energy*. United Nations Sustainable Development (blog). https://www.un.org/sustainable development/energy/

UNDP, (2005). *Energizing the Millennium Development Goals*. New York: United Nations Development Programme.

Walker, G., & Day, R. (2012). Fuel Poverty as Injustice: Integrating Distribution, Recognition and Procedure in the Struggle for Affordable Warmth. *Energy Policy*, 49, 69–75. http://dx.doi.org/10.1016/j.enpol.2012.01.044

Wilhite, H., Shove, E., Lutzenhiser, L., & Kempton, W. (2000). The Legacy of Twenty Years of Energy Demand Management: We Know More about Individual Behaviour but Next to Nothing about Demand. In: *Society, Behaviour, and Climate Change Mitigation*. New York: Kluwer Academic Publishers, pp. 109–126.

Zulu, L. C., & Richardson, R. B. (2013). Charcoal, Livelihoods, and Poverty Reduction: Evidence from Sub-Saharan Africa. *Energy for Sustainable Development*, 17(2), 127–137. https://doi.org/10.1016/j.esd.2012.07.007

Part I

Methodological Approaches

2 Identifying Energy-Poor Households, Experiences from the Global North

Niall P. Dunphy, Breffní Lennon, and Paola Velasco-Herrejón

2.1 Introduction

As we will see in later chapters, understanding the wicked problem of energy poverty is not an easy undertaking. Rather, it is replete with multiple layers of complexity across numerous intersecting societal and environmental scales. Experience working on the EnergyMeasures H2020 project[1] has demonstrated how difficult it can be to engage with hard-to-reach households vulnerable to energy poverty. Notwithstanding the additional barriers to engagement, put in place during the early stages of the project to counter the impact of the COVID-19 pandemic raging at the time, reaching the most vulnerable households has always proven difficult even for those organisations working at the coalface of energy poverty.

This chapter will present findings from our experiences identifying and recruiting energy-poor households in seven participating countries (BE, BG, IE, MK, NL, PL, and UK) to the project. While all are located in Europe, the range of experiences our consortium partners faced were often very different and reflected the historical, cultural, and environmental factors specific to each country. Consequently, we critique range of approaches to identifying and measuring energy vulnerability[2], especially indicators of energy poverty and so-called supporting indicators. As is noted throughout this book, the focus on energy poverty analysis has tended to stay at the macro-, or meso-, levels while understanding contexts at the local level continue to be rather underdeveloped. In keeping with the overall theme of the book, approaches on how to appropriately identify energy-poor households are drawn from both the literature and from experiences of some practitioners active 'in the field'.

2.2 Understanding energy poverty

Energy poverty can best be understood as arising out of number of intersecting factors, including high energy prices, low household incomes, specific household energy needs (e.g., health), and the reliance on inefficient, poorly performing buildings and appliances (Bouzarovski, 2014). While one's level of income can be considered an important factor, it is by no means the defining factor. One should note that not all those who suffer from monetary poverty are necessarily energy poor. Also, not all those in energy poverty are necessarily income poor[3,4] (Palmer et al., 2008). The very wicked problem energy-poor households face is the choice of spending an above-average portion of their income on heating, lighting, cooling, cooking, and appliance use; or going without so they can afford other essentials such as food[5]. The resulting scenario many households in energy poverty face therefore is a cold (or overly hot) and uncomfortable home and reduced living standards. As Thomson et al. (2016) suggest, the consequences can have significant impacts on people's physical health and mental well-being (due to more restricted

DOI: 10.4324/9781003408536-4

lifestyles and social exclusion) and can also lead to premature death related to severe seasonal weather conditions, especially in winter.

A key component of any effort to engage with energy-vulnerable households is being able to identify who is experiencing energy poverty or who might be energy insecure. Very often, shame, stigma, and indeed refusal to accept one is actually experiencing energy poverty can lead to 'geographic disparities in the risk and incidence of domestic energy deprivation' (Bouzarovski & Simcock, 2017, p. 640). As Cong et al. (2022) suggest, traditional income-based energy poverty metrics tend to minimise or ignore people's behavioural patterns, especially those energy-limiting behaviours practised in low-income households. Such behaviours often mask the real extent of the issue and instead create hidden energy poverties (e.g., see also Betto et al., 2020; Yip et al., 2020). Scholars rightly observe (e.g., Dubois, 2012, p. 107) that these difficulties in identifying those in energy poverty continue to be a 'major obstacle to an efficient implementation of fuel poverty policies. Like general policies of poverty reduction, fuel poverty policies have the specificity that the population who forms their target cannot be "found" easily'.

In addition, the multiple overlapping causes of energy poverty[6] can lead to greater challenges for practitioners when trying to accurately identify energy-poor households. Baker (2011, p. 15), for example, outlines the issue as such: '[it] is not possible to identify individual fuel poor households without obtaining information on both their income and housing circumstances'. A key, overarching objective of the EnergyMeasures project has been to engage with and assist energy-poor households, and as such, we were keen to incorporate these understandings into our planning for identification and engagement activities. Consequently, the first work package of the project concentrated on designing an in-depth, collaborative planning and preparatory work. This involved multi-stakeholder engagement to assess and improve the institutional context, while at the household level, engagements were organised to maximise the targeting of specific segments of the energy poor in each country.

As such, we were keen to explore the most appropriate methods for identifying energy-poor households that were inclusive and respectful of each householder's individual circumstances. To do this, we initiated a review of current best practice, academic discourse, and new thinking, from, for example, the European Energy Poverty Observatory (EPOV), to help us better identify those methods that would prove most effective in each of the different locations[7].

2.3 Measuring energy poverty

Thomson et al. (2017, p. 882) note that energy poverty '*is a private condition, being confined to the home, it varies over time and by place, and it is a multi-dimensional concept that is culturally sensitive*'. Therefore, measuring it is not an easy task. For those tasked with addressing energy poverty, it raises significant questions around privacy and personal dignity, and how one should approach the topic with householders. This has resulted in a tendency on the part of practitioners to rely on quantitative and geographical dataset that suggest an increased likelihood of energy poverty in what are usually already regarded as socially disadvantaged neighbourhoods. However, this approach is beginning to change with practitioners, opting for more nuanced means of identifying energy poor, for example, through a comprehensive income, energy expenditure, and property assessment of all households. This exercise provides all the data needed to measure the level of energy poverty, appreciate its severity, and to identify individual energy-poor houses for support. However, as Hills (2012, p. 70) observes this approach, when done correctly, can be '*prohibitively expensive – and intrusive – to carry out*'.

Therefore, if we acknowledge that not all households can be comprehensively assessed, then determining the number of, and which type of, households fall within established thresholds for energy poverty would in principle require 'a detailed household and building sample survey followed by sophisticated modelling work' (Hills, 2012, p. 70). However, more detailed surveys invariably come with higher administrative burdens, as Dubois (2012, p. 108) notes, the administrative overheads associated with such precise targeting of energy-poor households 'is a necessary step but, as it is also costly, being as precise as possible is not necessarily optimal'. Hills (2012) points out that many of those working to address energy poverty directly use less complicated proxy indicators for identifying households. The rest of this chapter will provide an overview of the indicators used for understanding energy poverty along with the types of supporting indicators that can help prioritise support 'on the ground'.

2.3.1 Energy poverty indicators

The three principal methods for measuring energy poverty, according to the literature, can be summarised as (1) expenditure-based (absolute or relative energy expenditure thresholds); (2) subjective-based (self-reporting of living conditions and financial distress); and (3) needs-based (direct measurement of energy use compared to a set standard) measurements (Bouzarovski et al., 2020; Thema & Vondung, 2020). Of these, metrics based on expenditure tend to predominate[8], with relative thresholds used as proxies for energy deprivation (Waddams Price et al., 2007, p. 35). Rademaekers et al. (2014, p. 25) outline three types of expenditure-based metrics used for defining and measuring energy poverty:

i A high share of energy costs: the proportion of income spent on energy being above an established threshold, that is, those who spend too much. The Belgium energy poverty barometer for example sets its threshold for measured energy poverty (mEP) at twice the median expenditure of equivalent households (Meyer et al., 2018);
ii Low income, high costs (LIHC): amount of income after energy costs being below the established threshold, that is, high energy expenditure and/or insufficient income. Hills (2011, p. 9) for instance suggests that the LIHC threshold should be a combination of expenditure over the median level, and residual income less than official poverty line;
iii Insufficient energy spending: where absolute energy spending is below established the threshold, that is, those who self-restrain, the hidden energy poverty (hEP). The threshold set for hEP in Belgium is set at half the median expenditure of equivalent households[9] (Meyer et al., 2018).

The second, subjective approach comprises primarily on self-reporting of one's inability to heat one's home sufficiently (often described as 'consensual' data). However, Thomson and Snell (2014) suggest that such reporting more often focuses on the consequences of energy poverty, (e.g., presence of damp, mould, and arrears) rather than its causes (*e.g.*, high costs, specific demands, poorly constructed built environments, etc.) (Dunphy, 2020). However, they do agree that this approach does provide '*an insight into the perceived affordability of heating homes across the EU'* (Thomson & Snell, 2013, p. 567). While such approaches have been criticised for their subjectivity and reliance on householders' perceptions, it could be argued that this is exactly why this approach should be used.

As such, expenditure-based measures and self-reporting of energy poverty do not distinguish between those households that really need to consume energy (essentially, providing for their basic household needs) and those that do not, though this can be addressed using a needs-assessment modelling approach(e.g., see SEAI, 2018, p. 37)[10]. As Meyer et al. (2018, p. 280) argue, there are '*people feeling energy poor who cannot be detected by neither (sic) forms of objective measurement*' (i.e., measured energy poverty mEP or hidden energy poverty hEP); therefore there is a clear need for subjective measurements of energy poverty. Fahmy et al.'s (2011, p. 4371) observation that there is little overlap in households identified as energy poor using expenditure-based, subjective-based, or needs-based measurements also supports the notion of using multiple complementary measures. For Bouzarovski et al. (2020, p. 41), '*energy poverty is a culturally sensitive, multi-dimensional concept that varies over time and by place and is thus not easily captured by a single indicator*' and accordingly they also suggest using a selection of consensual and expenditure-based indicators to be used in conjunction with each other. To meet this challenge, Sokołowski et al. (2020) propose a multidimensional energy poverty index comprising five key indicators of energy poverty. While the EU Energy Poverty Observatory (EPOV) have built on the work of Sareen et al. (2020), selecting four primary indicators of energy poverty, comprising:

1 **High share of energy expenditure in income** (2M) – those households with share of energy expenditure in income >2x the national median.
2 **Low share of energy expenditure in income** (M/2) – those households whose absolute energy expenditure is <1/2 the national median.
3 **Inability to keep home adequately warm** (Keep warm) – based on self-reported thermal discomfort.
4 **Arrears on utility bills** (Arrears) – based on households' self-reported inability to pay utility bills on time in the last 12 months.

(Bouzarovski et al., 2020, p. 41)

2.3.2 Supporting indicators

Complementary to the above, Rademaekers et al. (2014, pp. 38–39) suggest the importance of also considering supporting indicators that measure factors contributing to the social experience of energy poverty. Such additional indicators deepen our understandings of energy poverty and have a role in policy development. For Hills (2012, p. 82) '*understanding the characteristics of fuel poor households is a first step to identifying them on the ground*'. Dunphy (2020, p. 13) outlines the key types of supporting indicators, including:

- Demographic factors – households with vulnerable members, for example, young children, elderly (especially those living alone), those with health issues, etc., are more energy vulnerable.
- Energy prices – differential pricing may apply in many cases (for reasons of, e.g., geography, bill arrears consumption level, service bundling) meaning that householders will not have equal access to tariffs.
- Income levels – lower income households are more likely to be energy poverty, no matter which way it is measured.
- Household composition – the make-up of the household may render it more susceptible to 'financial problems in general and energy poverty particularly' (Rademaekers et al., 2014, p. 38).
- Heating system – what type and condition? inadequate and inefficient heating systems will provide poor heating at a higher cost.
- Supply choice – due to lack of choice (e.g., due to geographical[11] and/or socio-political considerations[12]) or supply lock in (such as with district heating systems[13], or tenancy agreements).

- Building efficiency – age and technical characteristics of a building directly influence its energy performance, and householders of poor performing buildings are inherently more prone to energy poverty.
- Level of social supports – the amount of social security support in a household is an indicator of its vulnerability to financial shocks, which could manifest itself as, or intensify existing, energy poverty;
- Tenure (and security of tenure) – different forms of tenure (e.g., ownership, long-term leasehold, monthly rental, etc.) are associated with varying levels of control and influence in decision-making and 'may limit the energy efficiency interventions and fuel switching measures' that can be implemented (Bouzarovski, 2014, p. 280).

2.3.3 Indicators or identifiers?

Energy poverty indicators are essential to quantifying the number of energy-poor households while also helping to understand the depth of energy poverty they endure and allows for greater efficacy and efficiency in supporting energy-poor households through more informed target policy measures. While there are clear benefits to collecting and analysing additional data, larger data sets directly increase the administrative burden and may lead to certain organisational biases to emerge. When discussing the use of proxy indicators, Hills (2012, p. 71) argues that the use of strict thresholds, for example, may do more harm than good and suggests instead that 'imprecision in targeting may be desirable rather than undesirable'[14].

The use of indicators can greatly enhance our understanding of who is suffering from energy poverty, while the supporting indicators offer greater insights into the characteristics energy-poor households are likely to exhibit and allow for better targeting of policy measures to support the identified groups (Dunphy, 2020). However, it should be noted that the adaptation and use of indicators do not in and of themselves constitute a method for identifying (specific) energy-poor households and it is this identification stage that is still not always clearly understood.

2.4 Approaches to identifying energy-poor households

Much of the discourse on identification focuses on a combination of defining energy poverty, outlining the characteristics of energy-poor households, and/or macro- and meso-level data collection for analysis, monitoring, and benchmarking. In contrast, the mundane activities used to identify specific households in energy poverty are usually not seen as interesting or novel and are therefore underreported (if at all). More often, the emphasis in reports and publications on household-level work tends to centre on the engagement activities themselves. However, several scholars do discuss the practises for identifying energy-poor households and the approaches taken. Sokołowski et al. (2020) note that proper identification of energy-poor households should lead to more efficient application of the resulting support measures. Using national-level data analysis and extrapolation, they propose a multidimensional index with five indicators of energy deprivation: (1) low income and high costs; (2) high actual cost; (3) insufficient warmth; (4) building faults; and (5) difficulties in paying. A household is considered energy poor if it displays at least two forms of deprivation. While Morrison and Shortt (2008) present a GIS-based modelling approach designed to predict areas where energy poverty is most likely to be prevalent. Combining census data with individual dwelling-level data from local housing service datasets, they were able to identify the physical risk factors for energy poverty with social risk factors sourced from national census returns. The resultant risk score then enabled them to map energy poverty risk ratings onto individual dwellings, allowing for more targeted identification of vulnerable households.

Another approach by Walker et al. (2012) involves a small-area energy poverty risk index using a range of environmental and socio-economic variables. Analysis is managed at the lowest level of geographical area for census statistics, Output Areas (OAs) (this approach has also been taken by Morrison and Shortt (2008) and Fahmy et al. (2011). The index offers a useful means for assessing energy poverty risks at a more granular level, i.e., c. 125 dwellings (Dunphy, 2020). Walker et al. (2012) also suggest identification should be targeted in areas of high risk, while households in more socially advantaged areas are encouraged to self-refer for support. Zooming out, Liddell et al. (2011) outline a zonal approach undertaken by the local council in Kirklees (pop. c. 400,000) in the United Kingdom, which developed the *Kirklees Warm Zone Project* initiative, offering loft and cavity wall insulation free-of-charge for every suitable household in the area. This so-called 'Zip-Up Method' for identification and implementation is seen as an exemplar for local authorities tackling energy poverty (Dunphy, 2020).

Another novel approach, described by Mohan (2021), is the application of public health referrals through the Warmth and Wellbeing Scheme in Ireland. This pilot government-funded scheme provides extensive energy efficiency upgrades to households with members suffering from chronic respiratory diseases. While in the United Kingdom, NHS doctors participating in a pilot scheme in Gloucestershire can prescribe heating to patients at higher risk of hospital admission during colder months (Lacobucci, 2022). Similarly, Scarpellini et al. (2017) describe a case study in Aragón, northern Spain, where social workers were equipped to identify energy poverty in the houses they visited. Subsequently, they were able 'certify' those households for access to public supports.

A more mixed-methods approach to identification outlined by Gouveia and Seixas (2016), involves the combining of daily electricity consumption data (using smart metres) with a 110-question survey of 265 households in Évora, Portugal. This combination of quantitative and qualitative data allowed for the grouping of consumers into three main groupings: those experiencing fuel poverty, those with a standard level of comfort; and those living in 'fat energy' households. While Spiliotis et al. (2020) offer a support framework for utilities to identify potentially energy-poor households. Using open-source weather data to determine heating and cooling degree days and national building standards to estimate the energy needs of 'reference' households, they combined this information with actual energy consumption data of those households (using the utilities' own billing databases) and area-level average data from national statistics to estimate household income. Applying this four-dimensional approach, they were able to identify energy-poor households whose energy expenditure is above 10% of household income, or those households consuming less energy than the relevant reference household (Dunphy, 2020).

However, such innovation is not new. Dubois (2012) outlines a delegated approach using local actors taken by the French programme *Habiter Mieux*, involving a mix of centralised and decentralised tasks. While determining and communicating the characteristics of energy-poor dwellings is centralised, actual household identification is decentralised through local actors (*e.g.*, medical and social centres, home-help assistance, credit unions, firefighters, and fuel suppliers) who use their local knowledge and social networks to reach out to, and identify, energy-poor households. Finally, Fischer et al. (2014) describe a variation of this approach, describing the work of energy advisory centres and their role in identifying energy poverty. Interacting with the public through telephone help lines, drop-in face-to-face 'surgeries' held in community spaces, etc. these NGOs are able to identify at-risk households who use their services, referring them for further assessment and supports[15].

2.4.1 Thoughts from the field

A key methodological approach undertaken by the EnergyMeasures H2020 Project was to conduct short, structured dialogues with researchers and practitioners working in the participating countries[16]. All the respondents have a track record of identifying and engaging with vulnerable energy consumers from across Europe. These households experience broad variations in seasonal extremes in climate, in addition to the socio-economic, political, and organisational structures available to them to cope with the challenges posed by energy poverty. Many respondents reported adopting a practice-based, mixed-methods approach to identifying and engaging energy-vulnerable households. While some do rely on data-driven analysis, especially for reporting, the task of identifying and recruiting energy-poor households is usually done through a combination of approaches.

These included more traditional marketing and advertising activities while maintaining a presence in the local community and promoting new or existing services. These include press releases, articles, news features in local newspapers, radio, parish and/or community bulletin boards, branded vehicles, etc., primarily designed to target older people with an established, more traditional network who might not engage with newer media. For younger people and smartphone users, similar results were accomplished through new content and promoting marketing material featured in traditional media outlets using popular social media platforms. Such activities helped with ongoing outreach work by the organisations involved, which targeted householders who may self-identify as being energy poor (*i.e.*, those already using the organisation's services or a partner organisation's services) often through advertised (printed material and through staff) at an organisation's drop-in facility in the target community, or at bespoke promotional events.

A number of organisations described the importance of *ad hoc* campaigning strategies with staff informing their friend and family networks of new or existing services, in the process helping to build trust through those staff members. Again, this word-of-mouth approach was especially useful when targeting those already aware of the organisation involved and/or its services and staff working there. A notable development over the course of the project was a move by some organisations to develop more collaborative projects with other NGOs, state bodies, etc. These partnership projects allowed two or more organisations to share resources for a specific campaign or service provision and maximise individual impact by using each organisation's database of existing users. Another notable development, as described above, is the growing emphasis on both direct and indirect referrals from social services, medical professionals, and other charitable organisations to those organisations specifically working to address energy poverty. This approach also allows for the referring organisation to assist with a problem that, while outside their remit, may impinge on their clients' social, medical, and financial well-being.

Some respondents also mentioned using mapping and engagement material generated from several European Horizon 2020 and older FP7 projects as part of their identification toolkit. None of the respondents mentioned using European or national datasets for identifying energy-poor households but instead relied on self-generated data. Most were unfamiliar with the four key indicators mentioned in Box 1 above or only had a tangential understanding of them, resulting in many organisations tasked with addressing energy poverty having to rely on mixed-methods approaches (Dunphy, 2020). While not necessarily a negative, it does result in unintended divergences between organisations and the communities they engage with and leads to those in most need of support not getting access to, or are made aware of, supports even when they are available. Also, the activities mentioned above require a degree of digital, financial, and civic/social literacy on the part of the citizen that they may not necessarily have.

In response to such challenges, actors working on energy poverty are starting to do more collaborative work with other local and state actors who share a common goal. An example given by an energy charity operating in a west European country described how it developed a collaborative programme with another charity whose remit covered topics other than energy. Pooling resources, including the dataset of participants already signed up to both charities, they co-devised a scheme whereby participants were able to receive vouchers to top up their pre-payment electricity meters[17]. This initiative resulted in novel synergies emerging between the charities and government bodies involved and had a greater beneficial impact on participants who were able to receive assistance from multiple organisations through referrals and cooperation (Dunphy, 2020).

Another respondent described their work communicating with households and implementing home metres to measure temperature and humidity in the home, with information collected and clear instructions relayed back to the occupant on how best to respond to sudden adverse changes in both[18]. Identification of appropriate participants is carried out in partnership with other energy charities working with individual landlords and landlord associations, rather than approaching tenants directly. This model prioritises engagement with the landlord/association, thus removing any potential for tension or negative responses if they were to go to tenants directly, who might struggle to encourage a landlord to make the energy-efficient investments necessary. Invariably, the issue of trust is key here, and the process results in a shift in focus away from the (potentially fraught) relationships between landlord and tenant onto addressing the actual condition of the property itself. Adopting a more tailored approach is very much seen as the best option by most respondents, with approaches to potential recruits framed by questions like ‘how can we help you?’, rather than appealing to the common good or on environmental grounds[19].

2.5 Conclusion

While a great deal of effort analysing energy poverty is carried out at both the macro-, and the meso-levels, a gap remains between such approaches and the actual identification of specific energy-poor households. In an effort to address this gap, this chapter has presented a number of potential approaches that may prove more beneficial to the local context. Most notably, the establishment of deeper relationships between organisations already working with households likely to be in energy poverty offers significant opportunities for amplifying the specific strengths of individual organisations beyond more traditional supports like social workers (Scarpellini et al., 2017) and health professionals (Mohan, 2021), to more nuanced and broad-ranging advice from, for example, debt advisors, social stores, housing associations, advice bureaux, consumer groups, NGOs, municipalities, and utilities (Dunphy, 2020). While the practical approaches forwarded in this chapter may rightly be seen as not entirely novel in their own right, it is through the application and synergies between organisations that one sees the real innovation.

Acknowledgement

This work has been conducted in the context of the EnergyMeasures Project funded by the European Union’s Horizon 2020 Programme for Research and Innovation under Grant Agreement No. 894759. Further information about the project can be found at https://doi.org/10.3030/894759.

Notes

1 An earlier version of the work presented in this chapter can be found in Dunphy (2020), a research report for the EnergyMeasures H2020 project.
2 For a more comprehensive treatment of the topic, especially the biases skewing approaches to measuring energy poverty, see Dunphy (2020).
3 Palmer, MacInnes, and Kenway (2008) highlight that the link between energy poverty and income poverty is somewhat weakened by rising energy costs.
4 Castaño-Rosa, Solís-Guzmán, and Marrero (2020) further emphasise the multi-dimensional nature of the phenomenon, observing that there is not a clear link per se between energy poverty and the energy (in)efficiency of buildings. The issue is more complex than that.
5 This example was demonstrated to us in an earlier Horizon 2020 project, ENTRUST [2015-2018], where a participant told us of their experience working with energy-poor households in a recently refurbished housing development. She explained: '*this young lad walked in from the flats with a voucher, and they gave him food and he turned around and said, "I don't want that. I don't want that. I don't want that." And she (a co-worker) said, "That's not on [and became frustrated with the young man].*"' When the participant approached the young man to ask him why he did not want the food package, he explained he could not afford the gas or electricity needed to cook the food being given to him. The young man could not afford to heat his apartment and cook a nutritional meal for himself; it had to be one or the other (Dunphy et al., 2017).
6 Including, for example, low income, poor quality of housing stock, social contexts, market conditions, individual financial circumstances, and health-related factors.
7 As a result, different candidate methods were tried and then modified by the partners based on local specificities (an additional challenge partners faced concerned the ongoing impacts from the COVID-19 pandemic in the different locations).
8 The pre-eminence of expenditure-based measures is typically explained by the perceived superiority and objectivity of such data compared to self-reporting (Sareen et al., 2020).
9 Rademaekers et al. (2014, p. 50) note that absolute monetary expenditure must be used to measure hEP rather than the proportion of income spent on energy. Otherwise, higher-income households could potentially also be classified as energy poor given the (very) low share of income they expend on energy.
10 The Irish Government's 2016 review of residential energy incorporated bespoke residential building stock model data from the national household budget surveys and the official building energy rating database to compare household income to the theoretical energy spend required to keep houses heated to WHO-recommended standards. As a result, the emphasis was less on the actual expenditure incurred and instead focused on the expenditure that would be needed to achieve acceptable levels of warmth.
11 The lack of choice in rural areas is well documented, where typically natural gas is not an option, there will likely be fewer electricity suppliers, and there will be a disproportionate dependence on fuel oil heating systems, or even solid fuels from a rather limited number of suppliers. This lack of choice can also be found in urban and peri-urban areas, where, for legacy (often socio-economic) reasons, the choice of energy sources and suppliers may also be restricted.
12 Even when there are multiple energy suppliers, they may not be equally available to all households. Requirements around a variety of issues (e.g., credit scores, minimum purchases, tenure status, contract length, payment methods, etc.) often combine to restrict actual choice for the less economically privileged.
13 See, for example, Bouzarovski et al.'s (2016) examination of the embeddedness of energy poverty in post-communist Hungary.
14 Hills (2012, p. 73) suggests that having households who are not strictly identified as energy poor still be eligible for policies is '*not necessarily undesirable since eligibility criteria which include households on the margins of fuel poverty are still likely to help address the problem of fuel poverty in the long term, either by helping those households whose energy costs are just below the threshold and who may become fuel poor over time, or by helping those who are just above the income threshold but who are still facing high costs*'.
15 These supports may even be carried out within their own organisation or by an external service provider. A key attribute of this approach is that the accessible nature of these organisations (via energy advice centres) means that the barriers to engagement are low.
16 Belgium, Bulgaria, Ireland, the Netherlands, Poland, North Macedonia, and the United Kingdom.

17 While these vouchers were only given to households that engaged with their services, participants received the vouchers as part of a wider package that included other supports, such as food parcels. Households were identified through local partners, government agencies, general practitioners, health, and addiction services, etc.

18 Usually advising manual ventilation by either opening or closing windows, *etc.*

19 This, and the previous example, feed into what Mullally, Dunphy, and O'Connor (2018) describe as a paternalistic perspective, which sees that armed with the 'correct' knowledge and given the 'facts', people will be 'persuaded' to change their behaviour. This approach ignores how behaviour is often locked into socio-technical systems and normative practises that individual citizens have little agency to change.

Bibliography

Baker, W. (2011). *Reaching the Fuel Poor: Making the Warm Home Discount work.* Hampton: Consumer Focus. https://doi.org/10.13140/RG.2.1.1780.7761.

Betto, F., Garengo, P., & Lorenzoni, A. (2020). A New Measure of Italian Hidden Energy Poverty. *Energy Policy, 138*, 111237. https://doi.org/10.1016/j.enpol.2019.111237

Bhatia, M., & Angelou, N. (2015). *Beyond Connections: Energy Access Redefined. ESMAP Technical Report 008/15.* Washington DC: International Bank for Reconstruction and Development.

Bouzarovski, S. (2014). Energy Poverty in the European Union: Landscapes of Vulnerability, *WIREs Energy and Environment, 3*(3), 276–289. https://doi.org/10.1002/wene.89

Bouzarovski, S., & Simcock, N. (2017). Spatializing Energy Justice. *Energy Policy, 107*, 640–648. https://doi.org/10.1016/j.enpol.2017.03.064

Bouzarovski, S., Thomson, H., Cornelis, M., Varo, A., & Guyet, R. (2020). *Towards an Inclusive Energy Transition in the European Union : Confronting Energy Poverty Amidst a Global Crisis. Third Pan-EU Energy Poverty Report of the EU Energy Poverty Observatory.* Luxembourg: Publications Office of the European Union. https://doi.org/10.2833/103649

Bouzarovski, S., Tirado Herrero, S., Petrova, S., & Ürge-Vorsatz, D. (2016). Unpacking the Spaces and Politics of Energy Poverty: Path-Dependencies, Deprivation and Fuel Switching in Post-Communist Hungary, *Local Environment, 21*(9), 1151–1170. https://doi.org/10.1080/13549839.2015.1075480

Castaño-Rosa, R., Solís-Guzmán, J., & Marrero, M. (2020). A Novel Index of Vulnerable Homes: Findings from Application in Spain, *Indoor and Built Environment, 29*(3), 311–330. https://doi.org/10.1177/1420326X18764783

Cong, S., Nock, D., Qiu, Y. L., & Xing, B. (2022). Unveiling Hidden Energy Poverty Using the Energy Equity Gap, *Nature Communications, 13*(1), 2456. https://doi.org/10.1038/s41467-022-30146-5

Dubois, U. (2012). From Targeting to Implementation: The Role of Identification of Fuel Poor Households. *Energy Policy, 49*, 107–115. https://doi.org/10.1016/j.enpol.2011.11.087

Dünnhoff, E., Eisenmann, L., & Schäferbarthold, U. (2010). *Guidelines: Introducing Advisory Services on How to Save Energy for Low-Income Households (translated for ACHIEVE).* Heidelberg/Frankfurt am Main: Institut für Energie- und Umweltforschung Heidelberg GmbH and Caritasverband Frankfurt e.V.

Dunphy, N. P. (2020). *Review of Methods of Identifying Energy Poor Households.* A deliverable of the EnergyMeasures H2020 Project. https://doi.org/10.5281/zenodo.6867434

Dunphy, N. P., Revez, A., Gaffney, C., Lennon, B., Aguilo, A. R., Morrissey, J., & Axon, S. (2017). *Intersectional Analysis of Energy Practices. A deliverable of the ENTRUST H2020 Project.*

EC-Link Project. (2014). *Energy Check for Low Income Households.* Berlin: Berliner Energieagentur GmbH.

Fahmy, E., Gordon, D., & Patsios, D. (2011). Predicting Fuel Poverty at a Small-Area Level in England, *Energy Policy*, 39(7), 4370–4377. https://doi.org/10.1016/j.enpol.2011.04.057

Fischer, J. E., Costanza, E., Ramchurn, S. D., Colley, J., & Rodden, T. (2014). Energy advisors at work. *Proceedings of the 2014 ACM International Joint Conference on Pervasive and Ubiquitous Computing*, 447–458. https://doi.org/10.1145/2632048.2636081

Gouveia, J. P., & Seixas, J. (2016). Unraveling Electricity Consumption Profiles in Households Through Clusters: Combining Smart Meters and Door-to-Door Surveys. *Energy and Buildings, 116*, 666–676. https://doi.org/10.1016/j.enbuild.2016.01.043

Healy, J. D. (2003). *Fuel Poverty and Policy in Ireland and European Union*. Dublin: Policy Institute, Trinity College Dublin.

Healy, J. D., & Clinch, J. P. (1999). Alleviating Fuel Poverty in Ireland: A Program for the 21st century. *International Journal for Housing Science*, 23(4), 203–215.

Hills, J. (2011). *Fuel Poverty: the Problem and Its Measurement. Interim Report of the Fuel Poverty Review*. London: London School of Economics and Political Science. Available at: https://sticerd.lse.ac.uk/dps/case/cr/CASEreport69.pdf

Hills, J. (2012). *Getting the measure of fuel poverty: final report of the Fuel Poverty Review. CASEreport (72)*. http://eprints.lse.ac.uk/id/eprint/43153

IEECP (2020) ENPOR – *Actions to Mitigate Energy Poverty in the Private Rented Sector*. Available at: http://www.ieecp.org/project/enpor-actions-to-mitigate-energy-poverty-in-the-private-rented-sector/.

Lacobucci, G. (2022). GPs Prescribe Heating to at-Risk Patients to Tackle Effects of Fuel Poverty. *BMJ*, o2835. https://doi.org/10.1136/bmj.o2835

Liddell, C., Morris, C., & Lagdon, S. (2011). *Kirklees Warm Zone: The Project and Its Health Impacts: A Cost-Benefit Analysis*. Coleraine: University of Ulster. https://pure.ulster.ac.uk/ws/portalfiles/portal/11261680/KIRKLEES_PROJECT_and_COST_BENEFIT_REPORT.pdf

Meyer, S., Laurence, H., Bart, D., Lucie, M., & Kevin, M. (2018). Capturing the Multifaceted Nature of Energy Poverty: Lessons from Belgium, *Energy Research and Social Science*, *40*(April), 273–283. https://doi.org/10.1016/j.erss.2018.01.017

Middlemiss, L. et al. (2018) 'Plugging the Gap Between Energy Policy and the Lived Experience of Energy Poverty: Five Principles for a Multidisciplinary Approach', in Foulds, C. and Robison, R. (eds) *Advancing Energy Policy*. Cham: Palgrave Pivot, pp. 15–29. https://doi.org/10.1007/978-3-319-99097-2_2

Mohan, G. (2021). Young, Poor, and Sick: The Public Health Threat of Energy Poverty for Children in Ireland. *Energy Research & Social Science*, *71*, 101822. https://doi.org/10.1016/j.erss.2020.101822

Morrison, C., & Shortt, N. (2008). Fuel Poverty in Scotland: Refining Spatial Resolution in the Scottish Fuel Poverty Indicator Using a GIS-Based Multiple Risk index, *Health & Place*, *14*(4), 702–717. https://doi.org/10.1016/j.healthplace.2007.11.003

Mullally, G., Dunphy, N., & O'Connor, P. (2018. Participative Environmental Policy Integration in the Irish Energy sector, *Environmental Science & Policy*. Elsevier, 83, 71–78. https://doi.org/10.1016/j.envsci.2018.02.007

Nussbaumer, P., Bazilian, M., Modi, V., & Yumkella, K. K. (2011). *Measuring Energy Poverty: Focusing on What Matters* (No. 42; Oxford Poverty & Human Development Initiative (OPHI) Working Papers). https://www.ophi.org.uk/wp-content/uploads/OPHI_WP_42_Measuring_Energy_Poverty1.pdf

Pachauri, S., & Spreng, D. (2011. Measuring and Monitoring Energy poverty, *Energy Policy*, 39(12), 7497–7504. https://doi.org/10.1016/j.enpol.2011.07.008

Palmer, G., MacInnes, T., & Kenway, P. (2008). *Cold and Poor: An Analysis of The Link between Fuel Poverty and Low Income*. London: New Policy Institute.

Pye, S., Dobbins, A., Baffert, C., Brajković, J., Grgurev, I., De Miglio, R, & Deane, P. (2015) *Energy poverty and vulnerable consumers in the energy sector across the EU: analysis of policies and measures*. INSIGHT_E, https://www.insightenergy.org/static_pages/publications#?publication=15

Rademaekers, K., Yearwood, J., Ferreira, A., Pye, S., Hamilton, I., Agnolucci, P., Grover, D., Karásek, J, & Anisimova, N. (2014). *Selecting Indicators to Measure Energy Poverty. A Report for the European Commission ENER/B3/2015-507*. Rotterdam: Trinomics.

Sánchez-Guevara Sánchez, C., Sanz Fernández, A., Núñez Peiró, M., & Gómez Muñoz, G. (2020). Energy Poverty in Madrid: Data Exploitation at the City and District Level. *Energy Policy*, 144, 111653. https://doi.org/10.1016/j.enpol.2020.111653

Sareen, S., Thomson, H., Tirado Herrero, S., Gouveia, J. P., Lippert, I., & Lis, A. (2020). European Energy Poverty Metrics: Scales, Prospects and Limits. *Global Transitions*, *2*, 26–36. https://doi.org/10.1016/j.glt.2020.01.003

Scarpellini, S., Sanz Hernández, M. A., Llera-Sastresa, E., Aranda, J. A., & López Rodríguez, M. E. (2017). The Mediating Role of Social Workers in the Implementation of Regional Policies Targeting Energy Poverty. *Energy Policy*, *106*, 367–375. https://doi.org/10.1016/j.enpol.2017.03.068

SEAI. (2018). *Energy in the Residential Sector – 2018 Report*. Dublin: Sustainable Energy Authority of Ireland.

Sokołowski, J. et al. (2020) 'A Multidimensional index to Measure Energy Poverty: the Polish case', *Energy Sources, Part B: Economics, Planning and Policy*. Taylor & Francis, 15(2), pp. 92–112. https://doi.org/10.1080/15567249.2020.1742817

Spiliotis, E., Arsenopoulos, A., Kanellou, E., Psarras, J., & Kontogiorgos, P. (2020). A Multi-Sourced Data Based Framework for Assisting Utilities Identify Energy Poor Households: a Case-Study in Greece, *Energy Sources, Part B: Economics, Planning, and Policy*, *15*(2), 49–71. https://doi.org/10.1080/15567249.2020.1739783

Thema, J., & Vondung, F. (2020). *EPOV Indicator Dashboard. Methodology Guidebook*. Wuppertal: Wuppertal Institut für Klima, Umwelt, Energie GmbH. Available at: www.wupperinst.org

Thomson, H., & Bouzarovski, S. (2018) *Addressing Energy Poverty in the European Union: State of Play and Action*. Brussels: EU Energy Poverty Observatory. Available at: www.energypoverty.eu/sites/default/files/downloads/publications/18-08/paneureport2018_final_v3.pdf

Thomson, H., Bouzarovski, S., & Snell, C. (2017). Rethinking the Measurement of Energy Poverty in Europe: A Critical Analysis of Indicators and Data, *Indoor and Built Environment*, *26*(7), 879–901. https://doi.org/10.1177/1420326X17699260

Thomson, H., & Snell, C. (2013). Quantifying the Prevalence of Fuel Poverty Across the European Union. *Energy Policy*, *52*, 563–572.

Thomson, H., & Snell, C. (2014). *Fuel Poverty Measurement in Europe: a Pilot Study*. York: University of York.

Thomson, H., Snell, C., & Liddell, C. (2016. Fuel Poverty in the European Union: a Concept in Need of Definition? *People, Place and Policy*, 10(1), 5–24. https://doi.org/10.3351/ppp.0010.0001.0002

Waddams Price, C., Brazier, K., Pham, K., Mathieu, L., & Wang, W. (2007). *Identifying Fuel Poverty Using Objective and Subjective Measures. Centre for Competition Policy. Working Paper 07–11*. Norwich: University of East Anglia.

Waddams Price, C., Brazier, K., & Wang, W. (2012) 'Objective and Subjective Measures of Fuel poverty', *Energy Policy*, Elsevier, 49, pp. 33–39. https://doi.org/10.1016/j.enpol.2011.11.095.

Walker, R., McKenzie, P., Liddell, C., & Morris, C. (2012). Area-Based Targeting of Fuel Poverty in Northern Ireland: An Evidenced-Based Approach. *Applied Geography*, *34*, 639–649. https://doi.org/10.1016/j.apgeog.2012.04.002

Yip, A. O., Mah, D. N., & Barber, L. B. (2020). Revealing Hidden Energy Poverty in Hong Kong: a Multi-Dimensional Framework for Examining and Understanding Energy Poverty, *Local Environment*, *25*(7), 473–491. https://doi.org/10.1080/13549839.2020.1778661

3 How Do We Measure Energy Poverty? The Limitations of Energy Expenditure as an Indicator

María María Ibáñez Martín, María Florencia Zabaloy, and Federico Dubois

3.1 Introduction

The discourse around energy poverty, like economic poverty, has developed from one-dimensionality to a multidimensional perspective. The most recent advances recognise that energy poverty refers not only to access and affordability of energy services but also to the quality of energy and the equipment available to satisfy different energy uses. In this way, Ibáñez Martín et al. (2019) define energy poverty as the deprivation of essential energy services for human life, arising from a lack of access, quantity, and quality not only of energy but also of appliances. This deprivation is caused by various factors (socioeconomic, geographic, buildings, cultural and so on), which ultimately affect the level of well-being of household members. Also, the most severe state of energy deprivation, called energy indigence, refers to the lack of access to energy services (*ibid.*).

Despite recognising the multidimensionality, most of the antecedents that empirically address the subject do so from a single indicator. The studies commonly resort to the indicator proposed by Boardman (1991), under which a household is energy poor if it allocates more than 10% of its income to energy expenditure. Without considering spending as an input, there are also the indicators proposed by Nussbaumer et al. (2011), which focus on the deprivation of access to modern energy services. Alternative approaches have been proposed since then. For instance, Healy (2004) measures fuel poverty with an index that considers three objective and three subjective indicators. The objective indicators address housing conditions and equipment, whilst the subjective ones capture if people experience any kind of deprivation in their energy needs. Dehays and Schuschny (2018) propose indicators that seek to assess the degree of social inequality with respect to household access to and use of energy. García Ochoa (2014) applies the Method of Satisfaction of Absolute Energy Needs. Under this approach, energy poverty occurs when a household does not have all the assets (appliances or equipment) considered essential to satisfy the needs of household members. There are several studies that apply this method in the literature (Alfaro, 2017; Schimer et al., 2023; Velásquez, 2023). There are also other advances in terms of multidimensional indicators to measure energy poverty, both for Argentina (Durán & Condori, 2016) and the rest of the world (Bollino & Botti, 2017; Hernández et al., 2018; Mendoza et al., 2019; Sokołowski et al., 2020).

The objective of this chapter is to carry out an application of the various indicators proposed by the literature for the measurement of energy poverty in Argentina and to show the dissimilar results that can be reached depending on the indicator used. On the other hand, a discussion regarding the usefulness of the self-reported energy expenditure indicator will be presented. The results obtained from the energy expenditure declared in the National Household Expenditure Survey 2017–2018, conducted by the National Institute of Statistics and Censuses (INDEC

DOI: 10.4324/9781003408536-5

to use its acronym in Spanish), will be compared to the expenditure calculated from a model, following Dubois (2022), to estimate consumption per household based on its equipment and declaration of use of this survey. The main objective of this analysis is to expose that empirically approaching a multidimensional phenomenon with a single indicator based on energy expenditure is an extreme simplification of a central and complex problem in the present world.

3.2 Energy poverty: a multidimensional phenomenon?

Energy poverty is a phenomenon that has gained attention in academic discussions since the 1980s. However, its definition is still under debate (Romero et al., 2018). The first academic contributions on energy poverty adopt a subsistence approach and take the definition of poverty proposed by Rowntree (1901), for whom a household is deprivedwhen its members can't meet basic satisfiers to maintain a certain level of physical efficiency with their income. Under this perspective, also adopted by the European Economic and Social Council (EESC), fuel poverty is defined by the estimation of a poverty line using a temperature threshold that guarantees thermal comfort and the percentage of income allocated to fuel costs required to reach said level (García Ochoa, 2014). Under this notion, the most widely used indicator places households as energy poor if they spend more than 10% of their income to satisfy energy services (Boardman, 1991; Foster et al., 2000).

Some authors (Healy & Clinch, 2002; Whykey & Callender, 1997) criticise the subsistence focus on fuel poverty. Their criticism is based on the methodological difficulty of obtaining precise data on the comfort temperature in the dwellings, the occupation time of the people who inhabit them, and the sources of household income. Therefore, broader definitions of energy poverty arise, considering the lack of access to modern and clean sources of energy (UNDP, 2005). Modi et al. (2005) define as energy poor as a household that cannot access the basic energy services necessary to combat poverty. From this perspective, energy and energy services are considered fundamental inputs for the development of personal capacities.

In line with the discussions on economic poverty, the debate around the definition of energy poverty transitioned from a one-dimensional to a multidimensional perspective. Under the last approach, energy poverty is evaluated based on the incidence that this state of deprivation has on people's development. Then, the conceptions advanced towards multidimensional definitions (García Ochoa, 2014). Along these lines, the European Observatory on Energy Poverty considers a household to be in a situation of energy poverty if it experiences inadequate levels of energy services due to a combination of high energy expenditure, low income, inefficient buildings and appliances, and specific household energy needs. Rojas and Ibáñez Martín (2016, p. 123) define access to energy as "the scope to clean, reliable and affordable energy services for cooking and for heating, lighting, communications and productive uses. The lack of access is recognized as energy poverty and the people who suffer from it belong to the so-called vulnerable sectors". Castelao Caruana et al. (2019) agree that this problem is the result of a lack of access to modern and reliable energy services, conditioned by the connection, the quality of the service, and the type of technology available. At the same time, energy poverty is the product of the interaction between energy prices, the energy efficiency of dwellings –its infrastructure and equipment – and household income. In this sense, the definition of the phenomenon has advanced towards positions that contemplate not only access and the relationship between energy expenditure and household income but also incorporate elements such as subjectivity, the temporality of satisfaction, and the various energy services and appliances.

3.3 How to measure energy poverty?

As mentioned in the previous section, the conceptual discussion of energy poverty has advanced towards multidimensionality; however, the empirical approach to energy poverty has prevailed mainly in one-dimensional measurements. This predominance finds various justifications, but mostly it is due to the lack of information sources that contemplate multiple aspects related to the satisfaction of energy services, affordability, disposition of appliances, building conditions, etc. Additionally, some definitions contemplate subjective dimensions to assess the state of the energy shortage, making it even more complex to have such information in the periodic surveys of developing countries.

It is worth noting that in recent years, measurements from a multidimensional perspective have increased (Abbas et al., 2022; Falak et al., 2014; Liu et al., 2022; Mendoza et al., 2019; Sokołowski et al., 2020; Wang & Lin, 2022). However, because multidimensional measurements are very demanding regarding the availability of information, their measurement is not usual in the economies of Latin American countries (with the exception of Chile). Based on a single indicator or dimension in developing countries, studies are mostly used to compare realities between economies (Calvo et al., 2021; Lampis et al., 2022; Rocha & Schuschny, 2018; Soares et al., 2023).

Within the studies that focus their efforts on measuring the phenomenon of energy poverty, there are a series of indicators that are mostly used. One of the most widely used indicators is the one that derives from the concept of fuel poverty, known in the literature as the "10% indicator" (Boardman, 1991). The calculation of the latter consists of classifying a household as poor if it allocates more than 10% of its total income to the payment of energy services (*ibid.*), considering any type of carrier/fuel to satisfy energy needs (e.g., natural gas, bottled gas, firewood, coal, kerosene, waste, electricity, etc.). Using household energy expenditure in relation to their income is intended to capture their ability to pay to satisfy energy services (for cooking, heating, sanitary hot water, food refrigeration, and space cooling, among others). Behind this indicator, just like the measurements of economic poverty by income, it is implicit that income is a good measure to capture the well-being of the population. In the work of Calvo et al. (2021), in order to analyse the ability of households to achieve an energy expenditure that allows for satisfying the needs within the home, the representative expenditure of each country is estimated, weighting the average energy consumption with the respective energy prices for each year. In order to evaluate the economic pressure that this expense means on the home, it must be compared with the income of the population, considering quintiles. In other words, what is usually criticised for this indicator is its inability to capture the real-world perspective of each country. In countries, such as Argentina, where the energy used at the residential level is heavily subsidised, the 10% threshold may be too high and only capture very severe deprivation situations (Zabaloy et al., 2022).

The 10% indicator is also criticised for other aspects. On the one hand, permanent surveys usually collect information on spending declared by the household referent. This type of information usually has problems in its collection because, in many cases, the respondent is not the one who pays the household expenses, or, with the advance of electronic payments, in many cases there is no detailed monitoring of this type of expenditure (Cecchini et al., 2021). In homes that use firewood, coal, and kerosene, this type of expense is usually not linked to a bill and is made intermittently throughout a month (e.g., depending on the needs at the time, when the fuel runs out, when income is received, etc.). This type of behaviour can cause underestimations or overestimations of energy expenditure, which is one of the two main inputs for measuring energy poverty with this indicator. Although the indicator proposed by Boardman is based on the required energy expenditure and is therefore modelled and not based on statistics, many empirical applications build this indicator from household survey statistics. The other criticism is associated with the other input of this indicator: household income. The important survey problem presented by the income variable is well known. According to various studies, lower-income households tend to declare more income than they actually have, whilst higher-income households underreport their values. This type of bias has been extensively studied by specialists in vulnerability and poverty measurements (Bonelli et al., 2021; Gasparini et al., 2022).

Another strategy to measure energy poverty is to analyse the fuel that households use to satisfy their energy needs. In this sense, there are various studies that characterise as energy poor those households that burn highly polluting fuels for cooking, heating, and obtaining domestic hot water (Castelao Caruana et al., 2019; Ibáñez Martín et al., 2019; Lampis et al., 2022). Ibáñez Martín et al. (2022) analyse this situation of energy deprivation and reflect on its use as an indicator of energy poverty. The authors conclude that the use of traditional fuels to satisfy energy service needs indicates a deprivation of access to other less polluting (and generally less expensive) sources and, therefore, could be associated with the notion of energy indigence (*ibid.*).

The European Union's Energy Poverty Observatory (EPOV) proposed a set of indicators that are divided into primary and secondary. There are four primary indicators: two relating to subsistence – double the median (2M) and half the median (M/2) – and two consensual – delay in utility bills and inability to keep the dwelling adequately warm. The suggested subsistence indicators complement each other. The 2M indicator consists of capturing those households in which the energy expenditure-income ratio is more than double the national median. On the other hand, the M/2 indicator captures those who spend on energy less than half the national median. The latter could be innovative since it takes into account those households that may be saving energy in order to be able to pay for the service. However, there are several weaknesses to this indicator, which are pointed out in the next section. In the next section, an alternative absolute threshold metric that addresses some of the M/2 shortcomings, that is, the hidden energy poverty (HEP) indicator, is proposed. The EPOV's secondary indicators pursue the objective of explaining the origin of deprivation, that is, why it is possible for households to be in energy poverty. To do this, they consider the prices of the different types of energy used to generate electricity (e.g., petroleum derivatives, mineral coal, and solid biomass) and the final prices of electricity and natural gas paid by users. Aspects of the household are also considered, such as the number of people who live in it, its floor area, the presence of water leaks or humidity, the home's energy efficiency rating, and access to equipment. Finally, other macro indicators are considered, such as the economic poverty rate, deaths during the winter period, and population density.

Of the indicators proposed by EPOV, the most used in empirical application studies are 2M and M/2 (Emre & Sozen, 2022; Longo et al., 2020; Stevens et al., 2022; Thema & Vondung, 2020). The 2M indicator is also adjusted to capture the fact that the attributes and consumption of a household do not change proportionally but rather depend on the size of the household. This modification gives rise to the relative 2M (2Mr). Its diffusion is mainly linked to the ease of its calculation, the potential for making comparisons and the low demand for information that it possesses. On the contrary, the secondary indicators do not find a high application, at least for countries that do not belong to the European Union, due to the important requirement that they imply in terms of data collection (Martín-Consuegra et al., 2020).

Leaving aside the unidimensionality, indicators that try to capture the complexity of energy poverty have been developed. On this path are the "Satisfaction of Absolute Energy Needs" (NAEs, to use its acronym in Spanish) elaborated by García Ochoa (2014). García Ochoa (2014) develops the method of NAEs and includes the energy services of cooking and refrigerating food, heating water for personal hygiene, adequate lighting, and entertainment activities. This methodology uses economic assets as a measurement instrument to determine if a household is in energy poverty, considering that a household does not have the economic assets determined to satisfy the NAEs. The other approach is the "Multidimensional Energy Poverty Index" (MEPI) method, developed by Nussbaumer et al. (2011). The MEPI proposes to capture, both in quantity and quality, the access to energy services considered and offers a greater analysis of the elements that determine energy demand. Consequently, dimensions are defined for particular energy services relating to cooking, lighting, food preservation, entertainment, education, and telecommunications. Then, the methodology assigns dichotomous variables for each dimension that inquire about the possession of the necessary equipment, access to electricity, and access to safe energy sources for cooking. Finally, it assigns a weighting for each dimension, and the index is calculated. The MEPI measures incidence – number of households in PE; and intensity – how poor are the energy poor (Nussbaumer et al., 2011). As mentioned above, the application of these multidimensional perspectives has increased in recent years. The growth of multidimensional measurements of energy poverty is accompanied by an improvement in the collection of primary data in different countries. For instance, national surveys are beginning to collect information on energy issues, as is the case of the National Household Expenditure Survey 2017–2018 in Argentina.

3.4 Energy poverty in the Argentinian context: an application for the province of Buenos Aires

As mentioned in previous sections, energy poverty indicators based on energy expenditure are the most widespread. However, the conclusions that can be reached with its use can vary. The objective of this section is to carry out the application of these indicators for a specific case, an area of the province of Buenos Aires, and to highlight the disparity of results that can be reached based on the information used for its analysis. Additionally, it is discussed whether the 10% threshold is representative of the real-world situation in Argentina.

3.4.1 Data and sample

Given the objective to apply energy poverty indicators based on energy expenditure, the National Household Expenditure Survey (ENGHo, to use its acronym in Spanish) for the year 2017–2018 is used as the main source of information. The ENGHo is carried out by the National Institute of Statistics and Censuses (INDEC, to use its acronym in Spanish), with a variable frequency. The first available information base is from the period 1985–1986, and there are also the 1996–1997, 2004–2005, 2012–2013, and 2017–2018 editions. The National Household Expenditure Survey appraises urban centres whose population is equal to or greater than 2,000 inhabitants and reaches approximately 45,000 households throughout the country in order to represent the regional and socioeconomic diversity of Argentina.

INDEC, as coordinator of the National Statistical System (SEN), oversees the ENGHo, and the Provincial Statistics Offices are responsible for the survey in their respective jurisdictions. The ENGHo 2017–2018 edition incorporates – based on a technical cooperation agreement between the Ministry of Energy and INDEC in 2017 – a Special Energy Module (INDEC, 2022). This special module aimed to expand and update the equipment section of the ENGHO 2012–2013 and add (in addition to ownership) questions about the use of the equipment, its age, the energy efficiency label (if applicable), and other questions related to the use of energy. The aforementioned difference that ENGHo 2017–2018 has makes it a fundamental input for the development of energy policies oriented to the residential sector, and therefore it will be used as a source of information. Additionally, the "energy consumption" block allows for estimating the energy matrix of households.

In order to limit the volume of data and also reduce the climatic heterogeneity that the Argentine territory presents, a determining aspect in energy consumption (OLADE, 2017), the sample has been limited to those households that comprise part of the interior of the province of Buenos Aires (Figure 3.1). The ENGHo 2017–2018 contains a representative sample of 1,725 urban households in the considered region, whose representativeness reaches the value of 1,633,388 households. Within the considered subregion, there are two different bioclimatic zones according to the bioenvironmental characterisation specified in the IRAM 11603 standard. These are the warm temperate zone (III) and cold temperate zone (IV). A brief characterisation of each zone is explained below. Zone III, classified as warm temperate, has relatively hot summers and average temperatures between 20°C and 26°C, with average maximums greater than 30°C. Winter presents average temperature values of between 8°C and 12°C and minimum values that are rarely less than 0° C. This area is subdivided into two subzones: a and b, depending on the thermal amplitudes: subzone IIIa with thermal amplitudes greater than 14°C and subzone IIIb with thermal amplitudes less than 14°C. Zone IV, classified as cold temperate, is divided from Zone III by the isoline of 1,170 heating degree days (HDD18) and, as a lower limit with Zone V, by the isoline of 1,950 HDD18. Within it, summers are not harsh and have average maximums that are rarely higher than 30°C. Winters are cold, with average values between 4°C and 8°C, and the average minimums often reach values lower than 0°C. This zone is subdivided into four subzones (IVa, IVb, IVc, and IVd) by means of the thermal amplitude lines of 14°C and 18°C. In any case, within the considered area, there are only subzones IVc, transition, and IVd, maritime.

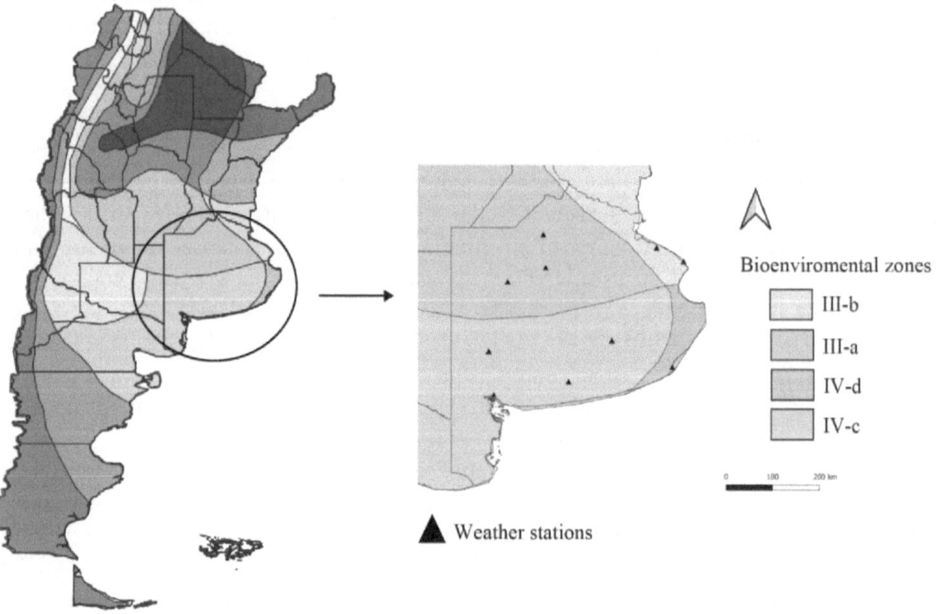

Figure 3.1 Geographic location and bioclimatic zones of the selected sample of households.

Source: Qwn elaboration.

3.4.2 Results

For this analysis, the indicators already mentioned are used: 10% and 2Mr (double the relative median). As developed, both indicators relate energy spending to household income. The 10% indicator considers in energy poverty the household that allocates more than this proportion of its income to pay for energy (Boardman, 1991). The 2Mr indicator determines that a household is energy poor if it has an "energy expenditure-income" ratio that exceeds twice the median of the "energy expenditure-income" ratio of the sample considered (EPOV).

For the estimation of the indicators, some methodological considerations were made, which is necessary given the Argentine context. Given that the survey is carried out for a full year and that, in addition, the increase in the Wage Index for said period was 28% (INDEC), income is deflated based on the quarter to which the declarations of each home correspond. For this, the last quarter (months September, October, and November) is considered the reference period, deflating the income of the declarations taken in the other quarters. Additionally, to estimate the 2Mr indicator, a "Modified Equivalence Scale" proposed by the OECD is used, which is based on the assumption (empirically validated) that certain economic variables of the household do not increase in the same proportion depending on the number of household members whilst also distinguishing between adult and minor members (Mancero, 2001). Thus, this equivalence expresses that both income and energy consumption, and consequently energy expenditure, do not increase in the same proportion according to the number of household members. The equivalence is applied using the following equation:

$$y = \frac{Y}{1 + 0{,}7\left(A - 1\right) + 0{,}5K} \tag{3.1}$$

Where y is the number of equivalent household members, Y is the number of members declared in the survey, A is the number of adults, and K is the number of children under the age of 14 years. Given the characteristics of the 2Mr indicator, using this equivalence allows for less distortion in the estimation of the median.

The indicators are calculated in two different ways: on the one hand, the information on income and expenses declared by the respondents; on the other hand, the estimated energy expenditure through the methodology proposed by Dubois (2022). The author proposes a model that allows for estimating energy consumption at the residential level for the selected sample. This model includes various energy services (cooking, food preservation, refrigeration, heating, lighting, domestic hot water production, laundry, and other uses), the equipment used to satisfy them (considering the use, the frequency of use, the number of hours of use, the level of efficiency of the devices, and the age), the number of rooms in the home, and the number of members who live there. With this information, the energy consumption of the home is obtained, which is then translated into energy expenditure. The second component – energy expenditure – is estimated using the electricity and natural gas tariff tables, in force for November 2018, and the Maximum Reference Price for the 10kg gas bottle, set by the National Secretariat of Energy. For electricity, the values of the EDEN distributor are taken, and for natural gas, those of the Camuzzi Gas Pampeana distributor. Likewise, the main taxes are considered: the value-added tax (IVA, for its acronym in Spanish) and the provincial taxes, which result in 32% for the electricity rate and 25% for the natural gas rate. A similar model of "required energy expenditure" (Barrella et al., 2022) is used to estimate the HEP indicator presented in Chapter 4 and the 2M indicator in the Spanish case study (Barrella et al., 2021).

The purpose of estimating energy poverty indicators considering two different sources of energy expenditure is to analyse whether self-declared expenditure is a good input to measure this phenomenon. It is worth clarifying that in Argentina, the energy market has a complex structure that differs between electricity and natural gas. This structure makes the estimation of spending through energy consumption (estimated by the Dubois (2022) methodology) laborious due to the diversity of taxes, tariffs, the application of energy subsidies, and other issues. Faced with this difficulty, it was decided to make the empirical application for the selected sample and not for the entire country.

When estimating both indicators using household income and the energy expenditure declared in the surveys as the main sources, it is found that 23.5% of the households in the sample spend more than 10% of their income on energy, which represents 375,600 families. This value is more severe if the 2Mr indicator is taken into account, indicating a third of the analysed population experiences energy deprivation. Although these results seem worrying, the situation of energy poverty in the selected regions is even worse when household energy expenditure is estimated based on their modelled energy consumption. In this case, the 10% indicator amounts to 28.3%, 4.8 percentage points more than in the estimate with declared energy expenditure (Figure 3.2). This change in the estimation methodology implies that 80,000 additional families experience energy poverty. In the case of the 2Mr estimator, the situation is similar: when contemplating the modelled energy expenditure, the indicator rises to 31.8%, distancing itself by almost two percentage points from the previous estimate (Figure 3.2).

The differences, beyond the result of the indicator, are also translated into the characteristics of households that are in energy poverty. When contemplating the results of the indicators estimated from the ENGho data, a certain parity is found in the gender of male and female heads of household who are in a situation of deprivation. However, when analysing the gender distribution (of the household leader) considering the modelled expenditure indicators, it is observed that energy poverty seems to have a higher incidence in households headed by women. Another disparity is observed when evaluating the size of the household. In the results obtained from the modelled energy expenditure, it is found that more than 45% of energy poor households have more than 5 members, whilst this value is reduced to 32% if they evaluate the indicators obtained with direct data from INDEC. The same trend is observed with regard to educational attainment and health coverage tenure; households with lower educational attainment and those without health insurance have a greater exposure to energy poverty.

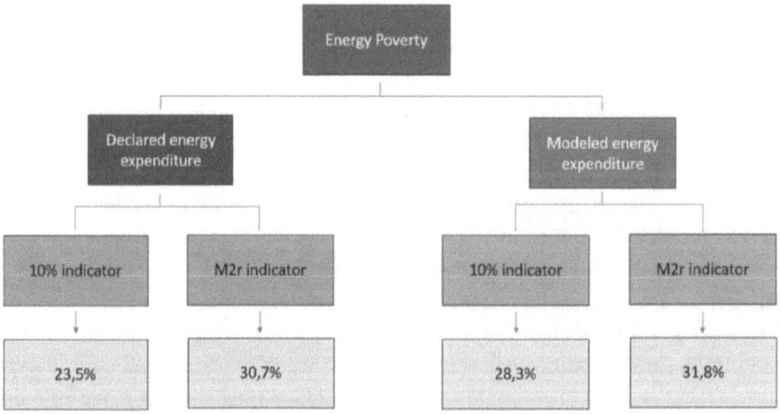

Figure 3.2 Energy poverty estimates for the selected sample – results.

Source: own elaboration based on own estimation.

These disparities are in line with studies that analyse the relationship between energy poverty and other multidimensional deprivations. In this sense, Ibáñez Martín et al. (2019) and Lampis et al. (2022) find that in Argentina, energy poverty can be promoted and generated by other deficiencies relevant to social life. In turn, these dissimilar results show that the way in which the estimators are calculated is not trivial and highlight the relevance of a good compilation of energy expenditure and income. Likewise, it allows us to visualise how a simple methodological change leads to a non-negligible measurement error.

Finally, as commented in previous sections, one of the great discussions around this type of indicator is "where to set the threshold". In Argentina, according to data from ENGho, on average households allocate 4.7% of their income to cover their energy expenses, whilst for the selected sample, this value is 5.4%. The difference can be explained by the variability of rates, subsidies, climates, and income that exist throughout the country. However, this value is well below the 10% threshold that the literature suggests for measuring energy poverty.

As previously mentioned, the calculations carried out up to now included subsidised energy expenditure, and only the subsidy known as the universal allowance per child was excluded. This methodological treatment is due to the fact that in Argentina, until 2022, a universal supply subsidy scheme was applied to natural gas and electricity services, so that the entire population connected to the distribution networks accessed this type of subsidy. In other words, the subsidy is not reflected in the energy bill, as the subsidy goes from the government to the companies at the wholesale level. Additionally, vulnerable families were favoured by an additional subsidy (known as Social Tariff)[1]. In Argentina, until mid-2022, a universal subsidy strategy for natural gas and electricity was applied, which in turn was complemented with targeted subsidies for vulnerable populations (which also affect bottled gas). Given this reality, a household in Argentina pays 45% of the total cost of natural gas and only 38% of the cost of electricity. Thus, energy expenditure in households in Argentina is very low due to the strong component of subsidies contained in energy[2]. Undoubtedly, given this situation, using the 10% indicator (which is the most widely used in research on the subject) underestimates a problem that affects a high percentage of the population.

3.5 Concluding remarks

The discussion about economic poverty has shifted from focusing on one-dimensionality to multidimensionality. The same has happened with the concept of energy poverty. Although energy poverty is defined from a multidimensional perspective, in practise, it is measured from a one-dimensional approach. Energy expenditure has gained prominence in energy poverty measurements; thus various indicators use it as the main input for measurement. Globally, there are studies that use the indicators of 10% and 2M proposed, respectively, by Boardman and EPOV. The 2Mr indicator, based on OECD suggestions, is calculated by modifying the 2M indicator in order to better capture the dynamics of household energy expenditure based on its composition and specificities.

The biggest criticism of one-dimensional indicators to measure energy poverty is their extreme simplification of a reality that is extremely complex and delicate. This debate is similar to the one in the literature regarding the measurement of economic poverty from indicators based only on income.

This chapter focused on the discussion around two indicators: 10% and 2Mr. From the application of these two indicators, the objective is to demonstrate that the way in which they are calculated and the data that feed their calculation can lead to dissimilar results. In this sense, it is found that the number of households that are captured in a state of deprivation is substantially different when the indicators are calculated from the energy expenditure declared by households versus the estimated expenditure from an energy demand model. In this sense, the modelling of the demand considering the hours of use, the equipment, and the essential energy services could incorporate (although in a reduced way) the intrinsic multidimensionality of energy poverty. This difference finds coincidence with evidence and papers that postulate that households do not always know their energy expenditure and that the permanent surveys on household expenditure have problems collecting these aspects.

Regarding the 10% indicator, one of the discussions that emerge from the analysis carried out is whether this value is really a valid threshold for the Argentine reality and the selected sample. In Argentina, subsidies have a strong impact on energy spending since until mid-2022, more than 50% of the national energy cost was covered by the National Treasury. Argentina began to change its energy subsidy policy. Indeed, it has begun to move from universal to targeted electricity and natural gas subsidies to reduce the burden that these represent on the public budget. The results of this work show that on average, the households in the interior of the province of Buenos Aires spend 4.7% of their income on energy bills. This finding allows us to question whether the 10% threshold is adequate to measure energy poverty in Argentina. Since universal energy subsidies are applied, another threshold could be considered in Argentina.

Finally, this application aims to highlight the need to complement the one-dimensional measurements of energy poverty. Reducing the measurement of the phenomenon to an indicator based on energy expenditure and household income is a simplification that leaves out many relevant aspects of energy deprivation. Being able to move towards measurements that include equipment, energy efficiency, the dwelling's envelope, occupant habits, and the macroeconomic conditions of the territory becomes essential if a conscious measurement of the phenomenon is to be sought. Additionally, this application has shown that the methodological decisions regarding the measurement of the variables used substantially change the characterisation of the energy poor. This aspect is important since when deprivations are analysed, it is not only important to know who is deprived, but also what other deprivations those groups suffer. Understanding these issues contributes to designing better public policies. Likewise, energy poverty can be reduced not only through energy policies but also through other economic policies that promote better education, health, and housing, among others.

Notes

1 A more exhaustive description of the tariff situation in Argentina can be found in Dubois (2022).
2 The subsidy is applied to the wholesale price because it is a supply side subsidy; that is to say, the government applies it directly to the supply. Therefore, final users do not visualise the subsidy in the energy bill. For example, in the case of electricity, the subsidy is in the difference between the monomic price (real cost of production) and the seasonal price (the one charged to the end user).

References

Abbas, K., Butt, K. M., Xu, D., Ali, M., Baz, K., Kharl, S. H., & Ahmed, M. (2022). Measurements and Determinants of Extreme Multidimensional Energy Poverty Using Machine Learning. *Energy*, 251, 123977.

Alfaro, Lara, S. (2017). Método satisfacción de necesidades absolutas de energía para un acercamiento al concepto de pobreza energética. *Avances en energías Renovables y Medio Ambiente*, 21. Disponible en: http://sedici.unlp.edu.ar/bitstream/handle/10915/140620/Documento_completo.pdf-PDFA.pdf?sequence=1

Barrella, R., Linares, J. I., Romero, J. C., Arenas, E., & Centeno, E. (2021). Does Cash Money Solve Energy Poverty? Assessing the Impact of Household Heating Allowances in Spain. *Energy Research & Social Science*, 80, 102216.

Barrella, R., Romero, J. C., Linares, J. I., Arenas, E., Asín, M., & Centeno, E. (2022). The Dark Side of Energy Poverty: Who Is Underconsuming in Spain and Why? *Energy Research & Social Science*, 86, 102428.

Boardman, B. (1991). *Fuel Poverty: from Cold Homes to Affordable Warmth*. Michigan: Pinter Pub Limited.

Bollino, C. A., & Botti, F. (2017). Energy Poverty in Europe: A Multidimensional Approach. *PSL Quarterly Review*, 70(283), 473–507.

Bonelli, Maldovan, Goren, J., & Corradi, N. (2021). ¿Nuevos problemas o profundización de desigualdades prexistentes?. *Trabajo y Sociedad*, 22(36), 55–79.

Calvo, R., Álamos, N., Billi, M., Urquiza, A., & Contreras, R. (2021). *desarrollo de indicadores de pobreza energética en América Latina y el Caribe*, serie Recursos Naturales y Desarrollo, N° 207 (LC/TS.2021/104), Santiago: Comisión Económica para América Latina y el Caribe (CEPAL).

Castelao Caruana, M. E., Méndez, F. M., Rosa, P. C., & Wild, G. (2019). Aportes para la medición de la Pobreza Energética: diagnóstico y propuestas para la intervención desde una Cooperativa de la Provincia de Santa Fe. *Revista de Ciencias Sociales*, 10(35), 45–62. Disponible en: http://ridaa.unq.edu.ar/handle/20.500.11807/1719

Cecchini, S., Villatoro, P., & Mancero, X. (2021). *El impacto de las transferencias monetarias no contributivas sobre la pobreza en América latina. Revista Cepal*, 134, 7–32.

Dehays, J., & Schuschny, A. (2018). Una propuesta de indicadores para medir la pobreza energética en América latina y el Caribe. *ENERLAC. Revista de energía de Latinoamérica y el Caribe*, 2(2), 106–124.

Dubois, F. (2022). *Pobreza energética en Argentina: el potencial de la Eficiencia energética como vector de mitigación*. Universidad Nacional de San Martín. Escuela de Ciencia y Tecnología.

Durán, R. J., & Condori, M. A. (2016). Índice multidimensional de pobreza energética para Argentina: su definición, evaluación y resultados al nivel de departamentos para el año 2010.

Emre, T., & Sozen, A. (2022). Energy Poverty Clustering by Using Power-Cut Job Order Data of the Electricity Distribution Companies. *International Journal of Energy Economics and Policy*, 12(3), 401–409.

Falak, S. H. E. R., Abbas, A., & Awan, R. U. (2014). An Investigation of Multidimensional Energy Poverty in Pakistan: A Province Level Analysis. *International Journal of Energy Economics and Policy*, 4(1), 65–75.

Foster, V., Tre, J. P., & Wodon, Q. (2000). Energy Prices, Energy Efficiency, and Fuel Poverty. *Latin America and Caribbean Regional Studies Programme*. Washington, DC: World Bank, vol. 131(42), pp. 1–7.

Gasparini, L. C., Gluzmann, P. A., & Tornarolli, L. (2022). Caracterización de la población vulnerable: una propuesta con estimaciones para Argentina. *Económica*, 68, 135–157.

Healy, J. D. (2004). *Housing, Fuel Poverty and Health: A Pan-European Analysis*. Londres: Ashgate.

Healy, J. D., & Clinch, J. P. (2002). Fuel Poverty in Europe: A Cross Country Analysis Using a New Composite Measurement. Environmental Studies Research Series Working Papers.

Hernández, M. F., Aguado, L. F., & Duque, H. (2018). Índice de pobreza energética multidimensional por regiones para Colombia, ipem_rc 2013. *Economía Coyuntural*, 3(3), 35–72.

Ibáñez Martín, M. M., Zabaloy, M. F., & Guzowski, C. (2022). Privación y pobreza energética: cocción limpia en Argentina. *SaberEs*, 14(1), 53–84.

Ibáñez Martín, M.M., Zabaloy, M.F., & Guzowski, C. (2019). Una Primera Exploración de la Situación de Pobreza Energética en Argentina: ¿Es la Pobreza Energética un Fenómeno Independiente de las Privaciones Multidimensionales? Anales de la Reunión Anual de la Asociación Argentina de Economía Política. Bahía Blanca, 2019. Available in: https://aaep.org.ar/anales/works/works2019/iban%CC%83ez.pdf

INDEC (2022). Innovaciones en el relevamiento de la Encuesta de Gasto de los Hogares. Disponible en: https://www.indec.gob.ar/engho/innovaciones.html

Lampis, A., Ibáñez Martín, M. M. I., Zabaloy, M. F., Soares, R. S., Guzowski, C., Mandai, S. S., & Bermann, C. (2022). Energy Transition or Energy Diversification? Critical Thoughts from Argentina and Brazil. *Energy Policy*, 171, 113246.

Liu, J., Jain, V., Sharma, P., Ali, S. A., Shabbir, M. S., & Ramos-Meza, C. S. (2022). The Role of Sustainable Development Goals to Eradicate the Multidimensional Energy Poverty and Improve Social Wellbeing's. *Energy Strategy Reviews*, 42, 100885.

Longo, D., Olivieri, G., Roversi, R., Turci, G., & Turillazzi, B. (2020). Energy Poverty and Protection of Vulnerable Consumers. *Overview of the EU Funding Programs FP7 and H2020 and Future Trends in Horizon Europe. Energies*, 13(5), 1030.

Mancero, X. (2001). *Escalas de equivalencia: reseña de conceptos y métodos*. Santiago de Chile: CEPAL.

Martín-Consuegra, F., Giménez, J. G., Alonso, C., Hernández, R. C., Aja, A. H., & Oteiza, I. (2020). Multidimensional index of Fuel Poverty in Deprived Neighbourhoods. Case Study of Madrid. *Energy and Buildings*, 224, 110205.

Mendoza, C. B. Jr, Cayonte, D. D. D., Leabres, M. S., & Manaligod, L. R. A. (2019). Understanding Multidimensional Energy Poverty in the Philippines. *Energy Policy*, 133, 110886. https://doi.org/10.1016/j.enpol.2019.110886

Modi, V., Mcdade, S., Lallement, D., & Saghir, J. (2005). *Energy and the Millennium Development Goals*. New York: Energy Sector Management Assistance Programme.

Nussbaumer, P., Morgan, B., Vijay, M., & Kandeh, Y. (2011). *Measuring Energy Poverty: Focusing on What Matters*. Oxford: University of Oxford. Oxford Poverty & Human Development Initiative (OPHI).

García Ochoa, R. (2014). Pobreza energética en América Latina. *Serie Documentos de Proyecto Cepal*, Santiago de Chile: Cepal-Naciones Unidas. 2014.

Organización Latinoamericana de Energía (OLADE). (2017). Manual de Balances de Energía Útil. Quito.

Rocha, J. D., & Schuschny, A. (2018). *Una propuesta de indicadores para medir la pobreza energética en América latina y el Caribe*. Quito: Organización Latinoamericana de Energía (OLADE).

Rojas, M., & Ibáñez Martín, M. M. (2016) "Planeamiento y gobernanza en el desarrollo de las energías renovables para la inclusión social". In Guzowski, Ibáñez Martín y Rojas (Eds.) "Los Desafíos de la política energética argentina". Ed. Dunken: Buenos Aires. ISBN 9789870291169

Romero, J. C., Linares, P., & López, X. (2018). The Policy Implications of Energy Poverty Indicators. *Energy Policy*, 115, 98–108.

Rowntree, S. (1901). *Poverty: A Study of Town Life*. Londres: Macmillan.

Schimer, R. S., Weiss, M., Lampis, A., Bermann, C., & Hallack, M. (2023) Pobreza energética en los hogares y su relación con otras vulnerabilidades en América Latina: el caso de Argentina, Brasil, Colombia, Perú y Uruguay. BID. NOTA TÉCNICA N° IDB-TN-02623. http://dx.doi.org/10.18235/0004702

Soares, R. S., Weiss, M., Lampis, A., Bermann, C., & Hallack, M. (2023). Pobreza energética en los hogares y su relación con otras vulnerabilidades en América Latina.

Sokołowski, J., Lewandowski, P., Kiełczewska, A., & Bouzarovski, S. (2020). A Multidimensional index to Measure Energy Poverty: the Polish Case. *Energy Sources, Part B: Economics, Planning, and Policy*, 15(2), 92–112.

Stevens, M., Raat, H., Ferrando, M., Vallina, B., Lucas, R., Middlemiss, L., & van Grieken, A. (2022). A Comprehensive Urban Programme to Reduce Energy Poverty and Its Effects on Health and Wellbeing of Citizens in Six European Countries: Study Protocol of a Controlled Trial. *BMC Public Health*, 22(1), 1578.

Thema, J., & Vondung, F. (2020). *EPOV Indicator Dashboard: Methodology Guidebook*. Wuppertal Institut für Klima, Umwelt, Energie GmbH.

United Nations Development Program. (2005). *Energizing the Millennium Development Goals*. New York: UNDP.

Velásquez, E. B. (2023) Pobreza energética en Colombia: un análisis departamental. En García Ochoa (coord.) Pobreza energética Visiones de América Latina. Tijuana: el Colegio de la Frontera Norte; Zamora, Michoacán: el Colegio de Michoacán. Disponible en: https://libreria.colef.mx/detalle.aspx?id=7919

Wang, Y., & Lin, B. (2022). Can Energy Poverty Be Alleviated by Targeting the Low Income? Constructing a Multidimensional Energy Poverty index in China. *Applied Energy*, 321, 119374.

Whykey, C., & Callender, C. (1997). *Fuel Poverty in Europe: Evidence from the European Household Panel Survey*. London: Policy Studies Institute.

Zabaloy, M. F., Ibáñez Martín, M. M., & Martinez, M. M. C. (2022). Una primera exploración de la situación de pobreza energética en Argentina: ¿Es la pobreza energética un fenómeno independiente de las privaciones multidimensionales de la población? LVII Reunión Anual de la Asociación Argentina de Economía Política. Noviembre, Córdoba. Disponible en: https://bd.aaep.org.ar/anales/buscador.php

4 Unveiling Hidden Energy Poverty in a Time of Crisis

A Methodological Approach for National Statistics

Roberto Barrella and José Carlos Romero

4.1 Introduction

The academic study of energy poverty dates back to the work of Brenda Boardman in the early 1990s (Boardman, 1991). In her pioneering studies, the definition of this issue identified as energy-poor households those that spent more than 10% of their income on required energy costs. That first step consolidated an approach to energy poverty in which this social problem was identified with excessive pressure of energy bills on the household budget. This initially identified issue was later recognised as a dimension of energy poverty due to disproportionate spending (Lowans et al., 2021).

However, over the years, it has become clear that there is another key dimension that is just as important or even more important than the previous one. Due to various circumstances linked to different vulnerabilities, many households extremely reduce energy expenses from their basket of basic needs (Anderson et al., 2012; Stojilovska et al., 2021). Of course, if they were to meet them, they would incur disproportionate expenses, but the reality is that they are pushed to decrease them to afford other basic needs. Such households 'choose' to deprive themselves of these energy services, further exacerbating their precarious living situation. This reality was called 'hidden energy poverty' (Meyer et al., 2018).

This is a more elusive dimension of energy poverty: how can we measure whether a household is underconsuming? It should be remembered that the main statistics used to calculate objective indicators come from the Household Budget Survey. This reflects the different consumption of goods and services by households. Thus, using these data, it is easy to identify whether a household is overstressing its economy because of energy bills. However, it is by no means easy to detect whether it is underconsuming.

The only EU official indicator to date – proposed in the Belgian barometer (Meyer et al., 2018) and introduced in both EPOV and EPAH dashboards – that attempts to measure this reality is the M/2. According to this indicator, a household is in energy poverty if it spends less than half the national median on energy expenditure. It is easy to understand the limitations of the indicator: how can we guarantee that this threshold is appropriate? Moreover, using a relative benchmark to identify a household in energy poverty is problematic and can be more a measure of inequality than poverty. Can we admit that what happens in an average household is the correct benchmark for others? We think not. A correct measurement of hidden energy poverty cannot take as a reference a value relative to the whole population. On the contrary, what is needed is an absolute reference. According to this, a household would be in hidden energy poverty if it spent less than it should to cover its energy costs. That 'should' can be defined based on a model that calculates the required energy needs of each household and later uses bills' prices to obtain its expenditure. In this chapter, we propose a set of energy poverty indicators that could be

DOI: 10.4324/9781003408536-6

applied at the national and regional levels, which puts particular emphasis on the hidden energy poverty dimension without neglecting the other main ones identified in the literature (Meyer et al., 2018), that is disproportionate expenditure, declared inadequate temperature and arrears on utility bills. This approach is then applied to the case study of an EU country: Spain.

The year 2021 got off to a frosty start within the Spanish Kingdom. *Filomena* snowstorm visited Spain during January of that year. The week of heavy rain and snowfall was followed by days in which people in many parts of Spain were forced to live with snow and ice, thus with exceptionally low temperatures. Luckily, the social shield introduced in 2020 was still in force, and by its mercy, the interruption of basic supplies to beneficiaries of the social tariffs was prohibited (a measure that would later be extended until the end of 2022). On the other hand, the second half of 2021 was characterised by exceptionally high energy prices. The take-off in prices began in June and reached unprecedented heights. In parallel and underlying, the increase in gas prices on the wholesale market began in the same summer months of 2021. In the case of electricity, the customers with a regulated tariff (PVPC is the Spanish acronym) suffered the most from the rise, as the price was directly indexed to the wholesale electricity market price. In the case of natural gas, the opposite occurred, with households on the regulated tariffs (TUR is the Spanish acronym) being the least affected, as prices are set quarterly by the Government and were well below free market prices.

Such was the pace of price increases in electricity and gas that the Spanish Government had to act. Firstly, Royal Decree Law (RDL) 12/2021, of June 24, introduced changes on taxation by reducing the value-added tax (VAT) and the generation tax, while RDL 17/2021, of September 14, acted on the special electricity tax, among other measures. All of this tarnished the proposed new electricity tariff that came into force on June 1, 2021.

From the point of view of vulnerable households, it is easy to imagine that the situation, far from improving, might have continued to worsen. Spain (like the rest of the world) was coming from the year of the pandemic, where, despite an exceptionally low energy price scenario, some energy poverty indicators, especially those of late payments and inadequate temperature in the home, had worsened considerably (Romero et al., 2023). It was therefore foreseeable that the situation, especially from the second half of the year onwards, would become tenser. This is precisely what we propose to analyse in this chapter, with particular emphasis, as highlighted above, on the hidden poverty dimension. To do so, we first propose a methodological approach to analysing the multiple dimensions of energy poverty through different metrics (Section 2). Secondly, we present the main results obtained for Spain (Section 3), starting with the EPOV primary indicators (also included in the indicators' review of EPAH; Gouveia et al., 2022) and continuing with two alternative indicators proposed, respectively, by Romero et al. (2018) and Barrella et al. (2022a). These are, on the one hand, the disproportionate expenditure indicator based on the Minimum Income Standard approach (MIS) and, on the other hand, the HEP indicator of Hidden Energy Poverty. The latter is analysed more deeply for its significance in objectively identifying people who are underconsuming because of the lack of energy affordability. We end the chapter with the main conclusions and lessons learned (Section 4).

4.2 Methodological approach to unveil energy poverty

As indicated in Section 1, the four primary indicators identified by the EPOV (and included in the new EPAH dashboard; EPAH, 2022) were selected as the first set of metrics of the proposed methodological approach:

- High share of energy expenditure in income (2M).
- Low absolute energy expenditure (M/2).

- Inability to keep the home adequately warm (Inadequate temperature).
- Arrears on utility bills.

The first two indicators can be classified as objective or income-expenditure-based indicators. In particular, the 2M attempts to detect the disproportionate expenditure dimension of energy poverty, while the M/2 detects the underspending one. On the other hand, the 'Inability to keep home adequately warm' and 'Arrears on utility bills' are subjective or consensual indicators. These indicators are also used by the Spanish Government to annually monitor energy poverty at the national and regional levels.

Regarding the 'alternative metrics', The MIS-based indicator is a disproportionate expenditure indicator that uses a different methodology than the 2M presented above. While the latter identifies a household as 'in energy poverty' when it spends more than twice the national median on energy, the MIS-based indicator sets the threshold of disproportionate expenditure at a basket of basic needs. This metric was proposed in the United Kingdom (Moore, 2012) (based on an investigation on the definition of the basket of basic needs in this country; Bradshaw et al., 2008) and has not been widely disseminated outside its borders. However, we consider it to be a very appropriate indicator of disproportionate expenditure (Barrella et al., 2022b; Romero et al., 2018) precisely because of the type of absolute threshold it uses. According to Romero et al.'s methodology, this threshold includes all non-energy household expenses to provide for their basic needs.

The MIS indicator considers 'in energy poverty' those households that, after subtracting the standard minimum income from their real income, do not have sufficient resources to cover their energy bill. The used formula is the one proposed by Romero et al. (2018). Regarding the absolute income threshold, Spain does not have a specific study for the quantification of equivalent MIS as other countries do, for example the UK, France and Portugal. In this chapter, a proxy for the equivalent MIS – income threshold – has been used, that is the minimum wage (SMI for its acronym in Spanish) set by the National Government, which is equal in all the regions. This wage reference was set at €900, €950 and €965 (in 12 payments), respectively, in 2019, 2020 and 2021. The last SMI value also corresponds to the maximum Minimum Insertion Income (RMI[1] is the Spanish acronym) available in 2021 for eligible households in the Community of Madrid.

On the other hand, the Hidden Energy Poverty (HEP) indicator seeks to complement the M/2 underspending indicator presented above by proposing an alternative absolute energy expenditure approach. The ultimate aim of this metric is to unpack the complex reality of hidden energy poverty, that is those households that consume less energy than necessary because they are financially unable to afford it. According to the selected HEP indicator – proposed by (Barrella et al. 2022a) – a household is in energy poverty if:

1 Its actual energy expenditure is less than half of its required energy expenditure (RENE/2).
2 The household belongs to one of the five lowest equivalised income deciles.

The first threshold is used to measure household energy underspending, and the second one is applied to filter households with low consumption due to affordability problems, that is households suffering from hidden energy poverty. The applied RENE model[2] was designed and validated for the Spanish case study (Barrella et al., 2022a). However, an adaptation of this model to other EU countries or the use of similar national models could make it possible to apply the HEP indicator to other European case studies. The former adaptation might be performed by taking into account the climate (Castaño-Rosa et al., 2021) and building stock (EU Building Stock Observatory (BSO), 2022) differences across Europe.

4.2.1 Data

The database used to calculate the objective indicators was the Spanish Household Budget Survey (HBS). This survey of the National Statistics Institute (INE) aims to obtain information on the nature and destination of consumption expenditure, as well as on various characteristics relating to the living conditions of households. It is carried out annually on a sample of approximately 24,000 households, which statistically represent all of the country's population.

On the other hand, the database used to obtain the subjective indicators was the Survey on Living Conditions (SILC). This is also an annual survey and is carried out on a sample of 15,000 dwellings distributed in 2,000 census sections throughout the national territory.

In this chapter, we have used the data from the HBS and SILC carried out in 2019, 2020 and 2021.

4.3 Energy poverty in Spain

4.3.1 EPOV primary indicators

Table 4.1 shows the results obtained for the four main EPOV indicators for 2019, 2020 and 2021 so that the evolution in the pre- and post-pandemic period can be seen.

First, we see that the 2M indicator of disproportionate expenditure decreases from 16.1% in 2020 to 15.2% in 2021, returning to 2019 values. In other words, just over 15% of Spanish households (almost 3 million) spent more than double the national median on energy in 2021, placing them in energy poverty according to this dimension of disproportionate expenditure. A relevant fact to understand this indicator is the value of the median household energy expenditure (M) itself, which, multiplied by two, constitutes the reference threshold. In 2020, this median was 4.5%, while in 2021 it stood at 4.4%. In other words, there is a slight decline. This may seem counterintuitive given the scenario of high energy prices that we outlined in the introduction. In our opinion, two phenomena explain this figure. On the one hand, it should not be forgotten that high energy prices started to escalate in the middle of the year, peaking just in December. This makes their impact on the survey much more moderate. On the other hand, although energy expenditure grew in 2021 compared to 2020, so did household income (as pointed out in Romero et al., 2022a), offsetting this growth to the point of marginally reducing both the median energy expenditure over income and the 2M indicator itself.

Secondly, the underspending indicator (M/2) was reduced from 11.2% to 10.1%. This is a relatively small but significant reduction: according to this indicator, in 2021, this energy poverty dimension affected around 200,000 fewer households than in 2020. What the M/2 tells us is that we detect a decrease in under-consuming households compared to the national median. This may be an indication that the social shield apparently continued to do its job of containing underspending. However, it is worth waiting for the rest of the results to see if this hypothesis is confirmed.

Table 4.1 EPOV primary indicators for Spain in 2019, 2020 and 2021

Indicator	*2019*	*2020*	*2021*
2M (households)	15.20%	16.10%	15.22%
M/2 (households)	11.50%	11.20%	10.12%
Arrears on utility bills (individuals)	6.60%	9.60%	9.50%
Inadequate temperature (individuals)	7.60%	10.90%	14.27%

Source: Own calculation from HBS and SILC databases.

Thirdly, the 'Arrears on utility bills' indicator, after a significant deterioration in the second studied year, remained almost constant in the last one: from 9.6% in 2020 to 9.5% in 2021. In other words, almost 10% of our country's citizens reported being late in paying their utility bills in 2021. This figure suggests two readings, one positive and the other somewhat less so. On the one hand, the indicator does not show a worsening of arrears during 2021, but at the same time, it should be noted that 10% of households with delayed bills is still a very worrying figure, especially when we note that the current situation does not suggest that the energy crisis will be resolved quickly and significantly in terms of a reduction in energy prices.

Finally, the 'Inadequate temperature' indicator had the most worrying development. It rose from 10.9% in 2020 to 14.3% in 2021. This metric denotes that more than 14% of the Spanish population stated that they do not have adequate thermal comfort conditions in the winter. Probably one of the causes of this increase has to do with the echoes of the *Filomena* storm. The extreme cold that settled over the peninsula at the beginning of January has no doubt engraved in our subconscious how fragile Spanish homes are to events of this nature. On the other hand, it can also be assumed that this indicator is showing what the official indicators based on income and expenditure, due to their limitations, do not manage to reflect. Namely, this metric could highlight that, in 2021, Spanish households continued and extended their dynamic of controlling energy expenditure even at the cost of maintaining an inadequate temperature inside their homes. To unpack this issue, Figure 4.1 shows the 2021 regional disaggregation of the 'Inadequate temperature' indicator. The map points out that the islands are the most affected by the subjective perception of inadequate temperature at home during winter, followed by southern regions like *Extremadura*, *Comunidad Valenciana* and *Región de Murcia*. Moreover, an autonomous city that performed as badly as the islands is Melilla, where the 18.91% of people could not adequately warm their home in 2021. In contrast, in the nearby city of Ceuta, the indicator was only 2.89%.

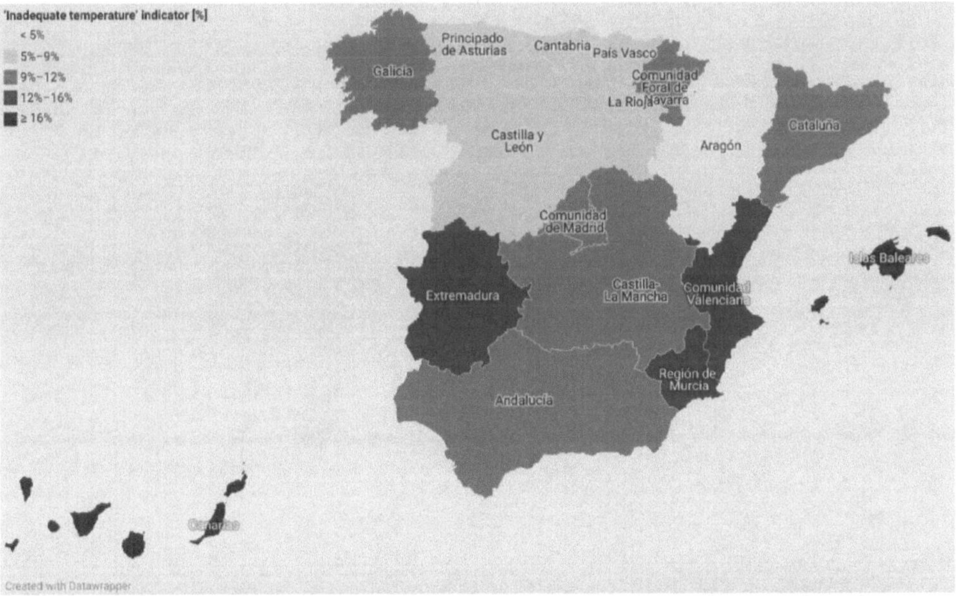

Figure 4.1 'Inadequate temperature' indicator by region for Spain in 2021 [% of population].

Source: Own elaboration from the calculation carried out using the SILC database.

Table 4.2 MIS-based indicator for Spain in 2019, 2020 and 2021

	2019	*2020*	*2021*
MIS_{SMI}	15.30%	16.60%	14.62%
SMI	€900	€950	€965

Source: Own calculation from HBS database and Ministry of Labour and Social Economy, https://www.mites.gob.es/estadisticas/bel/SMI/index.htm

4.3.2 Alternative indicators

Table 4.2 shows the results of the MIS-based indicator for 2019, 2020 and 2021. As with the 2M disproportionate spending indicator, the MIS also improved in 2021 compared to 2020, 14.6% compared to 16.6%. Again, we need to be cautious before celebrating a small improvement in these metrics. It is always positive that the incidence of some of the dimensions of energy poverty is decreasing. However, the extent of this issue remains very high. Therefore, we are still far from considering this problem of disproportionate spending under control in Spain.

Figure 4.2 shows the values obtained by the MIS indicator in the different Spanish regions. The MIS values in the cities of *Ceuta* and *Melilla* (not shown on the map) are, respectively, 20% and 18.5%. Figure 4.2 shows that the differences are very significant. The MIS ranges from 9% in the *País Vasco* to 22.7% in *Extremadura.* More generally, Northern Spain and the *Comunidad de Madrid* have a lower incidence of disproportionate spending in 2021, according to the MIS indicator. On the other hand, the Autonomous Communities with the highest incidence (*Extremadura, Canarias, Andalucía*) correspond to those with the lowest equivalent net income.

Table 4.3 shows the results obtained for the HEP indicator in 2019, 2020 and 2021. Before analysing the 2021 figure in comparison with the previous year, it is worth taking some time to

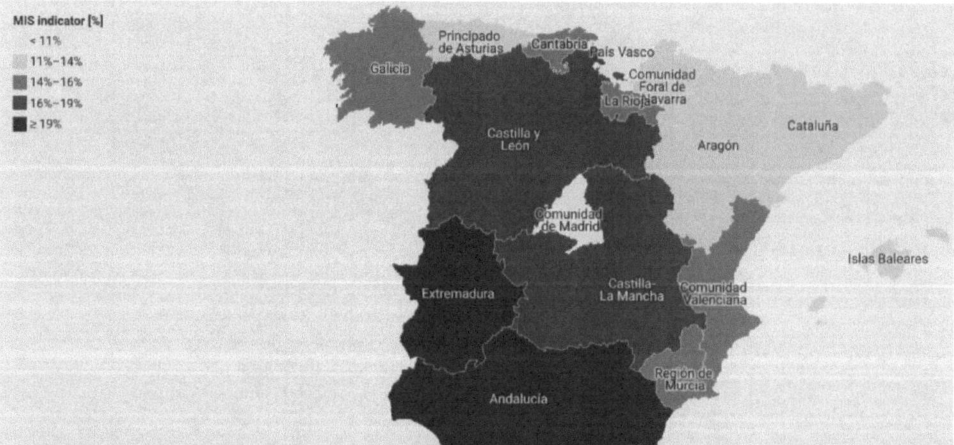

Figure 4.2 MIS indicator by region for Spain in 2021 [% of households].

Source: Own elaboration from the calculation carried out using the HBS database).

Table 4.3 HEP indicator for Spain in 2019, 2020 and 2021 [% of households]

Indicator	*2019*	*2020*	*2021*
HEP	25.20%	21.10%	31.21%

Source: Own calculation from HBS database.

understand these figures. The reader may be surprised by the magnitude of the result. In 2021, more than 31% of households were in hidden energy poverty, as defined by the indicator itself. What this is saying is that almost a third of households spent less than half of their theoretical expenditure on energy needed to cover their energy needs (thermal and electricity) once the income filter was applied. The key, as can be understood, is again the threshold used, which in this case is, as indicated above, an absolute threshold calculated from a theoretical model of household energy consumption and expenditure, that is the RENE model described in a previously mentioned study (Barrella et al., 2022a). It is a model that uses the Spanish building regulation as a reference with respect to the comfort temperature in the home, that is 20 degrees Celsius, and the operation hours and floor area, that is a 24h/7d occupancy and 100% climate-controlled area.

Regarding the evolution to 2021, Table 4.3 shows that HEP worsened extraordinarily. This contrasts with the evolution of the M/2 indicator (shown in Table 4.1), which improved slightly. The reason for this behaviour lies in the relative nature of the M/2. The rise in energy prices experienced in 2021 had a controversial effect on this indicator: the impact on the most vulnerable was masked by the increase in energy bills for the general population. This is not the case with the HEP indicator, as the threshold that determines the vulnerability of each household is absolute, that is it is not related to the behaviour of the rest of the households but only to its own benchmark of required expenditure. This observation points out the need to complement the vision offered by the EPOV/EPAH indicator of M/2 underspending with some other absolute metric, such as the HEP, when identifying hidden energy poverty in a country.

On the other hand, the HEP trend aligns with those of the subjective indicators of 'inadequate temperature' (shown in Table 4.1). This might confirm the abovementioned assumption that Spanish households reduced their energy consumption in 2021, even at the expense of their comfort, for reasons derived from price rises and the consequent 'fear of bills'. This result requires particular attention, given that the scenario of high prices that was outlined in the second half of 2021 went on long beyond that year and was even intensified by the Russian invasion of Ukraine in 2022.

Thereafter, an additional scenario is offered in which the threshold is set at one quarter of the required energy expenditure (RENE/4). In this way, we seek to measure 'extreme hidden energy poverty'. Table 4.4 shows the extreme HEP results in the past two years of the selected series. Although in logic with the previous HEP results (Table 4.3), we would expect a significant increase, the magnitude of the indicator is surprising and alarming. In 2021, more than 10% of Spanish households spent less than a quarter of their theoretical energy expenditure and belonged to one of the five lowest income deciles, that is the households in extreme HEP doubled with respect to 2020.

Table 4.4 Extreme HEP indicator for Spain in 2020 and 2021 [% of households]

Indicator	*2020*	*2021*
Extreme HEP	4.80%	10.32%

Source: Own calculation from HBS database.

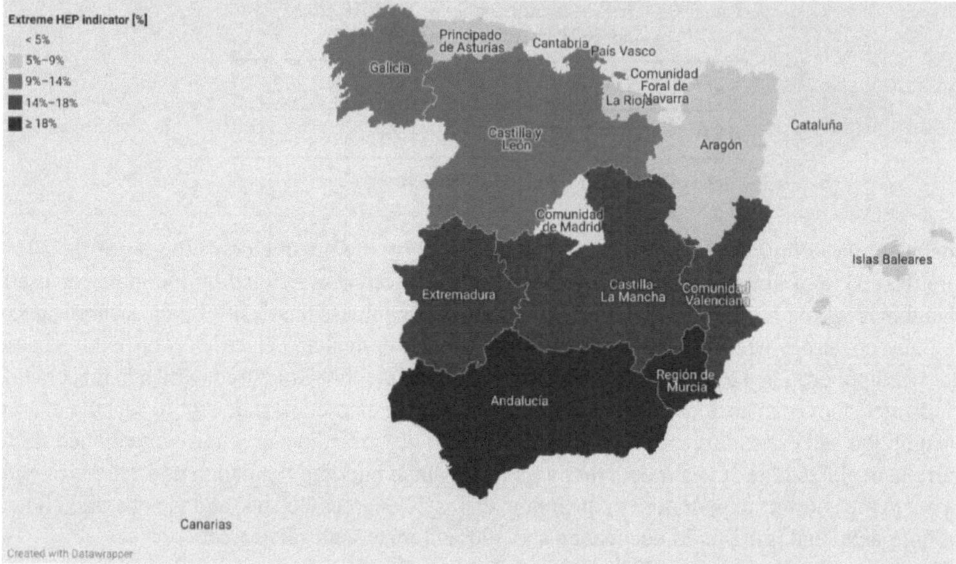

Figure 4.3 Extreme HEP (HEP-RENE/4) indicator by region for Spain in 2021 [% of households].
Source: Own elaboration from the calculation carried out using the HBS database.

Finally, Figure 4.3 shows the 2021 results for each Spanish region, thus pointing out a significant territorial disparity in the extreme HEP incidence. *Canarias* is, in this case, the region with the lowest incidence (3.2%), which is understandable considering that it is the territory with the lowest required demand for heating. At the other extreme is *Andalucía*, where the indicator rises to 19% of households. In the latter case, despite having a fairly favourable climate during the winter in most of the territory, the low penetration of air conditioning appliances in the lowest income deciles (both for heating – according to the HBS – and for cooling; Instituto de Estadística y Cartografía de Andalucía, 2019), the high demand for cooling and the insufficient level of income create a perfect breeding ground for the proliferation of hidden energy poverty. In general terms, it can be said that Figure 4.3 shows a higher incidence of the severe hidden poverty indicator in the southern regions of the peninsula compared to the north.

Finally, to check the impact of the government's measures on this indicator, a calculation of the HEP has been made based on a counterfactual scenario. This scenario is an alternative situation in which the National Government's electricity bill reduction measures[3] would not have been in place. In both cases (HEP and extreme HEP), the counterfactual scenario is 1% point above the actual scenario, that is 32.68% and 11.65%, respectively. That is, if the government's measures had not been implemented, the incidence of hidden energy poverty would have reached approximately 200,000 more households.

4.4 Conclusions

This chapter proposes a set of indicators to measure the different dimensions of energy poverty in a country, emphasising the underspending issue (hidden energy poverty), which has not been adequately addressed in official statistics. Moreover, this chapter compares the results obtained from the application of this methodological approach to the Spanish case study in 2021 with those from previous years.

Concerning subjective indicators, the results show that, in 2021, 9.5% (4.5 million people) and 14.27% (6.7 million people) of the population were, respectively, in arrears on utility bills and with a subjective perception of inadequate temperature at home, compared to 9.6% and 10.9% in 2020. The former indicator consolidates its value after the significant increase from 2019 to 2020, but the latter worsens very significantly: 1.6 million more people compared to 2020 and double compared to 2019.

Moreover, two indicators have been calculated to unpack the disproportionate expenditure dimension: 2M and MIS. The results obtained are, respectively, 15.22% (almost three million households) and 14.62% (2.7 million households) compared to the values obtained for 2020, 16.10% and 16.6%. The two metrics show the same trend: a slight reduction in the incidence of the disproportionate expenditure issue in 2021 compared to 2020.

Finally, two indicators of underspending were calculated: M/2 and HEP. The results for the M/2 and extreme HEP indicators in 2021 were, respectively, 10.12% and 10.32% (almost 2 million households), compared to 11.20% and 4.8% in 2020. In this case, there is a very noticeable trend discrepancy between the two metrics. While the former reveals an improvement in the last year, the latter indicates an extraordinary worsening. The analysis carried out in this chapter points out that the latter indicator deserves more credit for several reasons. The first is because it aligns with the subjective indicator of 'inadequate temperature'. It seems sensible to state that when a household declares that it cannot maintain its dwelling in minimum comfort conditions in the winter, it is in hidden energy poverty. On the other hand, it is crucial to bear in mind that 2021 was, on the one hand, the turning point of the COVID-19 pandemic (which started at the end of 2019 and affected the health and economy of the population all over the world) and, on the other hand, the year of the beginning of a global energy price crisis (that was lately exacerbated by the Russian invasion of Ukraine in March 2022). The latter has significantly impacted Spanish households' economies, especially from autumn onwards. In this sense, the results of the 'inadequate temperature' and HEP indicators seem to indicate that many families restricted their spending for fear of bills that became unaffordable overnight.

Several lessons can be drawn from this picture of hidden energy poverty in Spain. The first is that this might be the most critical dimension of energy poverty in the country. More than 30% of Spanish households did not cover half of their required energy expenditure, and around 10% did not meet even a quarter of it. Traditionally, in the fight against energy poverty, much weight has been given to the dimension of disproportionate expenditure, but this analysis highlights that it is urgent to balance the focus. A second lesson learned is that territorial differences are once again enormous, which might suggest to policymakers that specific local actions are needed. A third lesson comes from the counterfactual scenario. This tells us that the measures implemented to contain the increase in the bill, although they did not manage to stop the blow, at least brought relief to a non-negligible percentage of households. Finally, from a methodological perspective, using an absolute threshold might allow scholars and decision-makers to assess energy poverty rather than energy inequality, the latter being the most common outcome of relative threshold metrics such as the M/2. This issue is further analysed in Chapter 3. In the same line, according to this chapter's results, the HEP indicator seems to better capture the underspending behaviour of Spanish households (especially in an energy price hike scenario), which is misunderstood if using a relative energy threshold comparison.

Eventually, 2021, from an energy poverty perspective, was a year of some light and a big shadow for Spain. Fewer households spent a disproportionate share of their income on covering their energy costs, but, in contrast, many more households entered the dark abyss of extreme hidden energy poverty. The latter suggests the importance of using proper indicators to unveil this usually mismeasured and pressing dimension of hidden energy poverty, especially in a time of crisis.

Notes

1 The RMI was used as MIS income threshold in several studies carried out for Spain (for example, Romero et al., 2018; Rodriguez-Alvarez et al., 2019), but it is considered insufficient to meet household basic needs (Barrella et al., 2022a).
2 The RENE model is a bottom-up construct intended to estimate the theoretical energy costs that a Spanish household would have to pay to meet its energy needs, including both thermal (heating, cooling and DHW) and electricity (lighting, appliances and cooking) uses (Barrella et al., 2022a). These needs are estimated according to household characteristics, such as the number of members, the dwelling size, the location (climate zone) and the building age.
3 The Government reduced the VAT from 21% to 10%, the electricity tax from 5.11% to 0,5% and charges by 96%. Their impact on energy bills is shown in Barrella (2021). For a complete understanding of electricity bills in Spain, the curious reader may refer to Chaves Ávila et al. (2021).

References

Anderson, W., White, V., & Finney, A. (2012) 'Coping With Low Incomes and Cold homes', *Energy Policy*, 49, pp. 40–52. Available at: https://doi.org/10.1016/j.enpol.2012.01.002

Barrella, R. (2021) '2021 Energy Price Crisis impacts on Energy Poverty in Spain', *EP-pedia*. Available at: https://www.eppedia.eu/article/2021-energy-price-crisis-impacts-energy-poverty-spain

Barrella, R. et al. (2022a) 'The Dark Side of Energy Poverty: Who Is Underconsuming in Spain and Why?', *Energy Research & Social Science*, 86, p. 102428. Available at: https://doi.org/10.1016/J.ERSS.2021.102428

Barrella, R., Romero, J.C., & Mariño, L. (2022b). Proposing a Novel Minimum Income Standard Approach to Energy Poverty Assessment: A European Case Study, *Sustainability*, 14(23), p. 15526. https://doi.org/10.3390/SU142315526

Boardman, B. (1991). *Fuel Poverty: from Cold Homes to Affordable Warmth*. London: Belhaven Press. Pinter Pub Limited.

Bradshaw, J. et al. (2008) 'A Minimum Income Standard for the UK', p. 64. Available at: http://www.lboro.ac.uk/research/crsp/mis/

Castaño-Rosa, R. et al. (2021) 'Cooling Degree Models and Future Energy Demand in the Residential Sector. A Seven-Country Case Study', *Sustainability*, 13(5), p. 2987. Available at: https://doi.org/10.3390/su13052987.

Chaves Ávila, J.P., Gómez San Román, T., & Morell Dameto, N. (2021) *La electricidad en España : formación del precio, composición de la factura y comparativa con otros países*. Available at: https://www.fundacionnaturgy.org/publicacion/la-electricidad-en-espana-formacion-del-precio-composicion-de-la-factura-y-comparativa-con-otros-paises/

EPAH (2022) *National indicators*. Available at: https://energy-poverty.ec.europa.eu/observing-energy-poverty/national-indicators_en (Accessed: 15 December 2022).

EU Building Stock Observatory (BSO) (2022) *EU Building Stock Observatory | Energy*. Available at: https://ec.europa.eu/energy/topics/energy-efficiency/energy-efficient-buildings/eu-bso_en?redir=1 (Accessed: 13 March 2020).

Gouveia, J.P. et al. (2022) *Energy Poverty National Indicators. Insights for a more effective measuring*. Available at: https://energy-poverty.ec.europa.eu/discover/publications/publications/energy-poverty-national-indicators-insights-more-effective-measuring_en

Instituto de Estadística y Cartografía de Andalucía (2019) *Encuesta Social 2018. Hogares y Medio Ambiente en Andalucía*. Available at: https://www.juntadeandalucia.es/institutodeestadisticaycartografia/encsocial/2018medioambiente/ (Accessed: 22 September 2022).

Lowans, C. et al. (2021) 'What Is the State of the Art in Energy and Transport Poverty Metrics? A Critical and Comprehensive review', *Energy Economics*, p. 105360. Available at: https://doi.org/10.1016/j.eneco.2021.105360

Meyer, S. et al. (2018) 'Capturing the Multifaceted Nature of Energy Poverty: Lessons from Belgium', *Energy Research and Social Science*, 40, pp. 273–283. Available at: https://doi.org/10.1016/j.erss.2018.01.017

Moore, R. (2012) 'Definitions of Fuel Poverty: Implications for policy', *Energy Policy*, 49, pp. 19–26. Available at: https://doi.org/10.1016/j.enpol.2012.01.057

Rodriguez-Alvarez, A., Orea, L., & Jamasb, T. (2019) 'Fuel Poverty and Well-Being: A Consumer Theory and Stochastic Frontier approach', *Energy Policy*, 131, pp. 22–32. Available at: https://doi.org/10.1016/j.enpol.2019.04.031

Romero, J.C., Barrella, R., & Centeno, E. (2022a) *Informe de Indicadores de Pobreza Energética en España 2021, Informes Cátedra de Energía y Pobreza.* Available at: https://repositorio.comillas.edu/xmlui/bitstream/handle/11531/75498/Informe_Indicadores_2021_EyP_v4_completo.pdf?sequence=-1

Romero, J.C., Barrella, R., & Centeno, E. (2023) '*Understanding the Impact of COVID-19 Lockdown on Energy Poverty in Spain*', *Energy Efficiency* **16**, 56, Available at: https://doi.org/10.1007/s12053-023-10141-5

Romero, J.C., Linares, P., & López, X. (2018) 'The Policy Implications of Energy Poverty indicators', *Energy Policy*, 115, pp. 98–108. Available at: https://doi.org/10.1016/j.enpol.2017.12.054

Stojilovska, A., Yoon, H., & Robert, C. (2021) 'Out of the Margins, into the Light: Exploring Energy Poverty and Household Coping Strategies in Austria, North Macedonia, France, and Spain', *Energy Research & Social Science*, 82, p. 102279. Available at: https://doi.org/10.1016/J.ERSS.2021.102279

5 Reflections from Quantitative Characterisations of Energy Poverty in Mexico

Methods, Energy Justice, and Geographic Differences

Adolfo Mejía-Montero and Pável Soriano-Hernández

5.1 Introduction

Access to affordable energy services – for cooking, transport, heating, or cooling – is fundamental to the well-being of countries and communities around the world and contributes to collective efforts to overcome existing and/or potential cycles of poverty and underdevelopment (International Energy Agency, 2004; Sorrell et al., 2020). Such narratives linking energy use and access with well-being, mobilised by authors like Reynolds and Hughes (1984) or Nye (1999), have materialised in the United Nations Sustainable Development Goals (SDG7) (UN, 2020).

Within this context, issues related to Energy Poverty (EP) have been explored and analysed extensively over the last decades (Li et al., 2014) since the concept of *fuel poverty* was first coined in the 1980s (Day et al., 2016) in the UK. These concepts emerged in response to concerns about heating standards in British households (Ministry of Fuel and Power, 1946; Morris, 1961) during the oil crisis beginning in 1973 (Owen, 2010). This situation took Boardman (1991) to define a fuel-poor household as one spending >10% of its income on fuel costs. In recent years, fuel poverty evolved into the more widely defined concept of EP, later adopted by organisations like the European Commission and the International Energy Agency (IEA) (European Commission, 2021; IEA, 2020). Although the concepts of EP and fuel poverty are sometimes used as synonyms and descriptors of household energy consumption issues (Li et al., 2014), both concepts are typically used to represent different problems in different geographies. However, EP has become a more global term used to analyse the lack of access to modern energy services, an issue that intensifies in countries from the Global South. On the other hand, fuel poverty is mainly used in the UK to describe the inability to afford adequate heating services in households (Bouzarovski & Petrova, 2015).

Since Energy Justice (EJ) was endowed with its first conceptual frameworks (Jenkins et al., 2014; McCauley et al., 2013; Sovacool & Dworkin, 2015), it has become well-established in the energy literature as a scholarship aiming to have a positive impact on energy policy (Jenkins et al., 2016; Van Veelen et al., 2019). Within this context, there are increasing number of publications using both EP and EJ in combination, with different degrees of analysis of a wide variety of complex energy issues across different geographies. These include domestic energy injustice in the UK (Bouzarovski et al., 2022), issues of affordability in rural West Africa (Gafa et al., 2022), and solar power and EP in India (Akter & Bagchi, 2021).

Nevertheless, recent literature on EP and EJ still highlights the need for further research focusing more on frequently overlooked regions like Latin America to facilitate the development of alternative conceptualisations and methods (Jenkins et al., 2021; Thomson et al., 2022). This chapter contributes to energy research by drawing from the analysis of national characterisations of EP in Mexico (Soriano-Hernández et al., 2022) to further the theoretical, empirical, and

DOI: 10.4324/9781003408536-7

methodological exploration of EP and EJ, in turn elaborating on the Latin American context. Three distinctive discussion sections outline the main contributions from this chapter after the following section, which describes the methods and results. The first discussion section draws on theoretical contributions in the literature to analyse the conceptual links and tensions that arise between EP and EJ when examining the results of the case study. The second section illustrates how geographic differences influence the lived experiences of Mexican households under EP and how these might differ from the lived experiences of households in the Global North. The third and last discussion section makes a case for developing alternative methods for gathering and analysing data that facilitate a deeper, more robust, and complete understanding of EP. The concluding section of this chapter summarises the main takeaway messages and contributions.

5.2 Methods and results

The chapter draws from two main methodological approaches: accessibility to energy services (García, 2014; García-Ochoa & Graizbord, 2016b) and monetary expenditure in energy services (Boardman, 1991; Sovacool et al., 2017). The relevant data from Mexican households were collected from Mexico's National Household Income and Expenditure Survey (ENIGH) (INEGI, 2015), which is free and publicly accessible. The sample size included 19,124 households, representing a total of 31,128,396 households nationwide. A complete version of the methodology can be accessed in Soriano-Hernández et al. (2022).

5.2.1 Accessibility of energy services: meeting absolute energy needs

For García (2014), the lack of energy leading to the deprivation of households' needs and satisfiers implies one is living in EP. The Meeting of Absolute Energy Needs (MAEN) method operates this idea by building on Sen's (1981) theorisations which differentiate between categories of absolute needs (e.g., subsistence, entertainment, and freedom) from existential needs (e.g., being, doing, and interacting) and satisfiers (e.g., shelter, work, and privacy) (Max-neef et al., 1992). As these fundamental human needs are considered to be shared across human cultures, they are considered finite and, therefore, classifiable. Thus, it is possible to develop a Household EP (HEP) Index to measure the lack of economic goods, like appliances, in households (Table 5.1), causing the deprivation of energy services and absolute needs (García, 2014; García-Ochoa & Graizbord, 2016a; García-Ochoa & Graizbord, 2016b).

The HEP identifies six basic energy end-uses, listed in the third column of Table 5.1, linked to economic goods necessary for different final energy uses. Therefore, if a household does

Table 5.1 Selected six key energy services for the household EP index (HEP)

Energy Service	*Economic Good*	*Absolute Need*
Cooking	Stoves (gas or electric)	Subsistence
Refrigeration	Refrigerator (models later than 1998)	Subsistence and protection
Entertainment	TV or computer (with internet access)	Entertainment, leisure and creation
Lighting	One incandescent bulb or fluorescent lamp per room	Protection, entertainment, pleasure and creation
Water Heating	Water heater or electric/gas stove	Subsistence and protection
Air conditioning or ventilation	One fan (for every three people in a household) or air conditioning	Subsistence and protection

Source: Summary elaborated by the authors based on García-Ochoa and Graizbord (2016a).

Table 5.2 Estimated Mexican households in EP in areas where climatic conditions do not require thermal comfort, segmented according to HEP score

HEP Index	*Households*	*Percentage (%)*
0	37,516	0.3
0.2	162,962	1.1
0.4	367,557	2.5
0.6	455,911	3.1
0.8	2,842,711	19.6
1	10,646,465	73.4

Source: Modified from Soriano-Hernández et al. (2022).

not have access to economic goods or appliances, it cannot satisfy its human needs through an energy service, which can be considered in EP (García-Ochoa & Graizbord, 2016b). It is essential to highlight that air-conditioning or ventilation is added as an extra energy end-use for households in geographic locations where the average annual temperature is above 26°C.

Once the final energy uses, economic goods, and absolute needs have been defined, the HEP index can be constructed, assuming a household is living in EP if:

$$HEP(x_i) < 1 \tag{5.1}$$

where:

$$HEP(x_i) = \frac{1}{n}\sum_{i=1}^{n} x_i \tag{5.2}$$

Where xi corresponds to the different selected economic goods from Table 5.1. If a household has the selected economic good, then xi=1; otherwise, xi=0. Thus, lacking an economic good leads to HEP (xi)<1. Therefore, the household faces more severe degrees of EP the more HEP approximates 0. Tables 5.3 and 5.4 illustrate the results of access to or deprivation of the economic goods described in Table 5.1.

Under the MAEN method, an estimated 11,613,578 households suffer different degrees of EP in Mexico (38.84%). These results are close to those from García-Ochoa and Graizbord (2016a; 2016b), with only a slight increase of around 2% from the 2012 figures.

Table 5.3 Estimated Mexican households in EP in areas where climatic conditions need thermal comfort, segmented according to HEP score

HEP Index	*Households*	*Percentage (%)*
0	57,526	0.4
0.17	456,055	3.0
0.33	662,488	4.3
0.5	766,663	5.0
0.67	1,315,700	8.5
0.83	4,488,489	29.2
1	7,641,888	49.7

Source: Modified from Soriano-Hernández et al. (2022).

5.2.2 Approximation of household energy consumption based on monetary expenditure

The second method is based on Boardman's (1991) definition embedded in the EJ principle of affordability (Sovacool et al., 2017), stating that any household spending 10% or more of its income on energy and fuel services is considered to be energy poor. However, Boardman's initial approach was built for the UK and focused on heating to keep houses at 21°C. Building on earlier critiques of EP, notably those conceptualisations and methods traditionally limited to heating, this work embraces approaches like that of Sovacool et al. (2012), which include expenditure for mobility, specifically gasoline, given the high number of vehicles registered in Mexico. This work also provides an approximation of energy consumption based on work by Jiang and O'Neil (2004). Such an approach allows the transformation of monetary expenditure on fuels and services into energy values, considering unit prices and heat values of fuels and services (see Table 5.4).

Once the information on prices and calorific values is obtained, energy consumption in households can be calculated from their expenditure on these fuels or services using the following equation.[1]

$$HEC = \sum_{i=1}^{n} \frac{AFE_i}{AFP_i} \times CVF_i \tag{5.3}$$

where:

HEC = Total Household Energy Consumption in Joules.
AFE = expenditure on the ith fuel in monetary values.
AFP = price of the ith fuel in monetary values per unit ($/Kg or $/liter).
CVF = Heat value of the ith fuel in Joules/Kg or Joules/liter.

To compare the MAEN method and the monetary expenditure method, a filter was applied to use data from household reporting, comprising 1) quarterly expenditure on energy services; 2) responses to questions about the goods considered in the MAEN method; and 3) quarterly income. This filter was meant to reduce the sample to 18,102 households, representing 29,901,931 households in Mexico (96.15%). Using the criterion of 10% or more of the income spent on the energy services and fuels listed above, this work found that 10,164,200 (34%) Mexican households are living in EP based on expenditure. Additionally, the average energy consumption at each level of deprivation was approximated using Jiang and O'Neil's methodology

Table 5.4 Average price in Mexican Pesos and heat values per fuel or energy service in Mexico in 2014

Fuel/Energy Vector	*2014 Average Price (US$ per unit)*	*Heat Value*
Electricity	0.0036 per GJ	–
Petroleum Liquefied Gas (PLG)	13.762 per kg	0.046 GJ/kg
Natural Gas (NG)	13.94 per GJ	–
Coal	9.8 per kg	0.03 GJ/kg
Firewood	3.2 per kg	0.015 GJ/kg
Oil	13.94 per litre	0.038 GJ/litre
87 Octane gasoline	13.31 per litre	0.0311 GJ/litre
92 Octane gasoline	14.11 per litre	0.0314 GJ/litre

Source: Elaborated by the authors based on estimations based on Castillo et al. (2012), CFE (2014), CONAFOR (2013), and SENER (2014).

Table 5.5 Key indicators of households that do not require thermal comfort by deprivation level for households in temperate climate regions (not requiring thermal comfort)

HEP Index	*0*	*0.2*	*0.4*	*0.6*	*0.8*	*1*
Households	37,516	162,962	367,557	455,911	2,842,711	10,646,465
Households spending in energy poverty (spend >10% of income on energy services)	12,572	33,125	60,767	108,460	922,447	3,341,658
Average expenditure (% of total income)	10.52%	10.01%	16.31%	9.02%	9.40%	9.51%
Average quarterly energy consumption (GJ)	2.81	1.34	2.38	4.04	5.58	8.53
Total Households	14,513,122					

Source: Modified from Soriano-Hernández et al. (2022).

(see Tables 5.5 and 5.6). These results show that around 3,449,280 (11.54%) of Mexican households face EP through both accessibility and expenditure pressures.

5.3 Discussions

The discussions presented here build on the results reported by Soriano-Hernández et al. (2022), summarised in the previous section. The following sections provide deeper discussions within three main areas (geographic difference, energy justice, and alternative methods) where the chapter makes its contribution. Section 3.1 discusses essential differences emerging from geographic context that determine how EP is lived and experienced in Mexico, contrasting with traditional notions in EP developed mainly in the Global North. This section also explores intranational divides of EP lived experiences across regions with different urbanisation rates and climates. Such geographic differences are explored through the lenses of justice as distribution and recognition to explore the tensions between conceptualisations of EP in Mexico and EJ conceptual frameworks. Section 3.2 elaborates on the results and previous discussions to reflect on the need for alternative methods to measure and understand EP, calling for using qualitative and mixed methods to explore EP effectively.

Table 5.6 Key indicators of households that require thermal comfort by deprivation level for households in hot climate regions (requiring thermal comfort)

HEP Index	*0*	*0.17*	*0.33*	*0.5*	*0.67*	*0.83*	*1*
Households	57,526	456,055	662,488	766,663	1,315,700	4,488,489	7,641,888
Households spending in energy poverty (spend >10% of income on energy services)	12,258	68,027	99,880	151,273	388,727	1,591,744	3,373,262
Average expenditure (% of total income)	7.87%	6.28%	6.47%	7.92%	8.28%	10.26%	12.11%
Average quarterly energy consumption (GJ)	1.45	1.32	1.62	2.61	4.26	6.73	11.08
Total Households	15,388,809						

Source: Modified from Soriano-Hernández et al. (2022).

5.3.1 Reflections on geographic difference and energy justice implications

The existing conceptual frameworks on EJ comprise helpful analytical tools to conceptualise EP beyond a condition of material energy deprivation, enabling energy researchers to frame such conditions in terms of different justice tenets or principles. For instance, the three core tenets framework (distributive, procedural, and recognition) proposed by McCauley et al. (2013) aims to address social justice concerns within energy systems. Whilst distributive justice calls for the equitable distribution of costs and benefits coming from energy systems among all members of society (Jenkins et al., 2016), justice as recognition draws attention to the specific needs, conditions, and characteristics of particular social groups, avoiding forms of cultural domination, degradation, and devaluation (McCauley et al., 2013). Finally, procedural justice is concerned with fair access to decision-making processes involving all members of society in a participatory, impartial, and non-discriminatory manner to pursue social goals (Jenkins et al., 2016). Such an analytical approach has been mobilised to explore energy issues related to climate change (Newell & Mulvaney, 2013), energy resilience in Nepal (Underwood et al., 2020), or energy vulnerability within transitions to smart technologies (Sovacool et al., 2021). However, this research highlights important limitations of EJ conceptual frameworks for an in-depth analysis of EP in Mexico.

One of the main limitations lies in the ambiguity of some definitions and elements present in conceptual frameworks, with authors like Monyei et al. (2019) arguing that EJ frameworks are not yet complete and fail to provide ample exploration or tangible definitions for concepts like "sufficient" energy access. Indeed, such ambiguity is present in conceptual frameworks like the energy principles (Sovacool et al., 2017), which use loose terms such as "sufficient" and "high-quality" to describe energy resources in relation to the principle of availability. Within this context, these facilitate meaningful discussions about what is considered "sufficient" or "high-quality" energy resources. In this case, "sufficient" or "high-quality" energy resources are understood as specific economic goods (Table 5.1) or energy consumption levels, by households with different incomes, located in different climatic regions (Tables 5.5 and 5.6). Moreover, these same results enable discussions about the impacts of socio-economic disparity among households on accessing what they could consider "sufficient" energy resources. In this respect, it is necessary to recognise that frameworks like the EJ principles (Sovacool & Dworkin, 2015) have been modified over time by adding, replacing, or modifying different principles in later versions (Sovacool et al., 2017). Therefore, we call for energy researchers, especially those developing empirical research on EP and EJ in countries of the Global South, to engage in conceptual work that allows refining, expanding, or replacing EJ conceptual frameworks.

Concerning the three EJ tenets framework, the results show households suffering different degrees and combinations of EP due to accessibility or expenditure, highlighting the need to interlink justice through distribution and recognition (McCauley et al., 2013). Following a distributive approach to justice allows for differentiation among tiers or levels of households living under different degrees of EP. These households can be categorised into tiers of EP across geographies (temperate and hot regions) and social classes (income deciles). Therefore, the results help to identify the most vulnerable groups of households in distributive and geographic terms, particularly those living with hidden EP and experiencing issues of expenditure in combination with the lowest level of accessibility (HEP = 0). This means that households in this state of hidden energy poverty are locked into high rates of expenditure in relation to paying for energy services, whilst at the same time do not have all the necessary satisfiers to meet their energy needs. The double or hidden EP state experienced by these households can only be identified by combining the two methods. Therefore, even if the distributive results of EP from applying both methodologies individually lead to similar figures for Mexico, in isolation, they miss-recognise slightly more than one in ten Mexican households that live in this situation of double or hidden EP.

Moreover, the results from this research highlight how, in the Mexican case, EP by expenditure permeates all the tiers of EP by deprivation of energy services. Not only do these results highlight the relevance of an EJ approach that seeks to combine different EJ principles to analyse the multicausal nature of EP in Mexico, but they also raise important questions related to how different groups of people experience EP in the country. Based on the results in Table 5.5, it is possible to calculate the share of households (in temperate regions) living in EP according to expenditure across different tiers of EP by deprivation (different HEPs). Between 16.5% and 23.8% of the households suffering stronger deprivation (HEP = 0.2 and 0.4) also experienced poverty by expenditure. In contrast, households with moderated or no deprivation (HEP = 0.6, 0.8, and 1) presented higher shares of EP by expenditure (between 31.4% and 32.4%). On the other hand, data in Table 5.6 for hot climate regions also suggest differences between households living under intense situations of EP by deprivation (HEPs = 0, 0.17, and 0.33) and lower/non-existent (HEPs = 0.5, 0.67, 0.83, and 1). For hot regions, the share of households experiencing EP by expenditure decreases as HEP increases in the lowest tiers of HEP, going from 21.3% (at HEP = 0) to around 14.9-15.1% (at HEP = 0.17 and 0.33). Nevertheless, higher HEPs see a direct relationship, with the share of households in EP by expenditure consistently escalating from 19.7% at HEP = 0.5 to 44.1% at HEP = 1.

The last paragraph illustrates how, in Mexico, EP seems to evolve from an issue of accessibility/deprivation to modern energy services to an issue of affordability/expenditure for such services, with most households (61%) suffering either or both of these two types of EP in 2014. It could be argued that this is driven by the change in consumption habits when households access additional energy services or economic goods to satisfy previously unsatisfied energy needs, such as mobility (Sovacool et al., 2012). This new satisfier generates a new consumption habit. Examples include the use of gasoline for motor vehicles or the acquisition of home cooling systems in hot climate regions. This compromises the economic stability of households, as their income may not be sufficient to maintain the new levels of consumption they have acquired. As a consequence, they fall into EP for expenditure. However, the results of this research reveal how the progression of EP by accessibility to affordability unfolds differently depending on the geographic conditions (e.g., weather and average temperatures) that lead households to satisfy different energy needs through different economic goods.

On the other hand, the EJ tenet of recognition enables the exploration of the challenges faced by the "most vulnerable" households, allowing us to understand their lived experiences of EP from an analysis grounded in terms of their specific geographic locations and contexts. For instance, it is essential to recognise that two out of three households experiencing double or hidden EP are located in regions with hot climates. Within this context, our study recognises and differentiates the need for thermal comfort (in this case, cooling). However, our methodology fails to differentiate effectively and therefore recognises other faces of EP in the country, such as the lack of heating for households in cold regions. According to Ortiz Álvarez and Vidal Zepeda (2006), 69% of the national population was exposed to cold winters (12°C-cold and 5°C-very cold). This represents a critical gap in justice as recognition, hindering a deeper understanding of EP issues in countries like Mexico, where thermal comfort is approached from a hot climate perspective. Within this context, mobilising a tenet of justice as recognition in EP for Mexico would allow us to better analyse households by their geographical and socio-economic contexts, or by their consumption habits, to then design and implement concrete strategies targeted to minimise their vulnerability towards EP.

The above discussions around justice as recognition lead us to wonder how EP emerged as a concept mainly from countries in the Global North, notably the United Kingdom. A consequence of which has been that studies have traditionally been framed by issues of affordability of fuel and heating services (Li et al., 2014; Sovacool, 2015). Against this background, it is essential to recognise the need for contextual understandings of EP, as studies focusing on countries from the Global South may require different approaches or interpretations of EP based on more locally specific contexts or grounded understandings of justice (Castán Broto et al., 2018; Monyei et al., 2019). Here it is important to highlight that, according to Thomson et al. (2022), the study of EP in Latin American countries remains a persistent gap in academic literature. For Mexico, Garcia's works (García, 2014; García-Ochoa & Graizbord, 2016b) that other studies (six including the work of Garcia-Ochoa) have begun to address this gap.

The results based on EP by expenditure from our case study enable the identification of gasoline, electricity, and liquefied petroleum gas (LPG) as the most significant sources for expenditure in Mexican households vulnerable to EP, due to affordability. According to official figures, in 2014, Mexico had a vehicle fleet of more than 25 million cars (INEGI, 2021), almost one per household. Therefore, our work illustrates the relevance of contextual understandings of EP, arguing that, for Mexico, fuels used in private transportation have a critical role for in driving EP by monetary expenditure. Therefore, Mexican understandings and lived experiences of EP, seems to be fundamentally different from those depicted on literature from European or North American case studies, where energy services like thermal comfort seem to have the leading role (Kelly et al. 2020; Middlemiss, 2022). These results support Sovacool et al. (2012) in suggesting services like mobility to be included in the understanding of EP. Essential energy services must be provided in an affordable, reliable, adequate, quality-assured, safe and environmentally benign manner (Parajuli, 2011). Therefore, as our definition of household EP follows that of other authors like Robinson and Mattioli (2020) and Martiskainen et al. (2021) – by considering household expenditure on gasoline for private transport – this chapter calls for developing empirical research for other countries in the Global South, especially where heating (or thermal comfort) does not hold a central role in household energy services. This will further our understanding of EP and contribute to decolonising the concept. Such a process of decolonisation implies deconstructing EP research and understandings from all potential assumptions and biases generated as a result of its epistemic foundations in the Global North. Facilitating greater research in the Global South will complement, expand, refine or replace the way we conceptualise and work with energy-related concepts like EP or EJ (Tornel, 2022; Walker, 2022). This chapter makes a small contribution by briefly exploring how, in the Mexican case, access to and affordability of services like mobility or cooling have a prevalent role in driving households into different degrees of EP, in contrast to other energy services (e.g., heating) traditionally mentioned in academic literature.

5.3.2 Reflections on methodological diversity

The discussions above highlight how the situation in Mexico serves to reflect conceptual and empirical gaps within the existing EP literature. Within this context, Thomson et al. (2022) also call for greater methodological diversity, identifying only two publications in Latin America using qualitative methods to explore EP. Whilst methodological approaches like the one by Jiang and O'Neil (2004) and García-Ochoa and Graizbord (2016a, 2016b), both used in this work, illustrate an effective combination in studying this phenomenon, such methods continue in the main to be quantitative in approach. Since EP is such a complex societal phenomenon (influenced by age, geography, income, level of education, etc.), this chapter recognises the need for using qualitative and/or mixed methodologies under an interdisciplinary approach to facilitate more effective and robust analyses of EP complexities (Sovacool et al., 2017). Within this context, some noteworthy examples of the use of qualitative methods in examining EP in Latin America can be found, including Ricalde et al. (2021) and Brand-Correa et al. (2018) with both using participatory workshops in Mexico and Colombia, respectively.

An outstanding share of Mexico's household energy services focus on cooking and water heating either with electricity or with some fossil fuel like liquefied petroleum gas (García-Ochoa & Graizbord, 2016a). However, some households use some form of biomass (e.g., firewood) for the same purposes. This research illustrates that up to 33.5% of households in temperate regions suffering the worst deprivation issues (HEP = 0) also suffer from EP due to expenditure or affordability (higher than those with HEP = 0.2 or 0.4). Indeed, firewood and gasoline represented the most significant expenditure for these households, even without considering costs associated with working hours or risks related to recollecting wood, etc. (Santos González et al., 2012). Although official figures for the use of this type of fuel are debatable (Sánchez-Peña, 2012), the National Forestry Commission in Mexico (CONAFOR, 2013) argues that firewood satisfies around 80% of some rural households' energy needs. Nevertheless, official figures reported in the ENIGH appear to only include purchased firewood, which might underestimate the fact that much of the firewood is collected rather than purchased (Santos González et al., 2012; Silva Aparicio et al., 2018). The work from Pérez et al. (2022) estimates that around 26% of the Mexican population uses firewood, with 75% of wood users located in population centres with less than 2,500 inhabitants and 68% of the users not paying for the acquisition of firewood.

Consequently, although this work relies on the statistics from ENIGH on purchased firewood, it calls for the development of surveys that consider quantifying energy sources like recollected firewood. Such alternative methodologies could facilitate the assessment of collected firewood for satisfying energy needs in countries in the Global South and the Global North. Moreover, these alternative methods could open opportunities for EP researchers to collaborate with colleagues working on areas like forest management, conservation, and international development.

On the other hand, this research also limits its analysis to household expenditure on fuel for private vehicles. However, expenditure in relation to public transport has been highlighted as a relevant statistic for future research into understanding EP (Lowans et al., 2021; Sareen et al., 2022). Within this context, it is relevant to highlight that in Mexico, the level of data aggregation of the ENEIGH survey does not allow us to identify relevant information related to the type of transport, fuel, or the percentage of the fare that is directly used in fuel consumption necessary for a critical analysis of public transport. Solving such a lack of data aggregation represents a methodological challenge beyond the scope of this work. However, it highlights an opportunity to improve the statistical tools currently used by institutions like INEGI (the National Institute of Statistics and Geography) in Mexico as well as those in other countries.

5.4 Conclusions

This chapter illustrates the different ways in which Mexican households live and experience EP due to issues of accessibility, affordability, or both. According to our results, in 2014, approximately six out of ten Mexican households (61.3%) faced EP. In addition, this chapter discusses how 38.84% of Mexican households suffered from accessibility issues impeding them from meeting their absolute energy needs, whilst 34% faced a high economic burden purchasing energy services and fuels. This work also recognises that 11.54% of Mexican households live in both situations of EP, arguably comprising the most vulnerable groups. These results demonstrate and illustrate the multicausal origin of EP in Mexico. From the discussion of the results above, this chapter makes a triple contribution to the topic.

Conceptually, this chapter discusses issues of EP in Mexico in relation to conceptual frameworks relating to EJ. This chapter discusses the ambiguity present in some of the definitions and elements within those conceptual frameworks and calls for energy researchers – especially those developing empirical research on EP in the Global South – to engage with EJ to refine, expand, and/or replace these existing conceptual frameworks. Having said that, this chapter also demonstrates the usefulness of EJ frameworks to analyse the distributive justice dimension of the multicausal nature of EP in Mexico (by expenditure or deprivation). It also identifies a fundamental gap in EP literature by recognising that the most vulnerable groups live under two different kinds of EP. Finally, this chapter discusses the lack of recognition for people experiencing cold winters in countries that are traditionally labelled as having hot climates and in which thermal comfort is approached only through cooling. Therefore, addressing EP within an approach that involves EJ frameworks that facilitate identifying and potentially further remediating energy injustices across different geographies and income groups.

Empirically, this study illustrates how Mexican households have a significant burden of expenditure for gasoline for private transportation instead of heating, as is normally the case for countries in Europe or North America. Such recognition of transport in the Mexican EP experience also calls for future research that challenges traditional conceptualisations of household energy needs to advance the understanding of EP in overlooked regions like Latin America. A further complicating factor in Mexico involves the evolution of EP from issues of accessibility to modern energy services to affordability due to newly acquired energy services. However, this process seems to evolve differently depending on climate conditions (e.g., hot and temperate regions). Therefore, this chapter calls for analytic approaches that facilitate recognition and reflection on the relevance of geographic difference to understanding how EP arises and how it is experienced within households across different countries and regions. Such an understanding of the lived experiences of EP is especially relevant in countries frequently overseen and regions like Latin America, which can effectively contribute to decolonising our understanding of EP.

Methodologically, this quantitative research facilitates the recognition of the complexity and multifaceted nature of the EP problem in Mexico. However, the recognition of such complexity also leads to a call for an interdisciplinary approach based on the deployment of qualitative and/or mixed methods that facilitate a complete analysis and discussion of EP complexities. Finally, this chapter discusses the challenges related to the existing methods for capturing and analysing data related to variables like the recollection of wood or public transport, which apparently play a determinant role in measuring and understanding household EP in countries like Mexico. These alternative methods could open opportunities for EP researchers to develop collaborative work in areas like forest management, conservation, and international development, as well as provide helpful feedback to improve the statistical tools used by institutions like INEGI in Mexico as well as those in other countries.

Note

1 Energy values do not consider losses due to the efficiency of appliances or losses due to the storage, distribution, or transmission of energy. Therefore, the energy per household obtained represents an approximation of the actual energy consumption.

References

Akter, S., & Bagchi, K. (2021). Is off-Grid Residential Solar Power Inclusive? Solar Power Adoption, Energy Poverty, and Social Inequality in India, *Energy Research and Social Science*, 82, 102314. https://doi.org/10.1016/j.erss.2021.102314

Boardman, B. (1991). *Fuel Poverty: From Cold Homes to Affordable Warmth*. London, UK: Belhaven Press.

Bouzarovski, S. et al. (2022) The Diversity Penalty: Domestic Energy Injustice and Ethnic Minorities in the United Kingdom, *Energy Research and Social Science*, 91(July), p. 102716. https://doi.org/10.1016/j.erss.2022.102716

Bouzarovski, S., & Petrova, S. (2015). A Global Perspective on Domestic Energy Deprivation: Overcoming the Energy Poverty-Fuel Poverty Binary, *Energy Research and Social Science*, 10, 31–40. https://doi.org/10.1016/j.erss.2015.06.007

Castán Broto, V., et al. (2018). Energy Justice and Sustainability Transitions in Mozambique, *Applied Energy*, 228(July), 645–655.

CONAFOR (2013). *Bosques de Mexico frente al Cambio climatico*. Mexico City: Comision Nacional Forestal. Available at: http://www.conafor.gob.mx/biblioteca/Manual-de-la-Lena.pdf

Day, R., Walker, G., & Simcock, N. (2016). Conceptualising Energy Use and Energy Poverty Using a Capabilities Framework, *Energy Policy*, https://doi.org/10.1016/j.enpol.2016.03.019

European Commission (2021). What is energy poverty? Energy Poverty Advisory Hub. Available at: https://energy-poverty.ec.europa.eu/energy-poverty-observatory/what-energy-poverty_en (Accessed 15 October 2021).

Gafa, D. W., Egbendewe, A. Y. G., & Jodoin, L. (2022. Operationalizing Affordability Criterion in Energy Justice: Evidence from Rural West Africa, *Energy Economics*, 109(March), 105953. https://doi.org/10.1016/j.eneco.2022.105953

García, R. (2014). *Pobreza energética en América Latina*. Comisión Económica para América Latina y el Caribe (CEPAL).

García-Ochoa, R., & Graizbord, B. (2016a). Privation of Energy Services in Mexican Households: An Alternative Measure of Energy poverty, *Energy Research and Social Science*, 18, 36–49. https://doi.org/10.1016/j.erss.2016.04.014

García-Ochoa, R., & Graizbord, B. (2016b). Spatial Characterization of Fuel Poverty in Mexico. *An Analysis at the Subnational Scale*. Economía, sociedad y territorio.

IEA (2020). Defining energy access: 2020 methodology. Available at: https://www.iea.org/articles/defining-energy-access-2020-methodology (Accessed 19 January 2023).

INEGI (2015). Ingresos Y Gastos De Los Hogares Enigh 2014.

INEGI (2021). INEGI – Vehículos de Motor Registrados en Circulación (motor vehicles registered in circulation).

International Energy Agency (2004). World energy outlook chapter 10: Energy and development. World energy outlook.

Jenkins, K. et al. (2014) Energy Justice, a Whole Systems Approach, *Queen's Political Review*, II(2), pp. 74–87.

Jenkins, K., & McCauley, D., et al. (2016). Energy Justice: A Conceptual Review, *Energy Research and Social Science*, 11, 174–182. https://doi.org/10.1016/j.erss.2015.10.004

Jenkins, K. E. H., Sovacool, B. K., Mouter, N., Hacking, N., Burns, M. K., & McCauley, D. (2021). The Methodologies, Geographies, and Technologies of Energy Justice: A Systematic and Comprehensive Review, *Environmental Research Letters*, 16(4). https://doi.org/10.1088/1748-9326/abd78c

Kelly, J. A. et al. (2020) Enabling a Just Transition: A Composite Indicator for Assessing Home-Heating Energy-Poverty Risk and the Impact of Environmental Policy Measures, *Energy Policy*, 146(January), p. 111791. https://doi.org/10.1016/j.enpol.2020.111791

Li, K., Lloyd, B., & Liang, X. J., et al. (2014). Energy Poor or Fuel Poor: What Are the Differences? *Energy Policy*, 68, 476–481. https://doi.org/10.1016/j.enpol.2013.11.012

Lowans, C., et al. (2021). What Is the State of the Art in Energy and Transport Poverty Metrics? A Critical and Comprehensive Review, *Energy Economics*, 101, 105360. https://doi.org/10.1016/j.eneco.2021.105360

Martiskainen, M., et al. (2021). New Dimensions of Vulnerability to Energy and Transport Poverty, *Joule*, 5(1), 3–7. https://doi.org/10.1016/j.joule.2020.11.016

Max-neef, M. A., Hopenhayn, M., & Hamrell, S. (1992). Human scale development: Conception, application and further reflections, volume 1.

McCauley, D., et al. (2013). Advancing Energy Justice: The Triumvirate of Tenets, *International Energy Law Review*, 32, 107–116.

Middlemiss, L. (2022. Who Is Vulnerable to Energy Poverty in the Global North, and What Is Their Experience? *Wiley Interdisciplinary Reviews: Energy and Environment*, 11(6), 1–13. https://doi.org/10.1002/wene.455

Ministry of Fuel and Power (1946). Domestic fuel policy: Report by the Fuel and Power Advisory Council.

Monyei, Chukwuka G., et al. (2019). Response to Todd, De Groot, Mose, McCauley and Heffron's Critique of "Examining Energy Sufficiency and Energy Mobility in the Global South Through the Energy Justice Framework", *Energy Policy*, 133(August), 1–5. https://doi.org/10.1016/j.enpol.2019.110917

Morris, P. (1961). Homes for today and tomorrow (the Parker Morris report).

Newell, P., & Mulvaney, D. (2013. The Political Economy of the "just Transition, *Geographical Journal*, 179(2), 132–140. https://doi.org/10.1111/geoj.12008

Nye, D. E. (1999). *Consuming Power: A Social History of American Energies*. IEEE Transactions on Professional Communication. https://doi.org/10.1109/tpc.1999.768169

Owen, G. (2010). Review of the UK fuel poverty measure: Report for Ofgem. (London).

Parajuli, R. (2011). Access to Energy in Mid/Far West Region-Nepal from the Perspective of Energy Poverty, *Renewable Energy*, 36(9), 2299–2304. https://doi.org/10.1016/j.renene.2011.01.014

Pérez, G. et al. (2022) Socioeconomic and Environmental Aspects of Traditional Firewood for Cooking on the Example of Rural and Peri-Urban Mexican Households, *Energies*, 15(13). https://doi.org/10.3390/en15134904

Reynolds, T. S., & Hughes, T. P. (1984). Networks of Power: Electrification in Western Society, 1880–1930, *Technology and Culture*, https://doi.org/10.2307/3104214

Ricalde, K., Cedano, K.G., Thomson, H., & Robles-Bonilla, T. (2021). Contextualising Nussbaumer via Nussbaum: Unveiling a Multi-Disciplinary, Human Capabilities- Centred Approach to Energy Poverty from Mexico. In: Webb, J., Wade, F., & Tingey, M. (ed.) *Research Handbook on Energy and Society*. Edward Elgar Publishing.

Robinson, C., & Mattioli, G. (2020). Double Energy Vulnerability: Spatial Intersections of Domestic and Transport Energy Poverty in England, *Energy Research and Social Science*, 70(July), 101699. https://doi.org/10.1016/j.erss.2020.101699

Sánchez-Peña, L. (2012). Hogares y consumo energético en Mexico. *Revista Digital Universitaria*, 13, 1–8.

Santos González, A., Estrada Lugo, E. I. J., & Rivas Lechuga, G. (2012). Uso de la leña y conservación del bosque en el volcán Huitepec, Chiapas, México, *LiminaR Estudios Sociales y Humanísticos*, 10(1), 138–158. https://doi.org/10.29043/liminar.v10i1.41

Sareen, S., et al. (2022). Double Energy Vulnerability in the Norwegian Low-Carbon Urban Transport Transition, *People, Place and Policy Online* (September 2021), 1–20. https://doi.org/10.3351/ppp.2022.3953567323

Sen, A. (1981). *Poverty and Famines: An Essay on Entitlement and Deprivation: The Great Bengal Famine*. Oxford University. https://doi.org/10.1093/0198284632.001.0001

Silva Aparicio, M., Castro Ramírez, A. E., & Perales Rivera, H. R. (2018). Áreas Voluntarias de Conservación y la extracción de leña en núcleos agrarios de Oaxaca, *Ecosistemas y Recursos Agropecuarios*, 5(15), 435–449. https://doi.org/10.19136/era.a5n15.1743

Soriano-Hernández, P., Mejía-Montero, A., & van der Horst, D. (2022). Characterisation of Energy Poverty in Mexico Using Energy Justice and Econophysics, *Energy for Sustainable Development*, 71, 200–211. https://doi.org/10.1016/j.esd.2022.09.005

Sorrell, S., Gatersleben, B., & Druckman, A. (2020). The Limits of Energy Sufficiency: A Review of the Evidence for Rebound Effects and Negative Spillovers from Behavioural Change, *Energy Research and Social Science*, https://doi.org/10.1016/j.erss.2020.101439

Sovacool, B. K., et al. (2012). What Moves and Works: Broadening the Consideration of Energy Poverty, *Energy Policy*, https://doi.org/10.1016/j.enpol.2011.12.007

Sovacool, B. K., et al. (2017). New Frontiers and Conceptual Frameworks for Energy Justice, *Energy Policy*, 105(January), 677–691. https://doi.org/10.1016/j.enpol.2017.03.005

Sovacool, B. K., & Dworkin, M. H. (2015). 'Energy Justice: Conceptual Insights and Practical Applications, *Applied Energy*, 142, 435–444. https://doi.org/10.1016/j.apenergy.2015.01.002

Sovacool, B. K., Martiskainen, M., & Furszyfer Del Rio, D. D. (2021. Knowledge, Energy Sustainability, and Vulnerability in the Demographics of Smart Home Technology Diffusion, *Energy Policy*, 153(February), 112196. https://doi.org/10.1016/j.enpol.2021.112196

Thomson, H., et al. (2022). Understanding, Recognizing, and Sharing Energy Poverty Knowledge and Gaps in Latin America and the Caribbean – Because Conocer Es Resolver, *Energy Research and Social Science*, 87(December 2021). https://doi.org/10.1016/j.erss.2021.102475

Tornel, C. (2022). Decolonizing Energy Justice from the Ground up: Political Ecology, Ontology, and Energy Landscapes, *Progress in Human Geography*, 47(1), 43–65. https://doi.org/10.1177/03091325221132561

UN (2020). The sustainable development goals report 2020. New York. https://doi.org/10.4324/9781003099680-3

Underwood, G., Hill, D., & Lamichhane, S. (2020. Earthquakes, Blockades and Energy Crises: A Conceptual Framework for Energy Systems Resilience Applied to Nepal, *Energy Research and Social Science*, 69(May), 101609. https://doi.org/10.1016/j.erss.2020.101609

Walker, G. (2022). Whose Energy Use Matters? Reflections on Energy Poverty and Decolonisation, *People, Place and Policy Online*, 16(1), 6–12. https://doi.org/10.3351/ppp.2022.3833594884

6 Understanding the Lived Experience of Summer Energy Poverty through Participatory Action Research

Marta Gayoso Heredia, Daniel Torrego Gómez, Miguel Núñez Peiró, and Carmen Sánchez-Guevara Sánchez

6.1 Introduction

According to the Eurostat survey[1], more than 20% of households in southern European countries declared that they are unable to keep their homes in comfortable conditions during summertime. In the context of climate change, it is also known that extreme hot weather and heat waves are expected to increase. Different reports and studies predict that the impacts of these extreme events will worsen for those households suffering from Summer Energy Poverty (SEP) (Founda et al., 2019; Santamouris & Kolokotsa, 2015).

Compared to the literature focused on the study, energy poverty linked to winter conditions is still an overlooked issue (Thomson et al., 2019), and both the characterisation and measurement of the phenomenon are in the very initial stages. The review of methods, definitions (Middlemiss & Gillard, 2015), and characterisation of energy poverty points out a lack of summer dimension in the studies focused on the phenomenon (Gayoso et al., 2022). It is only possible to measure SEP by using secondary indicators that are only available for limited countries and annual periods whose consultation is freely accessible from the website of the Energy Poverty Advisory Hub. Although the influence of different features is well documented in the urbanscape with the experience of heat, there is a total absence of indicators that could guide policy makers to reduce the vulnerability related to heat exposure.

During summertime, cities located in Southern Europe such as Madrid experience an increase of social activities in public space. For vulnerable households, using public space may respond to a need to find a climatic shelter and avoid extreme indoor temperatures (Tsoulou et al., 2020). Although there is a growing interest in characterising the urban scale of the phenomenon by studying Urban Heat Islands, passive urban mitigation, or urban microclimates (Sanchez-Guevara et al., 2019), there is not yet a qualitative characterisation focused on relationships and community strategies to cope with excessive heat.

Qualitative research methods allow for the characterising of household strategies that cannot be reflected in quantitative databases and go beyond the dwelling usage and the individual scale. The few studies carried out exploring summer energy poverty (Gayoso et al., 2022) identify the phenomenon as more difficult to characterise from a quantitative perspective since qualitative aspects such as behaviour, adaptation strategies, and cultural factors have a great influence on how to perceive and cope with high temperatures. There is not yet sufficient methodological framework that integrates the lived experience into the study of energy poverty. This is especially relevant for summer conditions where thermal adaptation relies mainly on users' habits. Thus, sharing methods and insights derived from fieldwork are essential to develop a premature body of literature and raise awareness to make the adaptation process accessible to manage the impacts of climate change.

DOI: 10.4324/9781003408536-8

The use of public space integrates the dynamics and logics of the community that shares a location (Leivas, 2019), and it is diverse from one community to another. Therefore, delving into collective experiences becomes important to get a more complete picture of relations and resources that may alleviate energy poverty. Regarding qualitative research studies, participatory action research and focus groups are the most common methods used to characterise energy poverty from a collective perspective (Grey et al., 2017; Lorenc et al., 2013). However, only a few experiences focused on energy poverty have been conducted, and the application and integration of techniques into energy poverty studies have room to improve.

COOLTORISE is the first project to be funded by the European Commission to deepen and widen the understanding of SEP in the context of southern Europe. Four countries participate in the project: Spain, Italy, Greece, and Bulgaria. The main objective of the project is to raise awareness of summer energy poverty and reduce cooling needs by working together with households on integrating passive strategies to cope with the heat and improve indoor thermal conditions during summertime. For this aim, the project offers different workshops to increase summer energy culture. Understanding home usage patterns related to improving summer thermal comfort and reducing energy needs (e.g., optimising the use of solar shades, night ventilation, and cooking habits). Also, different workshops are focused on understanding energy bills and how to reduce their cost. Finally, another set of workshops were aimed to improve public space through collective outdoor intervention to obtain climatic shelters. The project's main activities will be developed during the summers of 2022 and 2023 reaching up to 7,240 beneficiaries.

This chapter presents insights regarding the implementation of Participatory Action Research and qualitative research methods for focus groups during the activities developed within the COOLTORISE project for the case of the city of Madrid. The chapter is structured as follows: in the first section, an approach to the phenomenon of SEP and qualitative research is introduced, in addition to the framework of the project in which the methodological proposals have been implemented; in Section 2, the scale of actions and the activities carried out are reviewed, together with the methodological proposal implemented in each activity to achieve the objectives described. Section 3 presents the results obtained in relation to the lived experience for each of the activities described; finally, Section 4 establishes the contribution of each methodological proposal in relation to the knowledge of the lived experience and how the information extracted from the experiments can be used to tackle Summer Energy Poverty.

6.2 Methods

Among the activities proposed within the COOLTORISE project in the city of Madrid, three different workshops have been designed that integrate qualitative research methods to explore and bring new insights into how vulnerable households cope with heat. A total of four collaborative mapping workshops have been conducted between June 2022 and September 2022, and 35 people have participated (Table 6.1).

The activities proposed have followed an approach to understand the lived experience of SEP involving different scales: the housing scale, where participation is based on the self-reported strategies to cope with the heat at home, and is materialised in energy culture workshops; the urban scale, where participation is based on sharing experiences of location-based knowledge, and is addressed with collaborative mapping workshops; and the body scale, where participation has been focused on testimonies of the lived experience using public space while taking a walking-tour workshop.

Participatory Action Research is an approach that is applied not only to understand the behavioural actions and habits of the participants but also to bring about change by actively involving

Table 6.1 Relation of type of workshops, number of participants, gender and profiles

Type of workshop	*Number of participants*	*Gender (non-binary was an included option, but the sample was 0)*		*Profiles*
Energy Culture Workshops	131 people	Women: 98	Men: 33	Older adults from municipal centres for older adults (ages > 55) 2022 The European researchers' night Municipal cultural centres Municipal equality community centre
Collaborative Mapping Workshops	35 people	Women: 25	Men: 10	Older adults from municipal centres for older adults (ages > 55)
Walking Tour Workshops	10 people	Women: 10	Men: 0	Older adults from municipal centres for older adults (ages > 55)

Source: Own elaboration.

them to take part in the solution and raising awareness to address the problems identified (De Haro & Koslowski, 2013). It differs from other methods because the aim is to enable actions achieved by participants: they collect and analyse data and decide the action that should follow. This structure changes the power relationship between participant and researcher and shares the power as researcher and participants become partners.

A focus group is a qualitative research method in which a small group of people, usually six to ten, are brought together to discuss a topic. The discussions are led by a researcher who asks open-ended questions to encourage dialogue and exploration of different perspectives.

From the perspective of those studies that integrate focus groups and Participatory Action Research (PAR) as methods within the research, it is possible to find some barriers when introducing focus groups in a conventional way regarding object-subject research relationships (Chiu, 2003). Also, focus groups allow participation in a very homogenous and limited way, as oral participation is the only way for expression instead of more visual or other methods of expression. Reformulating and transforming conventional methods allows other forms of participation to arise that attend better to diversity.

In the context of climate change and adaptation in cities, it is clear that effective mitigation cannot be achieved if different actors act independently: climate change has the characteristics of a global collective action problem, so cooperative principles and participative approaches are essential to address mitigation and adaptation (IPCC, 2022).

For each of these scales, participation implies different types of samples. Regarding the housing scale, the strategies described correspond to a reality that encompasses the cohabitation unit. Participants reflect the routines and consumption habits adopted by the dwellers. On an urban scale, the description of climatic shelters, paths, and routes transcends personal experience and entails a reflection of community networks and social interactions. At the body scale, however, the experience of each person is reflected in relation to the individual and local conditions at that moment.

6.2.1 Energy culture workshops

The aim of the workshop is to explore the strategies that participants already know and have incorporated in their dwellings to cope with heat and then to include new unknown strategies that can potentially improve their indoor thermal conditions and reduce cooling needs. The objectives in a more abstract sense are concerned with the revaluation of knowledge, empowerment, and

Figure 6.1 Energy culture workshops. As can be seen in the image, the participants are wearing masks. This is due to the fact that the workshops were held at a time when precautions related to COVID-19 had to be taken. This had no meaningful impact on the engagement of participants.

Source: Photograph by Marta Gayoso.

carefully listening to voices (H. Yoon & Saurí, 2019) that do not require technical training to know which strategies could be implemented in their dwellings. To this end, these workshops have been provided to groups of eight to ten people who belong to vulnerable neighbourhoods (Figure 6.1). The considerations relate to the climatology of the location, the type of construction, and socio-cultural factors and belong to the collective wisdom (Day & Hitchings, 2011; O'neill et al., 2006; Pelenur & Cruickshank, 2013). In the context of adaptation to climate change, it is essential to focus on local needs and specific conditions to identify suitable strategies.

At the beginning of the workshop, participants describe which strategies they already use during the summer to cope with high temperatures by writing these down on sticky notes. Secondly, a theory presentation is conducted by the workshop facilitators focused on passive conditioning and a brief review of what strategies are appropriate for their local climate to cope with excessive heat. This is followed by a discussion of the strategies that were reported at the beginning by the participants. These strategies are reviewed among all participants, and the strengths, weaknesses, and limitations encountered in implementing the strategies are discussed. Finally, those strategies that were not presented at the beginning of the workshop and that can be implemented to improve indoor thermal conditions are introduced.

To conclude, participants were asked to reflect on the strategies discussed on three different panel boards (see Figure 6.1) to facilitate better visualisation of results. On one panel, they locate the strategies to eliminate as they are contraindicated for combating heat, while on another panel they locate the strategies to maintain, which are those that are already incorporated in their day-to-day lives and are favourable to them. On the third panel, they locate the strategies to implement, which are those that the participants can incorporate after receiving the theory presentation.

6.2.2 Collaborative mapping workshops

The aim of this activity was to identify existing climate-related resources and shelters that exist (Hunt et al., 2018) in a shared location and to analyse the routes, paths, and public spaces from the perspective of their summer usage. The secondary objective was to generate a space for conversation and sharing of knowledge between different people from the same location in order to raise awareness of the importance of public space as a potential universally accessible resource for dealing with high temperatures (Yoon et al., 2022). These workshops complement the body/dwelling/building scales explored in the Energy Culture Workshops incorporating an urban scale in the dialogue. For vulnerable households, the identification of available climate shelters near their locations may reduce their exposure to extreme temperatures at home.

These workshops are carried out with groups of ten to 15 people who shared the same neighbourhood. To do this, large-format, unlabelled maps of the locations were prepared and printed to allow for additional information to be written directly onto them (Figure 6.2). The mapping codes used were shared with the participants and the person facilitating the workshop guided them on which elements are to be mapped, at each stage of the workshop. These codes included elements that describe key places, pathways and routes, sun/shade zones, greenery, blue infrastructures, and public facilities.

During the workshop, predetermined questions were asked about the use and perception of public space from a SEP perspective, to facilitate participation (Table 6.2).

The results of the mapping were photographed and transferred to a collaborative online map, to record the elements and comments that emerged during the workshops. Also, this was done in order to return the results to the participants so that they can include further comments and relevant information and continue to complete the map.

Figure 6.2 Collaborative mapping workshop.

Source: Photograph by Daniel Torrego.

Table 6.2 Predefined questions used in the mapping sessions. They encouraged labelling actions on the maps provided in the workshops

Collaborative mapping workshop questions script
Which places in your immediate surroundings are more pleasant during the summer?
Do you adapt the spaces and activities in which to meet and enjoy leisure activities according to the season? Are these options free, paid or linked to consumption?
Which places are not in your immediate surroundings and require transport but are also nice places to go to in summer or winter?
What is the existing vegetation, what are its characteristics?
What vegetation has existed in the past, and what were the causes of its disappearance?
Is it possible to identify areas and routes with specific uses that have shade for use during the summer?
Is it possible to identify areas and routes with specific uses that have sunlight for use during the winter?
What surfaces are in this space? What is the state of the street furniture or other equipment?
Are there any fountains, pools or water facilities that cool the environment?

Source: Own elaboration.

6.2.3 Walking tour workshops

The objective of this activity was to visit the points identified and described in the Collaborative Mapping Workshop activity to record how they actually are used and what is the embodied experience (Brager & de Dear, 1998; Shirani et al., 2020). In this activity, the information that emerges is related to the history and community codes that exist around the local microclimate. For instance, alternative routes to the ones described, usage schedules of common spaces, and descriptions of health impacts that relate to the public space were all considered. As a secondary objective, the aim was to have a more detailed record of the problems and improvements to be implemented in spaces that already serve as climate refuges and that can potentially serve as climate refuges. Also, it was hoped participants could create a space where the body experience is also expressive, going beyond the oral testimonies (Longhurst & Hargreaves, 2019; Pink, 2011).

In this workshop, participants belong to the same sample as in the collaborative mapping workshops, although the workshop was also open to new participants who come from the locale where the activity took place.

Initially, participants are informed of the route to be followed during the activity. This route has been previously designed based on the results recorded during the mapping session, ensuring accessibility, shade, and refreshment points were available. For each point along the route, a sheet with a special template was designed and printed to direct on-site recording of the problems detected, the potential for improvement, and the history or other uses that the place has had in the past (Figure 6.3).

The records made on the sheets were also collected and added to the collaborative online map. This enabled us to complete the registration of the resources to cope with heat.

6.3 Results

In this section, the results obtained in relation to the lived experience for each of the proposed activities are presented.

6.3.1 Results for energy culture workshops

Within this activity, the results collected for a sample of 13 workshops carried out between May 2022 and October 2022 are presented. They are based on a sample of 131 people. Results

Figure 6.3 Walking tour workshops.

Source: COOLTORISE project images (all participants).

are classified into three categories: strategies and resources that households use incorrectly and therefore need to be modified or eliminated; those already known and carried out by the participating sample; and those unknown to dwellers. In order to validate these last ones, after participation in the workshops, households express whether they will be prone to implement them (Table 6.3).

- Habits to be eliminated: in some workshops, households reported misuse of natural ventilation. Ventilation is used in some cases during the central hours of the day when the outside temperature is high, and some participants express a belief in forcing cross ventilation with the use of a fan as an air circulator. This is enhanced when the facade or envelope is not efficient enough to maintain a low indoor temperature. The use of evaporative coolers without proper ventilation of the dwelling was also reported as counterproductive.
- Habits to be maintained: in general, strategies that have to do with keeping the body at an adequate temperature are the most common and widespread actions expressed by the participants, positively responding to awareness campaigns by public health institutions. Strategies such as taking short, cold showers, eating cold meals, avoiding physical activity during central times of the day, or adapting clothing are known. At the dwelling scale, attempts are also made to block the entry of sunlight by using blinds.
- Habits to be implemented: in general, the specific bioclimatic criteria necessary for the climatic zone in which the activities have been developed for Madrid are not known. No awnings are installed for shading, delegating sun blocking only to the lesser effective blinds; strategies to favour evaporative cooling such as spraying water on curtains or mopping the floor are unknown; there is no housing design that incorporates cross-ventilation strategies.

Table 6.3 Summary of results for energy culture workshops

	Body scale	*Dwelling scale*
Habits to be eliminated		Misuse of ventilation. Placing water containers instead of choosing shallow containers, as they don't facilitate evaporative cooling. Evaporative coolers without proper ventilation.
Habits to maintain	Quick and cold showers. Cold meals. Avoiding outdoor activities during central hours.	Using blinds to block the sun.
Habits to implement		Installing awnings. Cross ventilation and evaporative cooling strategies such as mopping the floor or spraying water.

Source: Own elaboration.

For those households that can afford air conditioning, possible passive strategies to be implemented are not explored.

6.3.2 Results for collaborative mapping workshops

The results are summarised in four categories:

- Outdoor space: locations that have a recreational character and are not merely transit areas have been identified. For each location, the available sunlight/shading, the presence of blue infrastructure, the existence or absence of vegetation and the activities associated with it have been described. In general, the mapping activities described outdoor sites with no water infrastructure or drinking water points, little shading, and no strategically placed vegetation to provide relief in the summer. The mapped spaces do not incorporate a bioclimatic design that adapts materials and resources to reduce the effects of the urban microclimate.
- Indoor space: the climatic shelters available for summer use are mostly privately owned. On the one hand, there are swimming pools and residential facilities, which, although communal, are only available to households and their networks; on the other hand, there are spaces associated with commercial uses, such as the market, restaurants, or shopping mall; there are also communal places such as the residents' association or the church, but these are also places with restricted use. There are only indoor public places such as libraries, health centres, cultural centres, or municipal centres for older people, with limited opening hours and possibilities for use.
- Paths and routes: Through these mapping processes, it was possible to identify the most common existing routes and paths and how they are modified during the summer depending on the solar incidence or other overheating factors. During the rest of the year, the criteria for the layout of the routes are marked by the time variable, so that the most common routes are the shortest; in summer, these routes are modified, and the most common routes are not the shortest but those that incorporate the better shade. During the mapping sessions, reference was also made to the elimination of certain routes due to the impossibility of adapting

them to a path with more shade. Consequently, some places initially thought of as being useful as climatic shelters were discarded.
- Memories and disappeared places: in these mapping activities, those who had a more extended knowledge of the area described and identified old climatic shelters or elements that have disappeared and that served as a relief from the high temperatures. The disappearance of numerous water points, vegetation, and green spaces was noteworthy.

6.3.3 Results for walking tour workshops

This activity was carried out only once, with a sample of ten people. The results of this activity are classified into three categories:

- Strategies to cope with the high temperatures: some strategies that have to do with corporal expression and that did not emerge during the energy culture workshops could be registered during the walk. Actions such as fanning oneself or drinking water had not been mentioned among the strategies analysed in previous activities and could directly relate actions to places; looking for resting points and water refilling points conditioning the use of outdoor space were not aspects that received consideration for the routes described during the mapping, but they did during the walking tour activity.
- In-situ reporting of the experience of heat: during the walk, it was possible to collect the testimony of the people who participated in the activity in relation to their experience while experiencing high temperatures. Tiredness, annoyance, and overwhelming were sensations that were recurrently reported; thirst or the need to wear fewer clothes were also reported during the activity.
- Reporting resources and urban elements in poor conditions: this activity served to report numerous flaws or aspects to improve in terms of the urban design. The needs identified allowed numerous ideas and improvements for the space to be incorporated when using the space during the walking tour activity.

The latter can be exemplified in the more concrete results that were obtained in the mapping and walking sessions from knowledge of the participants' situations. Two specific locations can be singled out as particularly relevant for the participants in summer. The first of them is a location aggravating the experience of heat, and the second one is an attenuator of the same. The sidewalk space in front of the pharmacy was identified as a place prone to generating discomfort due to overheating. The pharmacy was considered a hotspot because of the participants' need to go to the pharmacy for medications and the absence of a shaded space for the waiting queues that form in front of the door (Figure 6.4). On the other hand, the local church emerged as a commonly agreed-upon place that could alleviate excessive heating conditions among the participants. The building's thick walls provided a cool space inside (Figure 6.5). This is a space open to visitors, and many of the people spoken to declared it as a space they would take refuge in during times of extreme heat in the neighbourhood.

6.4 Discussion and conclusion

The experiences shared in this chapter aim to contribute to the knowledge on the characterisation of the lived experience of energy poverty. The activities that are described here can be replicated in different settings with different groups of people to generate more insights.

Figure 6.4 C. Guetaria Pharmacy. As can be seen in the image, the absence of shading elements on the facade encourages situations of overexposure to the sun.

Source: Photography: Google Street View.

Figure 6.5 San Bartolomé Church. Several participants mentioned this place as a cool resource to go to on the hottest days.

Source: Photograph from Google Street View.

It has demonstrated the usefulness of PAR (Revez et al., 2020) and the adaptation of methods such as focus groups for the detection of EP (Energy Poverty Advisory Hub, 2023), the characterisation of SEP, and the in-depth study of the phenomenon from the body scale to the urban scale.

In particular, Energy Culture Workshops made it possible to characterise the existing deprivation of cooling and situations of energy poverty through the habits of household use (Gayoso Heredia et al., 2022). In addition, taking a deeper look at the gender perspective, it is possible to detect how household installations and resources are managed to cope with heat. For example, ventilation schedules or shading are added to all the caring tasks carried out by women. These also contribute to a greater understanding of the gendered nature of energy poverty, as the most vulnerable households depend on these passive cooling strategies to a greater extent than those households that can afford to use air conditioning (Nicholls & Strengers, 2018). On the other hand, the use of these workshops has achieved the intended secondary objective of intergenerational and intercultural knowledge sharing.

The mapping activities enabled participants to focus on their urban perceptions in relation to heat and how daily pathways related to it are enacted. These pathways are more complex than the home-work-home displacements of full-time employed workers, as they involve more stops and non-linear routes (Gil Sola & Vilhelmson, 2022). These daily care-related trajectories are affected by different factors such as affinities and security perception, points of relationship with the environment (water points, benches and other street furniture, etc.), the existence or not of climatic shelters, the public or private nature of the climatic shelters and their access rules, etc.

Through these mapping activities, it has been possible to detect how informal networks fulfil citizens' needs that are not covered by public resources. As an example of this, swimming pools in private communities and affinity or caring networks emerge as *cooling* resources.

The use of walking tours has allowed for other forms of knowledge expression to emerge, such as situated knowledge (Haraway, 1988), and to discover what is otherwise not accessible through traditional interviews. These types of understandings about specific geographic locations achieve an in-situ account of the experience of heat that is not possible by conducting traditional interviews alone.

In relation to the different results obtained according to the workshop modality, some interesting aspects emerged. In the energy culture workshops, the results mainly focus on the patterns of use, habits, and concrete actions aimed at reducing energy consumption at the household and building level. There are hardly any results for solutions outside the home or actions that have to do with the body in public spaces. Therefore, the results obtained in the collaborative mapping workshops are important to characterise the phenomenon of summer energy poverty as they contribute to an urban and microclimatic scale that complements the discourse on patterns of use inside the dwelling. The emotional and subjective aspects are only present during the walks, where the actions to combat the heat in situ allow for the construction of a personal heat perception narrative.

In addition to the secondary objectives pursued, such as the revaluation of knowledge, community work, and the strengthening of other methods of recording and expression, the implementation of the PAR and adaptation of the focus group method in the study of SEP is important in terms of its potential contribution to the design of public policies and urban design. This experiment makes it possible to address urban heat as a public matter of concern. The mixed methodologies implemented allow to evaluate the social and local dimension of urban heat, yielding fundamental information for climate change adaptation in the cities.

Acknowledgements

This research was conducted within the scope of the COOLTORISE project, funded by the European Union's Horizon 2020 research and innovation programme under the Grant Agreement No. 101032823. Participants provided informed consent for publication of the images included in this chapter.

It has also been granted by *Next Generation EU*-Margarita Salas grants, Universidad de Alicante (MARSALAS21-32).

Universitat d'Alacant Universidad de Alicante · Gobierno de España · Ministerio de Universidades · Plan de Recuperación, Transformación y Resiliencia · Financiado por la Unión Europea NextGenerationEU

Note

1 Graphical information at https://www.eea.europa.eu/data-and-maps/figures/percentage-of-households-unable-to. Regarding a more recent survey on winter conditions (2020), 8% of EU population declare the inability to keep home adequately warm. https://ec.europa.eu/eurostat/web/products-eurostat-news/-/ddn-20211105-1. There is no updated data for summertime conditions.

References

Brager, G. S., & de Dear, R. J. (1998). Thermal Adaptation in the Built Environment: A Literature Review, *Energy and Buildings*, 27(1), 83–96. https://doi.org/10.1016/S0378-7788(97)00053-4

Chiu, L. F. (2003). Transformational Potential of Focus Group Practice in Participatory Action Research, *Action Research*, 1(2), 165–183. https://doi.org/10.1177/14767503030012006

Day, R., & Hitchings, R. (2011). 'Only Old Ladies Would Do that': Age Stigma and Older People's Strategies for Dealing With Winter Cold, *Health & Place*, 17(4), 885–894. https://doi.org/10.1016/J.HEALTHPLACE.2011.04.011

De Haro, M. T., & Koslowski, A. (2013). Fuel Poverty and High-Rise Living: Using Community-Based Interviewers to Investigate Tenants' Inability to Keep Warm in Their Homes, *Journal of Poverty and Social Justice*, 21(2), 109–121. https://doi.org/10.1332/175982713X668917

Energy Poverty Advisory Hub. (2023). Energy Poverty Advisory Hub (EPAH) Handbook 1: A Guide to Energy Poverty Diagnosis. European Commission.

Founda, D., Pierros, F., Katavoutas, G., & Keramitsoglou, I. (2019). Observed Trends in Thermal Stress at European Cities with Different Background Climates, *Atmosphere*, 10(8), 436. https://doi.org/10.3390/atmos10080436

Gayoso Heredia, M., Sánchez-Guevara Sánchez, C., Núñez Peiró, M., Sanz Fernández, A., López-Bueno, J. A., & Gómez Muñoz, G. (2022). Mainstreaming a Gender Perspective into the Study of Energy Poverty in the City of Madrid, *Energy for Sustainable Development*, 70, 290–300. https://doi.org/10.1016/J.ESD.2022.08.007

Gayoso, M., Torrego, D., San Nicolás, P., Núñez-Peiró, M., & Sánchez-Guevara Sánchez, C. (2022). Methodological Action Framework: Energy Poverty Definition, Understanding and Policy Framework and Summer Energy Poverty Specificities. Universidad Politécnica de Madrid. http://cooltorise.eu/

Gil Sola, Ana, & Vilhelmson, Bertil. (2022). To Choose, or Not to Choose, a Nearby Activity Option: Understanding the Gendered Role of Proximity in Urban Settings, *Journal of Transport Geography*, 99. https://doi.org/10.1016/j.jtrangeo.2022.103301

Grey, C. N. B., Schmieder-Gaite, T., Jiang, S., Nascimento, C., & Poortinga, W. (2017). Cold Homes, Fuel Poverty and Energy Efficiency Improvements: A Longitudinal Focus Group Approach, *Indoor and Built Environment*, 26(7), 902–913. https://doi.org/10.1177/1420326X17703450

Haraway, D. (1988). Situated Knowledges: The Science Question in Feminism and the Privilege of Partial Perspective, *Feminist Studies*, 14(3), 575–599. https://doi.org/10.2307/3178066

Hunt, J, Aktas, C. R., Mahalov, Y. D., Moustaoui, A., Salamanca, M., & Georgescu, F. (2018). Climate Change and Growing Megacities: Hazards and Vulnerability, *Proceedings of the Institution of Civil Engineers – Engineering Sustainability*, 171(6), 314–326. https://doi.org/10.1680/jensu.16.00068

IPCC. (2022). *Climate Change 2022 Summary for Policymakers*.

Leivas, M. (2019). From the Body to the City: Participatory Action Research With Social Cartography for Transformative Education and Global Citizenship, *Educational Action Research*, 27(1), 40–56. https://doi.org/10.1080/09650792.2018.1519453

Longhurst, N., & Hargreaves, T. (2019). Emotions and Fuel Poverty: The Lived Experience of Social Housing Tenants in the United Kingdom, *Energy Research & Social Science*, 56, 101207. https://doi.org/10.1016/j.erss.2019.05.017

Lorenc, A., Pedro, L., Badesha, B., Dize, C., Fernow, I., & Dias, L. (2013). Tackling Fuel Poverty Through Facilitating Energy Tariff Switching: A Participatory Action Research Study in Vulnerable Groups, *Public Health*, 127(10), 894–901. https://doi.org/10.1016/j.puhe.2013.07.004

Middlemiss, L., & Gillard, R. (2015). Fuel Poverty from the Bottom-up: Characterising Household Energy Vulnerability Through the Lived Experience of the Fuel Poor, *Energy Research and Social Science*, 6, 146–154. https://doi.org/10.1016/j.erss.2015.02.001

Nicholls, L., & Strengers, Y. (2018). Heatwaves, Cooling and Young Children at Home: Integrating Energy and Health Objectives, *Energy Research and Social Science*, 39, 1–9. https://doi.org/10.1016/j.erss.2017.10.002

O'neill, T., Jinks, C., & Squire, A. (2006). Heating Is More Important Than Food, *Journal of Housing For the Elderly*, 20(3), 95–108. https://doi.org/10.1300/J081v20n03_07

Pelenur, M. J., & Cruickshank, H. J. (2013). Investigating the Link between Well-Being and Energy Use; An Explorative Case Study between Passive and Active Domestic Energy Management Systems, *Building and Environment*, 65, 26–34. https://doi.org/10.1016/j.buildenv.2013.03.020

Pink, S. (2011). Ethnography of the Invisible: Energy in the Multisensory Home, *Ethnologia Europaea*, 41(1). https://doi.org/10.16995/ee.1082

Revez, A., Dunphy, N., Harris, C., Mullally, G., Lennon, B., & Gaffney, C. (2020). Beyond Forecasting: Using a Modified Delphi Method to Build Upon Participatory Action Research in Developing Principles for a Just and Inclusive Energy Transition, *International Journal of Qualitative Methods*, 19, 160940692090321. https://doi.org/10.1177/1609406920903218

Sanchez-Guevara, C., Núñez Peiró, M., Taylor, J., Mavrogianni, A., & Neila González, J. (2019). Assessing Population Vulnerability Towards Summer Energy Poverty: Case Studies of Madrid and London, *Energy and Buildings*, 190, 132–143. https://doi.org/10.1016/j.enbuild.2019.02.024

Santamouris, M., & Kolokotsa, D. (2015). On the Impact of Urban Overheating and Extreme Climatic Conditions on Housing, Energy, Comfort and Environmental Quality of Vulnerable Population in Europe, *Energy and Buildings*, 98, 125–133. https://doi.org/10.1016/j.enbuild.2014.08.050

Shirani, F., Groves, C., Henwood, K., Pidgeon, N., & Roberts, E. (2020). 'I'm the Smart meter': Perceptions of Smart Technology Amongst Vulnerable Consumers, *Energy Policy*, 144, 111637. https://doi.org/10.1016/j.enpol.2020.111637

Thomson, H., Simcock, N., Bouzarovski, S., & Petrova, S. (2019). Energy Poverty and Indoor Cooling: An Overlooked Issue in Europe, *Energy and Buildings*, 196, 21–29. https://doi.org/10.1016/j.enbuild.2019.05.014

Tsoulou, I., Andrews, C. J., He, R., Mainelis, G., & Senick, J. (2020). Summertime Thermal Conditions and Senior Resident Behaviors in Public Housing: A Case Study in Elizabeth, NJ, USA, *Building and Environment*, 168(September 2019), 106411. https://doi.org/10.1016/j.buildenv.2019.106411

Wolf, J., Adger, W. N., & Lorenzoni, I. (2010). Heat Waves and Cold Spells: An Analysis of Policy Response and Perceptions of Vulnerable Populations in the UK, *Environment and Planning A*, 42(11), 2721–2734. https://doi.org/10.1068/a42503

Yoon, H., & Saurí, D. (2019). 'No More Thirst, Cold, or Darkness!' – Social Movements, Households, and the Coproduction of Knowledge on Water and Energy Vulnerability in Barcelona, Spain, *Energy Research and Social Science*, 58. https://doi.org/10.1016/j.erss.2019.101276

Yoon, S., Woo, S., Kim, J., Hwang, S. W., & Kweon, S. J. (2022). The Location Routing Problem for Cooling Shelters During Heat Waves, *Urban Climate*, 44, 101138. https://doi.org/10.1016/j.uclim.2022.101138

7 What Energy Poverty Looks Like

Methodological Insights from a Study in the Republic of Georgia

Nino Antadze

7.1 Introduction

The lived experience of energy poverty is a context-dependent phenomenon influenced by geographic, historical, social, cultural, and economic factors (González-Eguino, 2015; Sovacool et al., 2012). Understanding the complex nature of the lived experience of energy poverty is important for developing effective solutions to this problem. Therefore, the question of *how* to document and study the lived experience of energy poverty gains importance.

In studying and assessing energy poverty, quantitative methods seem to be the dominant methodological choice. For example, scholars recently noted the absence of detailed qualitative research to study energy poverty in Latin America and the Caribbean (Thomson et al., 2022). The studies that do use qualitative methods seem to be mostly using conventional research methods, such as interviews, document analysis, and observations (e.g., see McKague et al., 2016; Mould & Baker, 2017). Yet more recently, calls have been made to adopt relatively newer methods such as video ethnography since video footage serves as a record of a "multi-sensorial and multi-modal experience" (Pink, 2008, p. 190) that text alone is not able to convey.

This chapter outlines the process of investigating the lived experience of energy poverty in the Republic of Georgia using video ethnography. The methodological design of this research can be considered an experimental one. As restrictions due to the COVID-19 pandemic disrupted my travel plans, I decided to use the photovoice method (Wang & Burris, 1997) to engage study participants in data collection process. However, instead of taking photos, I asked study participants to depict on video energy poverty as they saw and experienced it. In this chapter, I will describe the research process, outline the methodological implications of this experience, and discuss the need for diversifying the methodological toolkit applied to studying and documenting energy poverty.

7.2 Methodological approaches to studying energy poverty

Discussions on the lived experience of energy poverty have spanned various aspects of this topic in different parts of the world (Ambrose et al., 2021; Cao & Frigo, 2021; Furszyfer Del Rio & Sovacool, 2023; Porto Valente et al., 2022; Yip et al., 2020). From the methodological perspective, these studies mostly rely on interviews, study visits, participant observations, and quantitative surveys (Ambrosio-Albala et al., 2020; Furszyfer Del Rio & Sovacool, 2023; Porto Valente et al., 2022; Yoon & Saurí, 2019). In this diversity of methodological approaches, the application of visual methods is relatively limited. A Scopus search with the combination of keywords "visual"/"video"/"visual ethnography"/"visual method" AND "energy poverty" resulted only in three results. Two papers were about modelling (Vasconcelos et al., 2021) and cost equation

DOI: 10.4324/9781003408536-9

(Majhi et al., 2021) and therefore were not relevant to this research. One more paper explored energy-vulnerable households in Greece; however, it did not use visual methods to explore the lived experience of energy poverty (Kaliampakou et al., 2021). In this study, video was used for survey interviews because of the inability to conduct them in person due to the on-going COVID-19 pandemic. I next searched for the papers with the keyword "video ethnography" without specifying "energy poverty" as a keyword. This search resulted in 42 results, but most were not about energy consumption or poverty. I found one study by Pink and Mackley (2012), in which the authors document the use of video ethnography to study domestic energy consumption. The lack of studies using visual methods to explore the lived experience of energy poverty, and video ethnography in particular, is a noteworthy gap as visual methods can contribute much to documenting, studying, and understanding the lived experience of energy poverty.

Scholars studying energy poverty can draw on literature discussing and describing video ethnography as one of the research methods (e.g., Vannini, 2020). Of particular interest for this book chapter are the studies in which videos are recorded not by scientists, but by citizens. Jones and Raymond (2012) refer to such videos *as third-party videos*, when recordings are made by nonscientists or nonresearchers, but are used for social science research. In their article, Jones and Raymond (2012) describe how they used video recordings of police-citizen interactions created by neighbourhood residents. In this case, the videos were not intended to be used in the research project initially; rather, they became data afterward. In our study, videos were created by nonscientists; however, they were recorded with the intention of being used as data. In the following sections, I describe how we engaged study participants in creating videos and how we used these videos to study energy poverty in the Republic of Georgia.

7.3 Using video to study the lived experience of energy poverty in the Republic of Georgia: the description of the process

The study aimed to investigate and document the manifestations of energy poverty in Georgia. According to the initial research design, my initial plan was to travel to Georgia during the summer of 2021 and winter of 2021–2022 to collect data through in-person video ethnography and interviews. While thinking about redesigning the study without me requiring to travel, I was inspired by the photovoice method (Wang & Burris, 1997), which enables study participants to identify and document issues of their concern and interest. As Sutton-Brown explains (2014, p. 170), "photovoice interrogates contextually based meanings from an insider perspective as a means to generate new insights into our socially constructed realities and cultures. It oscillates between private and public worlds in its attempt to publicize and politicize personal struggle via photography, narratives, critical dialogue, and social action." Instead of asking them to take photos of energy poverty around them, I asked study participants to depict on video energy poverty as they saw and experienced it. In what follows, I explain the key stages of conducting this study.

7.3.1 Recruiting study participants

The first step was to recruit study participants who would be interested and willing to collect the data through videos and engage in conversations about energy poverty. My collaborator, who lives and works in Georgia, helped reach out to possible participants. However, the most effective method was posting the request on my Facebook account. It should be noted that Facebook is very actively used in Georgia, and most of the societal and political discussions are unfolding on Facebook. After posting the invitation on my Facebook page, several people reached out to me. After initial exchanges, three of them decided to participate in the project. All three participants were female, in their 20s or 30s, and lived in different parts of Georgia. I had an online conversation, using the video conferencing platform Zoom, with each study participant to explain the expectations of participation and answer any questions they had. These conversations were very useful in getting to know each other and for establishing expectations and procedures relating to the research. In particular, we discussed the ethical implications of the research and the required ethics protocols to follow (please see the section below for more details). I also clarified that study participants would have the possibility to comment on the edited video. In fact, during the interviews that I conducted after the data was collected, I consulted with the study participants about which themes they thought were dominant in their videos and why. Study participants were offered GBP 50 honorarium for participating during the summer and winter (i.e., they were offered GBP 100 if they participated both in the summer and winter periods). The honorarium was offered to compensate for their time and efforts and to reimburse any costs they might have incurred (e.g., transportation).

7.3.2 Data collection

Study participants collected video data during two periods: August–September 2021 and January–February 2022. The study design gave them full freedom in choosing locations, topics, and how they wanted to record the videos. During our initial Zoom meeting, we discussed the concept of energy poverty and what it implies. For example, I highlighted that the lived experience of energy poverty could be visually depicted not only at home but also in such spaces as libraries, for example, as well as in relation to transportation and conditions of working and studying (especially given the need for teleworking and remote learning during the COVID-19 pandemic). The participants were asked to shoot videos with a total length of about 20 minutes per three-week period in winter and again in summer (a total of 40 min). It was up to the study participants how long each video would last (e.g., they could choose to create four 5-minute-long videos or several videos of various lengths). All participants drew on their personal experiences and immediate surroundings. As a result, the produced video essays depict energy poverty experiences in different parts of the country. Participants also had different approaches while editing (or not editing) their videos.

7.3.3 Debrief and production of video essays

After the data collection was completed, I conducted a second Zoom interview with each participant. During these interviews, we discussed their impressions/observations of the energy poverty around them, why and how they collected the data (e.g., why they chose to record the specific video footage), and anything beyond the collected data that they thought was important to consider in this study. As this process was new and interesting for me, I asked the study participants about their experiences and impressions of the data collection process. All of them noted that they found the process not only interesting but also enjoyable. They noted that the videos provide an immersive experience because the viewer can imagine themselves in the depicted places and can see what the lived experience looks like. Some also noted that videos helped them remember details that they may have forgotten or pay attention to nuances that they may have otherwise ignored.

Next, I edited collected video footage and added titles and captions where needed. I shared edited video essays with study participants to receive their comments and approval prior to uploading them online and sharing them with a broader audience. In the information letter and consent form, it was clarified that the videos could be shared through social media after the study participants reviewed and approved the videos. As a result, eight video essays were produced.[1]

7.3.4 Receiving approval from the research ethics board to conduct research

Two separate applications were submitted to the University of Prince Edward Island Ethics Review Board to receive approval to conduct the research. The first application was related to the recruitment of the study participants and their engagement in the data collection process. Below is an excerpt from the information letter to the study participants that details what participants are expected to do should they choose to collaborate on this project:

> Participation in this study is voluntary. During the two three-week periods in summer 2021 and winter 2021, you will be asked to create video diaries (using a smartphone or any other video recording device of your choosing) that depicts your observations of energy poverty around you. The total length of the video diaries per three-week period should be about 20 min. It is up to you how long each video diary may last (for example, you may choose to create four 5-min-long video diaries, or several video diaries of various lengths). The locations, settings, and any other details relating to the video recordings are up to you. If you decide to record certain people or/and their private property, you will need to ask for written informed consent from them. I have enclosed these forms in this information letter. If you decide to participate in this study, we will set up a Zoom meeting where I will provide more information about this process and will answer any questions that you may have.

Study participants could depict energy poverty in their immediate surroundings (e.g., in their homes) or in other spaces (e.g., on the street). In case study participants decided to record certain people and/or their private property, they needed to obtain written informed consent.

The second application was submitted to receive permission to conduct interviews with the study participants after the data collection phase was completed. The interviews were conducted individually with each study participant via Zoom and lasted for about 40–60 minutes.

In both applications, recruitment and information letters, as well as consent forms, were produced both in English and Georgian.

7.4 The lived experience of energy poverty as depicted in the videos

The study participants decided to depict the manifestations of energy poverty in different locations. One study participant documented energy poverty-related issues in a rural area; one participant focused on a city; and the third participant shot videos both in urban and rural locations. Two study participants depicting rural areas took videos in two different villages. Although these villages are located in different parts of the country, they are both located at high elevations and share some similarities concerning the lived experience of energy poverty. For example, in both places, the electricity supply is not reliable, there is a lack of access to the natural gas network, and as a result, residents heavily rely on firewood for cooking and heating. The images below are taken from the video essays depicting energy poverty during the winter months in the village of Ushguli in Western Georgia and the village of Patara Mitarbi in Eastern Georgia.

The first image (Figure 7.1) shows local residents using a wood stove to heat water in Ushguli. The amount of water heated on the wood stove suggests that water is heated for multiple

Figure 7.1 Heating water on a wood stove in Ushguli.

purposes: washing dishes, doing laundry, and cooking. The second image (Figure 7.2) depicts a similar woodstove in Patara Mitarbi, where firewood is also used for heating and cooking.

Other important challenges depicted in the videos relate to mobility issues and the lack of access to the Internet and telecommunication networks. Regarding mobility, rural areas lack safe and well-maintained roads, which makes it very difficult—and at times impossible—to travel to and from these villages, especially in winter. The lack of transportation is particularly important,

Figure 7.2 Wood stove used for heating in winter in Patara Mitarbi.

as the residents of rural areas must travel to the nearest towns and cities for various purposes, including medical care.

These rural areas also lack access to the internet and phone connectivity year-round. For example, during the COVID-19 pandemic when most higher education institutions and schools transitioned to remote learning, the lack of access to the internet put students from rural areas at a disadvantage.

The study participant who depicted energy poverty in the urban area paid attention to the lived experience of those who work outdoors year-round. She chose to depict the working conditions of street vendors and small traders in open markets and street shops. These settings are usually not equipped with air conditioning in the summer or heating in the winter. When asked why she chose to focus on street vendors, the participant explained that these people earn very little and are often ignored by the authorities and the public. According to the study participant, street vendors are everywhere and you can see them as you walk; on the other hand, their problems and harsh working conditions remain beyond the attention of many (see more on street vendors in Georgia in Polese et al., 2016; Rekhviashvili, 2015). The images below are taken from the video essay depicting the experiences of street vendors and small traders during the summer months in Tbilisi. The first image (Figure 7.3) shows a small shop where the only cooling method is a fan. The second image (Figure 7.4) is taken from the video footage of a flea market with little shade.

Another study participant documented the electricity infrastructure in the capital city of Tbilisi, especially the networks of illegal electricity cables, which can be noticed on almost any street. Figure 7.5 shows one such example.

Finally, study participants paid attention to the travel conditions relating to public transit (mainly intercity transit), especially during the summer months, when the temperatures may reach 40 degrees Celsius and when most of the vehicles either do not have functioning air conditioning systems or do not turn them on to save fuel.

Figure 7.3 A small shop selling food in Tbilisi in summer.

Figure 7.4 A flea market in summer in Tbilisi.

Table 7.1 summarises the key topics raised in the video essays.

Most of the topics raised in the video essays are discussed in energy poverty literature. For example, studies explore the lack of access to basic energy services such as cooking, heating, and lighting (Harrison & Popke, 2011; Jiang et al., 2020). This is particularly relevant to rural areas—a conclusion that was confirmed in this study as well. In relation to other topics raised in the video essays, there is discussion about mobility and specifically the lack of well-maintained road networks and limited options for transportation (Sovacool et al., 2012). Our findings reveal

Figure 7.5 Electricity cables on Tbilisi street.

Table 7.1 The summary of topics depicted in video essays

Location	*Season*	*Topics depicted in the video essays*
Rural		
Patara Mitarbi	Summer	• Damaged and dangerous road network • Lack of access to natural gas • Lack of access to communication services (Internet, TV networks, etc.) • Use of firewood for cooking • Poorly developed water supply network
	Winter	• Not maintained (snow is not cleaned) and dangerous road network • Lack of access to communication services (Internet, TV networks, etc.) • Use of firewood for cooking, heating • Electricity with low voltage, inability to use home appliances (e.g., washing machine) • Poorly developed water supply network
Ushguli	Summer	• Electricity supply is not reliable • Lack of access to communication services (Internet, TV networks, etc.) • Use of firewood for cooking • Working outside on the farm without mechanical power (e.g., tractor)
	Winter	• Road and street network not cleaned after heavy snowfall • Roads that are not maintained and cleaned after a rock slide • Electricity supply is not reliable • Use of firewood for cooking, heating, washing
Urban		
Rustavi	Summer	• Insufficient intercity transportation • Lack of air conditioning in intercity transport
Tbilisi	Winter	• Problems with electricity cable network, especially illegal electricity extensions • Natural gas network in the city, especially makeshift natural gas networks in houses that may not meet safety standards • Working conditions of street vendors and small trades in winter, lack of heating
	Summer	• Working conditions of street vendors and small trades in summer, lack of cooling

that mobility problems are relevant to both rural and urban areas, although they manifest themselves differently. There are several topics in video essays that are not as actively discussed in energy poverty literature, for example, the working conditions of those who work outdoors year-round. There is also relatively limited discussion on access to communication services such as the internet, TV networks, and so forth (Ochoa & Graizbord, 2016).

Overall, there is very limited literature on energy poverty in Georgia. It is also noteworthy that in consulted reports about energy poverty in Georgia (Kvaratskhelia et al., 2019; WeResearch, 2020), the issues associated with mechanical energy, mobility, or access to the internet and telecommunications are not discussed. The lack of literature on energy poverty in Georgia aligns with a more general trend in academic literature: studies that explore energy poverty in Central and Eastern Europe are relatively rare, and Georgia is not explored in any of them (Buzar, 2016; Karpinska & Śmiech, 2020; Mazurkiewicz & Lis, 2018). Going beyond documenting the Western European experiences of energy poverty is an important scholarly endeavour, as these experiences vary and are embedded in broader historical, political, cultural, socioeconomic, and geographical contexts.

7.5 Considerations for future research to study the lived experience of energy poverty

In this section I discuss some of the key considerations relevant to studying the lived experience of energy poverty using video ethnography and close collaboration with study participants. The discussion below draws on my experience of conducting this research and hopefully will motivate scholars to pursue this line of inquiry.

7.5.1 Research participants as collaborators

In our research, study participants were coproducers of knowledge. They were not treated as subjects of inquiry but rather as full-fledged collaborators who had control over data collection choices and who largely shaped the study findings.

Research designs similar to the one described in this chapter may change researcher-study participant relationships and their roles during the research process. Such methodological approaches encourage a more inclusive, participatory research process and, to some extent, challenge the superiority of the researcher's "expert" knowledge. In this case, study participants are free to collect data on the topics they consider important. Therefore, asking participants to depict the lived experience of energy poverty the way they see and experience it has epistemological implications, as "the acknowledgment of, and willingness to share, authorship gives agency to your project participants" (Garrett (2011, p. 529). The knowledge is generated with the participation of those for whom energy poverty is part of their everyday lives, either directly experienced, directly observed, or both. Such an approach aligns with what Pink et al. (2017, p. 372) describe as creating *empathetic encounters* with the experiences of others by using video "to access and understand participants' experience within the unspoken, mundane, routine, and unstopping flow of everyday life (by inviting them to attend to and, in collaboration with the researcher, account for this through co-documentation)." This research process is also similar to what Garrett (2011) describes as the creation of a *participatory video*, which is when study participants are actively coproducing research. Depending on the level of engagement in the research project, study participants can become coauthors of written outputs or academic presentations.

It is noteworthy that all three study participants approached the task of shooting and editing videos differently (they were also given full freedom in this regard). Two participants decided to edit their videos. One of them created video stories with extensive narration and interviews, while the second participant edited videos without narration and used background music. The third participant collected video clips, which I edited after the data collection was completed. All study participants used their cell phones to record and edit video footage. During interviews, those study participants who edited videos noted that they found the editing process creative and enjoyable, especially when they saw their stories coming together. Two study participants decided to shoot videos in colour, and one study participant decided to use a black-and-white filter.

From a practical perspective, participatory research design may require more time to recruit and engage study participants, as well as discuss study aims and procedures. In addition, study participants may incur additional costs, such as transportation expenses, if they want to travel to specific locations to shoot videos. Such involvement also requires time and commitment. The question of who can effectively record their experiences and observations on video is also relevant. The ability to produce video recordings will depend on access to video recording devices (e.g., cell phones, video cameras) and the knowledge required to use them. These issues need to be considered while designing a research project (e.g., consider including honorarium expenses in the budget) and clearly articulated and discussed when recruiting study participants.

7.5.2 A trade-off between structure and flexibility during the data collection process

I found it important to provide full freedom to the study participants during the data collection process. As discussed above, I explained general expectations (e.g., timeframes, length of videos, etc.); otherwise, more creative details were up to the study participants. For all study participants in our project, shooting videos for a research project was a new experience, and they all approached it in a way that made them feel comfortable and confident. This point speaks to the need for built-in flexibility in designing similar research. Certain expectations will be in place, for example, in terms of timeframes and procedures; however, within these expectations, study participants can be given room for creativity and experimentation.

7.5.3 Ethical considerations

One of the aspects of the research that requires more structure relates to ethical procedures. As described above, I sought ethics approval for data collection and interviews in two separate applications. Once I received approval from the Research Ethics Board for the data collection stage and after I recruited study participants, I discussed with them the importance and need to follow ethics procedures. I provided information letters and consent forms in the native language of the study participants and answered questions they had. This is an important stage of the research and may require slowing down to make sure that study participants are aware of and fully understand ethics-related procedures.

7.6 Conclusion

In this chapter, I described the use of the video ethnographic method to study the lived experience of energy poverty. Visual methods more generally, and video ethnography in particular offer much promise in producing a rich and nuanced understanding of the lived experience of energy poverty in different parts of the world. This is particularly true in the countries that are understudied in the current academic literature on energy poverty, such as Georgia. This chapter paid particular attention to how study participants were actively involved in the research process, especially during the data collection method.

To conclude, this chapter calls for scholars to diversify the methodological toolkit applied to studying and documenting energy poverty. The lived experience of energy poverty is a complex phenomenon, and having multiple ways of studying and analysing it is important, both from a scholarly and a practical standpoint. Hopefully, this chapter will provide some ideas and encourage scholars to explore the possibility of using video in their research.

Acknowledgements

I am grateful to the Fuel Poverty Research Network for financially supporting this research. My special gratitude goes to the project collaborator Kety Gujaraidze and three study participants for their time and willingness to contribute to this research.

Note

1 The video essays can be viewed at this link: https://drive.google.com/drive/folders/17pdM1TVNTNdvG4Jkc3DOXib6DS-yH1Yb?usp=sharing

References

Ambrose, A., Baker, W., Sherriff, G., & Chambers, J., 2021. Cold Comfort: COVID-19, Lockdown and the Coping Strategies of Fuel Poor Households. *Energy Reports* 7, 5589–5596. https://doi.org/10.1016/j.egyr.2021.08.175

Ambrosio-Albala, P., Middlemiss, L., Owen, A., Hargreaves, T., Emmel, N., Gilbertson, J., Tod, A., Snell, C., Mullen, C., Longhurst, N., & Gillard, R., 2020. From Rational to Relational: How Energy Poor Households Engage with the British Retail Energy Market. *Energy Research & Social Science* 70, 101765. https://doi.org/10.1016/j.erss.2020.101765

Buzar, S. (2016). *Energy Poverty in Eastern Europe: Hidden Geographies of Deprivation.* New York: Routledge.

Cao, U., & Frigo, G., 2021. Of Social Movements, Human Rights and Electricity Access: Exploring an Indigenous Civil Resistance in Chiapas, Mexico. *Energy Research & Social Science* 75, 102015. https://doi.org/10.1016/j.erss.2021.102015

Furszyfer Del Rio, D.D., & Sovacool, B.K., 2023. Of Cooks, Crooks and Slum-Dwellers: Exploring the Lived Experience of Energy and Mobility Poverty in Mexico's Informal Settlements. *World Development* 161, 106093. https://doi.org/10.1016/j.worlddev.2022.106093

Garrett, B.L., 2011. Videographic Geographies: Using Digital Video for Geographic Research. *Progress in Human Geography* 35, 521–541. https://doi.org/10.1177/0309132510388337

González-Eguino, M., 2015. Energy Poverty: An Overview. *Renewable and Sustainable Energy Reviews* 47, 377–385. https://doi.org/10.1016/j.rser.2015.03.013

Harrison, C., & Popke, J., 2011. "Because You Got to Have Heat": The Networked Assemblage of Energy Poverty in Eastern North Carolina. *Annals of the Association of American Geographers* 101, 949–961. https://doi.org/10.1080/00045608.2011.569659

Jiang, L., Yu, L., Xue, B., Chen, X., & Mi, Z., 2020. Who Is Energy Poor? Evidence from the Least Developed Regions in China. *Energy Policy* 137, 111122. https://doi.org/10.1016/j.enpol.2019.111122

Jones, N., & Raymond, G., 2012. "The Camera Rolls": Using Third-Party Video in Field Research. *The Annals of the American Academy of Political and Social Science* 642, 109–123. https://doi.org/10.1177/0002716212438205

Kaliampakou, C., Papada, L., & Damigos, D., 2021. Are Energy-Vulnerable Households More Prone to Informative, Market, and Behavioral Biases? *Societies* 11, 126. https://doi.org/10.3390/soc11040126

Karpinska, L., & Śmiech, S. (2020). Invisible energy poverty? Analysing housing costs in Central and Eastern Europe. *Energy Research & Social Science*, *70*, 101670. https://doi.org/10.1016/j.erss.2020.101670

Kvaratskhelia, T., Margvelashvili, M., Sinea, A., Balan, S., Nazarenko, D., & Cenusa, D., 2019. Energy Poverty. Guidance for State Policy and Public Discourse in the Time of Reform. Georgia, Moldova, Ukraine, Romania: World Experience for Georgia, Tbilisi, Georgia.

Majhi, C.K., Nanda, S., Pradhan, R.C., & Mohapatra, B.G., 2021. An Approximate Cost Equation of Offshore Wind Turbine Blade, in: Das, B.B., Barbhuiya, S., Gupta, R., Saha, P. (Eds.), *Recent Developments in Sustainable Infrastructure, Lecture Notes in Civil Engineering*. Singapore: Springer Singapore, pp. 973–983. https://doi.org/10.1007/978-981-15-4577-1_81

Mazurkiewicz, J., & Lis, P. (2018). Diversification of Energy Poverty in Central and Eastern European Countries. *Virtual Economics* 1(1), 26–41.

McKague, F., Scott, M., Wooliscroft, B., & Lawson, R. (2016). Understanding the Energy Consumption Choices and Coping Mechanisms of Fuel Poor Households in New Zealand. *New Zealand Sociology* 31, 106–126.

Mould, R., & Baker, K.J., 2017. Documenting Fuel Poverty from the Householders' Perspective. *Energy Research & Social Science* 31, 21–31. https://doi.org/10.1016/j.erss.2017.06.004

Ochoa, R., & Graizbord, B., 2016. Privation of Energy Services in Mexican Households: An Alternative Measure of Energy Poverty. *Energy Research & Social Science* 18, 36–49. https://doi.org/10.1016/j.erss.2016.04.014

Pink, S., 2008. An Urban Tour: The Sensory Sociality of Ethnographic Place-Making. *Ethnography* 9, 175–196. https://doi.org/10.1177/1466138108089467

Pink, S., & Mackley, K.L., 2012. Video and a Sense of the Invisible: Approaching Domestic Energy Consumption Through the Sensory Home. *Sociological Research Online* 17, 87–105. https://doi.org/10.5153/sro.2583

Pink, S., Sumartojo, S., Lupton, D., & Heyes LaBond, C., 2017. Empathetic Technologies: Digital Materiality and Video Ethnography. *Visual Studies* 32, 371–381. https://doi.org/10.1080/1472586X.2017.1396192

Polese, A., Rekhviashvili, L., & Morris, J. (2016). Informal Governance in Urban Spaces: Power, Negotiation and Resistance Among Georgian Street Vendors. *Geography Research Forum* 36, 15–32.

Porto Valente, C., Morris, A., & Wilkinson, S.J., 2022. Energy Poverty, Housing and Health: the Lived Experience of Older Low-Income Australians. *Building Research & Information* 50, 6–18. https://doi.org/10.1080/09613218.2021.1968293

Rekhviashvili, L., 2015. Marketization and the Public-Private Divide: Contestations between the State and the Petty Traders Over the Access to Public Space in Tbilisi. *International Journal of Sociology and Social Policy* 35, 478–496. https://doi.org/10.1108/IJSSP-10-2014-0091

Sovacool, B.K., Cooper, C., Bazilian, M., Johnson, K., Zoppo, D., Clarke, S., Eidsness, J., Crafton, M., Velumail, T., & Raza, H.A., 2012. What Moves and Works: Broadening the Consideration of Energy Poverty. *Energy Policy* 42, 715–719. https://doi.org/10.1016/j.enpol.2011.12.007

Sutton-Brown, C. A. (2014). Photovoice: A Methodological Guide. *Photography and Culture* 7(2), 169–185. https://doi.org/10.2752/175145214X13999922103165

Thomson, H., Day, R., Ricalde, K., Brand-Correa, L.I., Cedano, K., Martinez, M., Santillán, O., Delgado Triana, Y., Luis Cordova, J.G., Milian Gómez, J.F., Garcia Torres, D., Mercado, C., Castelao Caruana, M.E., & Pereira, M.G., 2022. Understanding, Recognizing, and Sharing Energy Poverty Knowledge and Gaps in Latin America and the Caribbean – Because Conocer Es Resolver. *Energy Research & Social Science* 87, 102475. https://doi.org/10.1016/j.erss.2021.102475

Vannini, P. (ed.) 2020). *The Routledge International Handbook of Ethnographic Film and Video, Routledge International Handbooks*. London; New York: Routledge/Taylor & Francis Group.

Vasconcelos, P.N., Zambroni de Souza, A.C., Ribeiro, P.F., & Balestrassi, P.P., 2021. Modeling the Power Sector of Wonderland: an integrated system dynamics model to assess a sustainable regional development, in: 2021 IEEE Mexican Humanitarian Technology Conference (MHTC). Presented at the 2021 IEEE Mexican Humanitarian Technology Conference (MHTC), IEEE, Puebla, Mexico, pp. 38–45. https://doi.org/10.1109/MHTC52069.2021.9419950

Wang, C., & Burris, M.A., 1997. Photovoice: Concept, Methodology, and Use for Participatory Needs Assessment. *Health Education & Behavior* 24, 369–387. https://doi.org/10.1177/109019819702400309

WeResearch. (2020). *Monitoring the Implementation of the Sustainable Development Goals in Georgia.* Tbilisi, Georgia: Women Engage for a Common Future (WECF).

Yip, A.O., Mah, D.N., & Barber, L.B., 2020. Revealing Hidden Energy Poverty in Hong Kong: a Multi-Dimensional Framework for Examining and Understanding Energy Poverty. *Local Environment* 25, 473–491. https://doi.org/10.1080/13549839.2020.1778661

Yoon, H., & Saurí, D., 2019. 'No More Thirst, Cold, or Darkness!' – Social Movements, Households, and the Coproduction of Knowledge on Water and Energy Vulnerability in Barcelona, Spain. *Energy Research & Social Science* 58, 101276. https://doi.org/10.1016/j.erss.2019.101276

Part II

Lived Experiences

8 Not Equal Before the Energy System

Can Energy Justice Principles Lift Women in Rural West Africa Out of Energy Poverty?

Laurent Jodoin and Carelle Mang-Benza

8.1 Introduction

Sub-Saharan Africa's (SSA) countries are facing severe impacts of climate change, despite bearing the least responsibility for historic greenhouse gas emissions. These countries are being more severely confronted with the harmful effects of climate change due to their limited economic development and weaker capacity to cope with the damage caused by extreme geoclimatic phenomena, thus increasing their vulnerability. This vulnerability varies across and within each country but hits women in rural areas particularly hard due to the intersecting burdens of poverty, limited access to essential services, and normative gender roles (e.g., Energia 2017). Since energy production and its use have significant impacts on both climate change and human development, and in a context where energy deprivation disproportionately befalls the weakest communities, it is now increasingly accepted that energy issues impinge on matters of justice and fairness (Galvin, 2020; Jodoin, 2021; Jones et al., 2015; Sovacool et al., 2017).

The concept of energy justice (EJ) is gaining traction in academic and policy settings as a possible decision-support tool for policymakers. The ideal of an energy-just world promotes "happiness, welfare, and due process for both producers and consumers, distributing environmental and social hazards associated with energy production without discrimination" (Sovacool & Dworkin, 2015, p. 437). However, there is no clear path to materialize this lofty ideal into energy interventions. In light of the growing socio-economic inequalities between and within countries, also reflected in energy systems, it is opportune to accelerate the usability of energy justice by energy planners. This is particularly germane for rural SSA, where the nexus energy-poverty-gender is a daily reality and where electricity poverty and energy poverty (including, e.g., wood fuel) are bedfellows. In this context, the capability approach can provide the necessary space for a coherent interpretation of multiple justice principles and for identifying how certain groups or people (such as those in our survey) may be more vulnerable than others to certain impacts of our energy systems (Jodoin, 2021; Mang-Benza et al., 2023; Melin et al., 2021).

This chapter synthesizes findings from a three-year project, funded by the International Development Research Centre (IDRC, Canada), aiming at generating new evidence that can help improve energy efficiency, access to energy, energy justice, and climate change mitigation and adaptation in SSA. This project took place over the period 2017–2020 and focused on engaging both women and men through surveys in 2,290 rural households in Benin, Senegal, and Togo.

DOI: 10.4324/9781003408536-11

8.2 Energy in rural West Africa

All conventional fossil fuels and renewable energy sources are abundantly available on the African continent (IEA, 2022). However, for the majority of the 1.1 billion Africans, accessing safe and reliable energy is a daily challenge. The continent is fraught with dichotomies between resource potential and access to these resources; geographic disparities between regions north and south of the Sahara; between urban and rural areas; as well as gender inequalities between men and women, even within the same household. In rural areas of sub-Saharan Africa, energy infrastructure is cruelly lacking, and women are particularly disadvantaged, as this chapter intends to show. Our results show that the vast majority of households' heads (HH) are men (81.2 to 90.9%) with no schooling (33.3 to 73.9%) and working in agriculture (47.8 to 72.7%). The overall portrait of women heads is similar where the majority (58.9 to 78.2%) have no schooling and a main occupation in small business (23.8 to 67.9%). Dividing the average household income by the average number of people yields an average monthly income per person of approximately 15USD in the three countries. Despite the prolific body of work on gender inequalities in energy systems and the gendered face of energy poverty (Clancy et al., 2002; Lee et al., 2021; Mang-Benza et al., 2023; Oparaocha & Dutta, 2011; Pachauri & Rao, 2013), most solutions advanced to address energy deprivation in rural Africa emphasize technological fixes, cost-benefit analyses, and economic modelling. The concept of energy justice offers an opportunity to challenge this approach (Guruswamy, 2010; McCauley et al., 2013; Wilkinson et al., 2007).

8.3 A principled approach to energy justice

Incorporating many dimensions of justice, such as distributive, recognition, and procedural, *energy justice* is articulated in two main ways (Wood & Roelich, 2020). There is, on the one hand, the conception outlined in McCauley et al. (2013) and called the *triumvirate conception of energy justice* and on the other hand, the approach developed by Sovacool and Dworkin (2015), later expanded as the *principled approach* (Jodoin, 2021; Sovacool et al., 2016; Sovacool et al., 2017). We will focus on the principled approach and the operationalization of its principles presented in Table 8.1.

The EJ framework (EJF) can be seen as aiming at "a global energy system that fairly disseminates both the benefits and costs of energy services, and one that has representative and impartial energy decision-making" (Sovacool & Dworkin, 2015: 436). It includes ten principles or criteria of justice and equity, and thus several values (see Table 8.1). Whilst presenting an arguably comprehensive normative framework, the EJF still needs work before it can be operationalized, especially in developing countries. For instance, justice principles tend to be rather vague in terms of the proper metric to adopt, for example, primary goods as advocated by Rawls (1971). This study makes a contribution to the literature based on the capabilities approach (as also proposed by Jodoin, 2021). The capabilities approach (CA) offers the benefit of having been operationalized both by the UN and several local and national governments (Robeyns & Brighthouse, 2010). This approach, developed most prominently by Amartya Sen and more recently also by Martha Nussbaum, centres around liberty (Nussbaum and Sen, 1993; Sen, 1999, 2009). It sees Rawlsian ideas on primary goods as a means towards an end and considers social factors such as institutions and social norms as factors influencing the realization of capabilities.

Capabilities can be seen as a person's real freedoms or opportunities to achieve functioning, i.e., various states of human beings and activities that a person can undertake (Sen, 1999, 2009). While functionings are actual achievements, capabilities are "freedoms in terms of

Table 8.1 Energy justice principles

Principe	*Description*
Availability	People deserve sufficient energy resources of high quality.
Affordability	The provision of energy services should not become a financial burden for consumers, especially the poor.
Due process	Countries should respect due process and human rights in their production and use of energy.
Transparency and accountability.	All people should have access to high-quality information about energy and the environment, and fair, transparent, and accountable forms of energy decision-making.
Sustainability	Energy resources should not be depleted too quickly.
Intragenerational equity	All people have a right to fairly access energy services.
Intergenerational equity	Future generations have a right to enjoy a good life undisturbed by the damage that our energy systems inflict on the world today.
Responsibility	All nations have a responsibility to protect the natural environment and reduce energy-related environmental threats.
Resistance	Energy injustice must be actively, deliberately opposed.
Intersectionality	Expanding the idea of recognitional justice to encapsulate new and evolving identities in modern societies, as well as acknowledging how the realization of energy justice is linked to other forms of justice, e.g., socio-economic, political and environmental.

Source: From Sovacool et al. (2017).

[a person's] actual ability to do the different things a person has reason to value doing or being" (Sen, 2009: 16). Justice is thus a matter of being in a position to enjoy and develop our own capabilities but also of devising the appropriate institutions and social arrangements that can secure this position (Jodoin, 2021). Applied to energy, the focus on individuals' capabilities within their socio-ecological environment can highlight (as we will show) how certain demographics and groups might be more vulnerable than others to adverse impacts of our energy systems (Day et al., 2016; Velasco-Herrejon & Bauwens, 2020). Consequently, what is sought is not energy *per se* but rather the services *and* global opportunities energy can provide for people striving to achieve what they value.

This project is born from a partnership between the IDRC (a public funding agency in Canada) and Econoler (a private firm), in which the first author (L. Jodoin) was the principal investigator. The main goal was to improve energy justice and capabilities in rural SSA, especially for women. The methodology mainly consisted of a quantitative survey of 2,291 rural households (640 in Benin, 1,001 in Senegal, and 650 in Togo, randomly chosen through two strata of geo-socio-economic factors) and qualitative surveys (i.e., focus groups with 8–10 persons, both males and females; 24 in Benin, 106 in Senegal, and 10 in Togo), while engaging with local governments and NGOs (e.g., Energy4Impacts). For the quantitative survey, a questionnaire of about a hundred questions was developed in French and then translated into local languages; questions were asked in local languages in the three countries to the head of the household and his wife (if applicable, or the woman head). Each interview lasted approximately 45 minutes and was held within the household in local languages. Surveyors were mainly students and were selected for both their skills (e.g., academic background in social sciences) and their knowledge of local culture and languages. As the next sections will show, the data were collected in order to assess the EJF criterion as well as the sex-disaggregated capabilities of the adults in the households (except the elders and children). These capabilities have been addressed with both quantitative and qualitative data.[1]

8.4 From principles to lived experiences

As mentioned, the quantitative results (Jodoin et al., unpublished) show that the vast majority of heads of households are men with no schooling and working in agriculture. The average household monthly income for the three countries is 145 USD, with Togo's being half of Senegal's. The differences between these numbers are in line with national statistics such as the GDP per capita (World Bank, 2023). At the same time, the household average number of people is the lowest in Togo, and this may affect average monthly income (as well as the individual energy poverty level). For reasons of space here and because the research project did not investigate the global governance issues (e.g., in international forums like the Conference of the Parties, which are relevant to criteria such as "responsibility"), the following sections illustrate only five of the principles listed in the EJF, namely, availability, affordability, intra-generational equity, transparency/accountability, and due process. In accordance with the CA, the analysis acknowledges that different people need different amounts and different kinds of goods to reach the same levels of well-being, but also that the focus should be on individuals (rather than institutions) and the scope of real (rather than merely formal) opportunities. We thus believe that the results from the analysis of these five principles are sufficient to show that the bottom-up, individualistic lens of the CA reveals situations of injustice that could otherwise be overlooked.

8.4.1 Availability

Availability of energy is obviously an important principle in the EJF, but its description referring to "sufficient energy resources of high quality" is quite ambiguous and expresses the need for operationalization. Criteria can apply quite differently to different energy sources. Yet, the rural areas probably show the greatest diversity of energy sources, with an amalgam of electricity grids, solar PV, gasoline, liquid petroleum gas (LPG), charcoal, wood, dung, and so on. Therefore, electricity and cooking fuel should be kept separate as distinguished sub-criteria of availability. Additionally, this distinction brings more insight into the different issues related to electricity, poverty, and energy poverty.

Overall, our results are consistent with the 2022 Energy Progress Report (IEA et al., 2022), which indicates rural access to electricity rates of 18% for Benin, 44% for Senegal, and 22% for Togo. Not surprisingly, the proportion of households (H) connected to the national electric grid is quite low, with an average of 24.9% for the three countries, as well as the presence of individual solar photovoltaic (ISPV) installations with an average of 16.3%. Furthermore, the nominal power of electricity is, for the vast majority, less than 10 kW, and the main source of electricity (not energy) is generally dry cells for about half of the population. Finally, the majority of households have light bulbs (fluocompact or LED), cell phones, and a radio, but rarely a refrigerator (7 for every 100 H), a television (44) or a computer (4) (with Senegal doing slightly better). This lack of availability of electricity services impacts women and men differently since energy services, through technology accessibility, are unequally distributed, with men generally having more economic means and more power in the households' decisions (see Section 4.4.), while women generally face more inconvenience and risk at night from the absence of light (public or else).

For cooking fuel, the quasi-totality (over 89% in the three countries) uses biomass (wood or its derivatives or organic waste), and the vast majority uses solid wood (either wood branches collected in nearby forests or wood faggots bought at the market), with at least 75% in the three countries (see Table 8.2). Only Senegal shows a significant usage of another cooking energy source than traditional solid fuel, with (only) 3.8% of households using LPG. Our results are aligned with country-specific studies. The cookstoves mostly used are for wood and

Table 8.2 Availability of energy (cooking fuel)

Items	*Sub-items*	*Benin*	*Senegal*	*Togo*	*Total*
Cooking fuel	Wood Branches	76.1%	18.5%	70.6%	49.4%
	Wood Fagots	3.0%	57.1%	13.2%	29.5%
	Wood Charcoal	18.3%	12.6%	10.8%	13.7%
	Agro-waste	1.3%	8.0%	4.5%	5.1%
	Liquid Gas Petroleum	0.2%	3.8%	0.9%	2.0%
Cookstoves	"Three stones"	66.3%	73.8%	79.1%	73.2%
	"Cloporte"	17.0%	11.1%	5.2%	11.1%
	Improved (wood)	3.4%	7.2%	2.9%	4.9%
	Ceramic (traditional)	10.8%	2.0%	11.4%	7.1%
Reason for choosing a cookstove	Availability	68.1%	70.3%	34.6%	59.6%
	Price	21.4%	12.4%	14.6%	14.3%
	Facility to obtain	18.6%	12.3%	22.8%	17.0%
	Better results	3.0%	0.0%	2.0%	1.4%
	Other	7.0%	5.0%	25.8%	11.5%
Perceptions of improved cookstoves	Available	72.2%	69.9%	54.9%	66.3%
	Well adapted	76.9%	64.8%	72.6%	70.4%
	Consider having sufficient financial resources?	63.0%	38.3%	46.0%	47.4%
	Amount ready to pay (USD)	7.06	19.43	3.66	11.50
	Amount ready to pay (USD)—if sufficient resources	8.05	19.56	5.58	12.38
Prevalence of health problems and accidents related to cooking	Respiratory	0.90	1.03	0.39	0.81
	Cataract	0.92	0.86	0.51	0.78
	Skin	0.40	0.97	0.18	0.59
	Heart	0.12	0.42	0.12	0.25
	With women	55.3%	39.3%	41.4%	44.3%
	With children	21.3%	24.4%	11.2%	19.8%
	With men	12.3%	22.9%	9.5%	16.2%
	Accidents	-[3]	26.2%	15.8%	

biomass consumption; at least 84% of households in the three countries use traditional, inefficient cookstoves (i.e., either "three-stones" or "cloporte"). Consequently, the penetration rates of improved cookstoves are very low, with only Senegal above 5%. These figures are even more staggering that here the category "improved cookstove" also includes a cookstove for solid fuel (i.e., without gas or electricity).[2]

Cooking with solid fuel on such inefficient cookstoves is really the *worst-case scenario* because it impedes several capabilities: it requires a lot of fuel due to inefficiency (only 5–20% overall thermal efficiency according to ESMAP), thus leading to more expenses and time poverty; it causes indoor air pollution and the diseases associated with it; the combined prevalence of health problems and accidents related to cooking is more than two per household. It is also an obstacle to social and economic progress for women since women and girls who must collect wood are excluded from schooling or revenue-generating activities, and traditional cooking methods are often unsustainable since uncontrolled cutting of wood for fuel can be a contributing factor in land degradation and deforestation. Furthermore, even firewood can be lacking, and this gives rise to other even more damaging practises such as using plastic as fuel, as explained by this mother in Benin: "*at certain times during the year, this is the only option we have to avoid starving our families for lack of cooking energy.*"

The main reason invoked for the choice of a cookstove was its availability, whereas price was the main factor for only 14.3% of households on average. For the vast majority of respondents, the main factor is the ease of procurement of traditional cookstoves (because of their availability) and their afferent cooking fuel (i.e., wood or plastic bags). Still, according to our survey, the majority of households admit that improved cookstoves are available and (relatively) affordable. An explanation of this situation is multifactorial: (1) low, global purchasing power (and thus low incentives in the supply chain); (2) the entry cost of improved cookstoves; (3) a misunderstanding of the comparative advantages of alternatives (especially improved and clean cookstoves); (4) an overlook of opportunity costs (especially of women's working time and health); and (5) a lack of procedural justice (particularly women's participatory power, discussed later). Consequently, the energy services that are available to rural people are inefficient and of low quality (high entropy/low exergy and present health risk), and thus provide *ceteris paribus*[4] less capabilities (Jodoin, 2021).

8.4.2 *Affordability*

Affordability, which is of course important for energy poverty and is also part of the EJF, is usually determined by a simple calculation of the proportion of the household's income that goes to energy services where a threshold is established beyond which energy is considered unaffordable. However, this way of assessing affordability, despite its simplicity, has two major shortcomings. First, it does not assess energy services but rather energy supply. Thus, it poorly assesses how energy can be beneficial in terms of capabilities. Second, it evaluates expenses or direct costs more than the real capacity of a household to truly acquire energy services. In effect, a poor household can avoid paying for energy (e.g., electricity) because it is too expensive and/or spend a lot of time and effort acquiring energy by other means (e.g., collecting solid fuel such as wood). It is thus important to assess the *opportunity costs* by including the time spent collecting fuel and the distance travelled. Our findings show that taking into account these opportunity costs decreases affordability from 51% to only 22% of households with a threshold of 10% of revenues. Also, non-internalized externalities, such as effects on human health from indoor smoke exposure, forest degradation, and the impacts of greenhouse gas emissions, could also be included, but they required more empirical data. Consequently, most orthodox, expenditure-based definitions of energy poverty fall short of identifying energy-related, capability-based injustices, whereas our findings show that women suffer the most from the opportunity costs (see Gafa et al., 2022 for more details).

8.4.3 *Equity*

Our findings highlight many forms of intragenerational inequity. From the obvious global inequity in electricity access between developed and developing countries, to the one between urban and rural areas, and finally to inequality within villages as well as households. As expressed by Mang-Benza et al. (2023, p. 6), "intra- and intergenerational equity challenges in the studied rural areas have a definite feminine face, which translates into restrained capabilities for women, both within and outside the household." In effect, women are solely responsible for cooking with all the afferent requirements (cooking fuel and technology). This 50-year-old female respondent in Baatonu, Benin, candidly lays out her perception of the gendered roles in her community: "*Here we, women, are the ones in charge of collecting firewood. But at least people acknowledge what we are doing. Everything related to cooking at home is on the woman. On the other hand, men are the ones working in the fields*." And sometimes women have to do both as noted by a 32-year-old female retailer from Natantchiogue, Togo: *"Back in the days, the field was for men, nowadays, men rest and women work.*" Consequently, due to gender norms, women suffer from time poverty and the arduousness of wood fetching, which (as mentioned) can result not only in exclusion but also in the risk of being attacked by men and animals, and they have almost three times the risk of having health problems due to air pollution.

Moreover, when available and relatively affordable, electricity services are far from being equally distributed between households' members. This 64-year-old head of household from Djougou, Benin, explains how he manages electricity at home: "*At home, I am the head of the household. Therefore, I am responsible for buying the pre-paid electricity card. My wives have nothing to do with electricity charges*." This situation is echoed by two female respondents, who stated: "*We are not equal regarding electricity*" (52-year-old woman in Moba, Togo) and *"Since it's the man who brought electricity at home, he installed a light bulb in his own bedroom but not in mine"* (32-year-old woman in Agbandi, Togo). A situation resulting from as well as causing a lack of empowerment for women and procedural injustice, both of which negatively impact negatively their capabilities (discussed next; see also Table 8.3).

Table 8.3 Multi-level issues and women's capabilities

Level	*Description*	*Increase in women's capabilities*	*Data and references*
Global	High disparities in energy access between Global North and Global South.	No	e.g., Sovacool (2016).
National	High disparities in energy access between urban and rural areas.	No	e.g., Jodoin (2021).
Regional/ local	Even when a village is connected to the grid, many households may not have access.	No	Our findings show that high disparities in energy access between same village's households, but also that people perceive arbitrariness and indifference from energy distributors.
Household	Even when a household is connected to the grid, it may not be able to afford the tariff and/or the technology granting energy services.	No	The average monthly income per person is approximately 15 USD while the electricity tariff is 0.16–0.19 USD/kWh.
Within household (members)	Even if a household can afford the tariff and/or the technology, it may not be equitably distributed; HH's heads generally decide who get what (ref. qualitative data).	No	Ex. "Since it's the man who brought electricity at home, he installed a light bulb in his own bedroom but not in mine" (32-year-old woman in Agbandi, Togo; *qualitative results*).
	Even if a household can afford the tariff and/or the technology, women will still be responsible for collecting wood and cooking food (with the inherent risks) *and* they will still have less power in the household.	Maybe	E.g., "At home, I am the head of the household. Therefore, I am responsible for buying the pre-paid electricity card. My wives have nothing to do with electricity charges. They don't contribute. Since I paid for the electricity connection, I take care of it and don't expect anything from them, not even to top up the account." (64 y.o. man, Djougou, Benin; *qualitative results*).

8.4.4 Transparency and accountability, due process, and procedural justice

The EJF includes criteria that are based on procedural justice with "due process" and "transparency and accountability" (see Table 8.1). Procedural justice deals with recognition, participation, and power; it is concerned with how decisions are made in the pursuit of social goals, and who is involved and has influence in decision-making. It requires fair, respectful, and unbiased behaviour by authorities or those in power. Though this kind of justice is usually applied and assessed at higher levels, such as government and institutions (e.g., Paavola, 2005; Sovacool et al., 2016), we assessed it at the level of the household according to our assessment stratification and the individualistic ethics of the capability approach.[5]

Being patriarchal societies, the three surveyed countries present asymmetrical gender roles and implicit biases against women's leadership and decision-making capacities. Not only are women not represented in decisions related to politics and society, but men are also often responsible for making household purchasing decisions, including those regarding the sources and type of energy in the household. That is why this project has moved away from the assumption that households are units without any power structure and rather attempted to assess intrahousehold decision-making processes with sex-disaggregated data collection (Alderman et al., 1995; see also Table 8.3). Our findings show that, while financial decisions are exclusively made by men in an average of 61%, tasks-dispatching decisions are more evenly distributed, with 33% and 32% for men and women exclusively, respectively.[6] However, energy decisions are (again) fixed by gender norms since men are responsible for buying electricity and gas (with the afferent technology) and women are responsible for cooking with the associated technology and energy supply (mostly biomass). Of course, this intra-household decision-making pattern is detrimental to women's capabilities to control their environment since they are the primary energy users due to their domestic duties and are therefore more vulnerable to risks related to energy supply and use. Consequently, this should be seen as an underachievement in the "due process" (local governance) criterion of the EJF.

Also, beyond the power relations between men and women within households, rural households are rarely consulted in decisions to build energy facilities and infrastructure. Indeed, to the question: "Have you been consulted in the past, on behalf of the government, for the consideration of your energy needs in a project/program/national law?", the answers show that in general the answer is: "no." Therefore, rural population is generally not consulted by authorities. Moreover, the qualitative results show that this population often feels that the whole process of grid connection lacks transparency from the energy distributors. Finally, procedural justice concerns are, as mentioned, more often than not addressed at the level of policy and international agreements. For instance, an energy policy could be assessed in order to determine whether or not it addresses the needs of all the population and especially women, who tend to be left behind in terms of advantages and opportunities of energy access and use, mainly because they are underrepresented in STEM[7] employment and decision-making processes (this would be the topic of future work, however).

8.5 Proposed interventions

The primary focus of the CA, as mentioned, is on individuals rather than institutions; it is thus more concerned with micro-level interventions but recognizes the importance of mezzo- and macro-level factors such as social norms and energy policy. Of course, energy interventions can be performed at different levels but those able to increase energy services and the capabilities of individuals are not necessarily evident. Top-down energy or resource-based interventions achieving grid connection for each rural village, and even each household could still be far from sufficient. In effect, our results show that even when each and every household has access to affordable energy as well as energy services, it does not necessarily follow that each and every household member see their capabilities increased, and women are generally the most disadvantaged. Table 8.3 shows different levels at which energy injustices can manifest and how supposedly fixing these injustices with macro level, resource-focused interventions can be misleading. For instance, in a house with grid electricity and electric appliances, a woman can still have less capabilities because gender roles impose wood-fetching for cooking and/or drudgery without electric lighting (because the man keeps it for himself). Furthermore, these gender roles can even reverse the beneficial factor of energy access when women are expected to work longer hours because "now they have electric lighting." Yet, our findings show that women would prefer more time for personal activities, socialization, and remunerative opportunities.

With the empirical evidence gathered from Benin, Senegal, and Togo, we showed that rural women experience the limited availability and affordability of both traditional and modern energy services with more difficulty than men. Yet, it is important that energy interventions focus on the most vulnerable, those with less capabilities—women in rural areas. An EJF informed by the CA thus calls for a bottom-up, more individualistic approach with differentiated measures for cooking fuel and electricity targeting women more directly. First, the focus should be on eliminating the *worse-case energy scenario*—inefficient, three-stone cookstoves with fetched fuel wood or plastic packaging. The prohibition of inefficient, three-stones cookstoves in SSA (through a standard and labelling programme) should be accompanied by the promotion of women-led design and local fabrication of improved cooking solutions, as well as the implementation of information campaigns, knowing that improved cookstoves are available but also that men have the purchasing power.[8] Fostering sustainable, community-based or women-led governance on local forests while imposing restrictions on access and collection should contribute to achieve sustainability for all while improving women's empowerment. Also, temporarily subsidizing LPG will help with the conversion away from biomass as cooking fuel.[9] Second, besides increasing the power of the national grid, improving women's capabilities through modern, electric energy services could be achieved with national energy efficiency programmes, decentralized renewable energy measures, and global improvement in the cold supply chain.

An EJF informed by the CA brings home the need to focus on the least advantaged—women—with a bottom-up, more individualistic approach. While acknowledging the need to increase grid connection, it calls for the elimination of the worst-case energy scenario for women. With the challenges imposed by climate change, SSA countries should at least have the right normative tools to face them.

Notes

1 Basic capabilities, such as maintaining good health and being educated, have *not* been determined empirically i.e. by systematically asking the surveyed population (quantitative data). However, the qualitative data reveal that many women wanted more time to maintain social relationships, personal leisure and financial opportunities. Moreover, the material (e.g. technology) and social (e.g. norms) environment has been addressed with the survey through various EJF principles. The capabilities of the surveyed population have thereby been analyzed with (1) the scope of choice (range of possible levels of functioning that an individual can enjoy considering all the constraints and opportunities of acquiring commodities) and (2) the act of choice (the ability for people to choose among opportunities). For instance, having some air pollution-related illness or a lack of time due to drudgery has been interpreted as negatively impacting the basic capabilities.

2 World Bank (2014) defined clean cookstoves as stoves running on LPG, electricity, kerosene, liquid and gel biofuels, biogas, and solar energy, as well as retained-heat cookers and biomass gasifiers (ACS).
3 This question was not part of the survey in Benin.
4 All things being equal.
5 The capability approach adopts an ethical individualism but rejects ontological individualism. Accordingly, it seeks to improve the well-being (through capabilities) of individuals while recognizing that social norms can influence their capabilities.
6 The percentages associated with decisions taken by both men and women are lower in Senegal than in Benin and Togo. Religious background can justify this result: Benin and Togo are animist whereas Senegal is Islamic-oriented where patriarchy is more pronounced.
7 Science, Technology, Engineering, Mathematics.
8 As the creation of a new agency is always time- and resources-consuming, a modification of rural electrification agency mandate to incorporate the promotion of clean cooking devices is a simple and quick avenue for SSA countries to consider.
9 It seems as a justified trade-off between intra- and intergenerational justice since SSA countries typically produce 10 to 20 times less GHG per capita than developed countries (IEA, 2022).

References

Alderman, H., Chiappori, P.-A., Haddad, L., Hoddinott, J., & Kanbur, R. (1995). Unitary Versus Collective Models of the Households: Is It Time to Shift the Burden of Proof? *The World Bank Research Observer*, 10, 1–19.

Clancy, J. S., Skutsch, M., & Batchelor, S. (2002). The Gender-Energy_poverty Nexus: Finding the Energy to Address Gender Concerns in Development. *DFID Project CNTR998521*. Internet: UK Department for International Development (DFID).

Day, R., Walker, G., & Simcock, N. (2016). Conceptualising Energy Use and Energy Poverty Using a Capabilities Framework. *Energy Policy*, 93, 255–264. https://doi.org/10.1016/j.enpol.2016.03.019

Energia (2017). Energy access and gender equality: What we know so far and knowledge gaps. *Gender-Energy Programme*.

Gafa, D. W., Egbendewe, A. Y. G., & Jodoin, L. (2022). Operationalizing Affordability Criterion in Energy Justice: Evidence from Rural West Africa. *Energy Economics*, 109, 105953.

Galvin, R. (ed.) (2020). *Inequality and Energy: How Extremes of Wealth and Poverty in High Incomes Countries Affect CO_2 Emissions and Access to Energy*. London/San Diego/Cambridge/Oxford: Elsevier.

Guruswamy, L. (2010). Energy Justice and Sustainable Development. *Colorado Journal of International Environmental Law and Policy*, 21, 231–276.

IEA (2022). *Africa Energy Outlook*. Paris: International Energy Agency.

IEA, IRENA, UNSD, WORLD BANK, & WHO (2022). *Tracking SDG 7: The Energy Progress Report*. Washington, DC: World Bank.

Jodoin, L. (2021). Let Capabilities Ring: Operationalizing Energy Justice in Guinea. *Energy Research & Social Science*, 72, 101894.

Jodoin, L., Gafa, D. W., Egbendewe, A., Domegni, L. K., Gaye, I., Ague, V. J., Lo, K., Ouendo, E.-M., Fagbémi, L., Diop, M., & Ndiaye, M. L. (Unpublished). Making Energy Justice Work for Women in Rural Sub-Saharan Africa: A Quantitative Diagnostic from Benin, Senegal, and Togo.

Jones, B. R., Sovacool, B. K., & Sidortsov, R. V. (2015). Making the Ethical and Philosophical Case for "Energy Justice". *Environmental Ethics*, 37, 145–167.

Lee, J., Kim, H., & Byrne, J. (2021). Operationalising Capability Thinking in the Assessment of Energy Poverty Relief Policies: Moving from Compensation-Based to Empowerment-Focused Policy Strategies. *Journal of Human Development and Capabilities*, 22, 292–315.

Mang-Benza, C., Jodoin, L., Onibon Doubogan, Y., Gaye, I., & Kola, E. (2023). Making Energy Justice Work for Women in Rural Sub-Saharan Africa: A Qualitative Diagnostic from Benin, Senegal, and Togo. *Energy Policy*, 173, 113–345.

McCauley, D., Heffron, R. J., Stephan, H., & Jenkins, K. (2013). Advancing Energy Justice: the Triumvirate of Tenets. *International Energy Law Review*, 32, 107–116.

Melin, A., Day, R., & Jenkins, K. E. H. (2021). Energy Justice and the Capability Approach—Introduction to the Special Issue. *Journal of Human Development and Capabilities*, 22, 185–196.

Monkelbaan, J. (2019). *Governance for the Sustainable Development Goals: Exploring an Integrative Framework of Theories, Tools, and Competencies*. Singapore: Springer.

Nussbaum, M., & Sen, A. (eds.) (1993). *The Quality of Life*. Oxford: Clarendon Press.

Oparaocha, S., & Dutta, S. (2011). Gender and Energy for Sustainable Development. *Current Opinion in Environmental Sustainability*, 3, 265–271.

Paavola, J. (2005). Seeking Justice: International Environmental Governance and Climate Change. *Globalizations*, 2, 309–22.

Pachauri, S., & Rao, N. D. (2013). Gender Impacts and Determinants of Energy Poverty: Are We Asking the Right Questions? *Current Opinion in Environmental Sustainability*, 5, 205–215.

Rawls, J. (1971). *A Theory of Justice*. Cambridge: Harvard University Press.

Robeyns, I., & Brighouse, H. (2010). Introduction: Social Primary Goods and Capabilities as Metrics of Justice. In: Robeyns, I. & Brighouse, H. (eds.) *Measuring Justice: Primary Goods and Capabilities*. Cambridge: Cambridge University Press.

Sen, A. (1999). *Development as Freedom*. New York: Knopf.

Sen, A. (2009). *The Idea of Justice*. New York: Allen Lane.

Sovacool, B. K., Burke, M., Baker, L., KotikalapudI, C. K., & Wlok, H. (2017). New Frontiers and Conceptual Frameworks for Energy Justice. *Energy Policy*, 105, 677–691.

Sovacool, B. K., & Dworkin, M. H. (2015). Energy Justice: Conceptual Insights and Practical Applications. *Applied Energy*, 142, 435–444.

Sovacool, B. K., Heffron, R. J., McCauley, D., & Goldthau, A. (2016). Energy Decisions Reframed as Justice and Ethical Concerns. *Nature Energy*, 1, 16024.

Velasco-Herrejon, P., & T. Bauwens. (2020). Energy Justice from the Bottom Up: A Capability Approach to Community Acceptance of Wind Energy in Mexico. *Energy Research & Social Science*, 70, 101711.

Wilkinson, P., Smith, K. R., Joffe, M. & Haines, A. (2007). A Global Perspective on Energy: Health Effects and Injustices. *Lancet*, 370, 965–978.

Wood, Nathan and Roelich, Katy (2020), Substantiating Energy Justice: Creating a Space to Understand Energy Dilemmas. *Sustainability*, 12 (1917).

World Bank. (2014). *Clean and Improved Cooking in Sub-Saharan Africa: A Landscape Report*. Washington, DC: The World Bank.

World Bank. (2023). *GDP Per Capita (current US$) – Benin*. New York: World Bank.

9 Connected Yet Suffering

The Lived Experiences of Czech Households through Energy Service Provider Bankruptcies

Dominik David and Hedvika Koďousková

9.1 Introduction

The issue of energy poverty has been making headlines across Europe since the Russian invasion of Ukraine in February 2022. In the Czech Republic, however, it was already a pressing concern several months prior. In Autumn 2021, in the aftermath of the COVID pandemic and shortly after elections to the Czech Chamber of Representatives, household energy poverty became a hot topic across national media for obvious reasons. Over the course of a few autumn weeks, several alternative energy providers went out of business, led by the largest alternative energy supplier—Bohemia Energy Group (Glaserová, 2021; Lopatka & Hovet, 2021).

Over a million customer sites suddenly found themselves without their electricity and/or gas supplier and were immediately transferred to the so-called 'supplier of last instance' (SLI) regime. Thus, in practice, the law protected them from disconnection (Moravec, 2021; Veselovský, 2021). However, many customers suddenly faced much higher deposit payments than before because the SLI may not have had the opportunity to buy enough energy in advance for their new customers and thus had to purchase energy at a higher price on the wholesale market to meet the new demand. The prices charged by the SLI also mirrored expectations of household consumption for the winter period, as the alternative suppliers went bankrupt just before winter. (Kulig, 2022; Veselovský, 2021). While this represented a merely unpleasant situation for some households, for specific population groups, such as pensioners, single parents, and single-income households, the prices became practically unaffordable, leading to concerns about how to stay warm through the upcoming winter and how to quickly get out of the regime, among other issues.

In this chapter, we draw from these and other recent events to provide an updated perspective on households' perceptions and behaviour under unexpected circumstances and thereby reveal what it means to be energy vulnerable in the Czech Republic. Through semi-structured interviews with affected households, we examine the interplay among their material culture, practises, norms, and external influences. We ask the following questions: How did energy service provider bankruptcies impact these factors? Which factors prevented households from meeting the newly encountered challenges, such as high deposit payments and the need to quickly get out of the SLI regime, successfully? Our goal is to highlight the need to focus not just on energy poverty alleviation but also on its prevention. We argue that the understanding of energy vulnerability in the Czech Republic is too shallow and does not reflect underlying factors that increase the likelihood of a household falling into energy poverty.

DOI: 10.4324/9781003408536-12

9.2 Examining the impact of energy supplier bankruptcies in the Czech Republic

It is not widely known that the Czech Republic experienced a first wave of alternative supplier bankruptcies well before 2021. In relation to the transposition of European liberalization packages, many alternative energy players have emerged in the country in recent decades. The modus operandi of many has relied on offering lower prices than traditional suppliers to attract customers. The new actors usually purchased energy on the wholesale market, hoping to take advantage of decreasing energy prices. However, when energy prices on the wholesale market started to rise constantly, such alternative suppliers were exposed to the risk of paying much higher amounts for energy without the benefit of stable long-term contracts. This sudden increase in energy costs posed an existential financial burden for these suppliers. Although such practises by energy companies were legal, some energy market experts criticized such actions as "gambling" (Kubátová, 2021; Moravec, 2021).

Unfortunately, such risky operations in the energy market and a lack of sufficient financing have driven some alternative players into bankruptcy, as exemplified by the collapse of Moravia Energo in 2009 (Kadlecová, 2021). The subsequent transfer of the company's customers to the SLI regime was unprecedented at the time. However, in terms of the number of consumers affected, this first wave of transfers was only a sign of things to come. At the time of its collapse in 2009, Moravia Energo was supplying about 1,000 customer sites (ČTK & Horáček, 2009). Moreover, most of the SLIs were able to provide favourable energy prices at that time in an effort to attract and keep new consumers. In autumn 2021, however, the collapse of Bohemia Energy Group was absolutely devastating, affecting some 900,000 customer sites and taking place against the backdrop of steeply rising European energy prices (OTE, 2023a, 2023b).

Regardless of previous experience, however limited in scope, the end of Bohemia Energy and other minor players revealed a certain unpreparedness of the state's decision-making and bureaucratic institutions for what might be the consequences of these bankruptcies. The state lacked practical tools that could be quickly applied to help affected households as well as a clear knowledge of who actually were the most affected consumers (Kabrhelová et al., 2021). It also became clear that there were hidden sources of energy vulnerability at play, which needed to be taken into account to successfully prevent households from falling into energy poverty. Aside from the focus on energy market rules, the cause of vulnerability was found to be the inadequately defined SLI regime. The relevant legislation protected households from being physically disconnected but not from being unable to afford energy services.

There may be two reasons for the Czech state's lack of preparedness. Firstly, energy poverty has not yet been significantly addressed in the country. There are no official definitions of energy poverty, tools for its regular monitoring, or policies specifically designed to tackle the issue. Energy poverty is only partially addressed, mostly by state social benefits and subsidy programmes (The Ministry of Industry and Trade, 2019). However, as previous research has demonstrated, these policies have not fulfilled their potential in dealing with the issue (Koďousková & Lehotský, 2021). This lack of focused attention may be caused by the relatively favourable results the Czech Republic as a whole has obtained in pan-European energy poverty assessments (Bouzarovski et al., 2020), although specific strata of the population may face energy deprivation quite extensively (Koďousková & Lehotský, 2021). Secondly, part of the problem may also be the prevailing narrow understanding of energy vulnerability in the country, which depicts energy vulnerable customers as already being in a situation of energy poverty and at the same time having specific needs that require permanent access to energy (e.g., due to a health condition or the presence of children in the household) (Vysoká škola ekonomická v Praze, 2021). Such a definition protects the most vulnerable but excludes households that could easily and unexpectedly fall into the condition. We thus argue for a broader definition of energy vulnerability in line with how it is usually conceptualized in the academic literature (Allen, 2003; Bouzarovski & Petrova, 2015; Bouzarovski & Thomson, 2018; Robinson et al., 2019; Walker et al., 2013). While energy vulnerability thinking typically "highlights the factors that affect the likelihood of becoming poor" (Bouzarovski & Petrova, 2015), policies in the Czech Republic only target a narrow group of people already struggling with energy poverty, making it difficult to prevent impending problems.

Applying interpretative phenomenological analysis, we offer insight into what it was like for households to experience energy service supplier bankruptcy and the SLI regime (Griffin & May, 2012). This research method derives its theoretical assumptions from exploring the lived experiences of (marginalized) groups and the situated nature of their experiences. As such, the method is inductive in nature but is often driven by a predetermined theoretical perspective. We derive our findings from 15 stories of 18 people gained through semi-structured interviews and analysed through the energy cultures framework (see Table 9.1) (Stephenson et al., 2010, 2015). The framework includes several interconnected elements, including material culture (have), practises (do), norms (think), and external influences, which together form a contextual soup of wider factors beyond a household's direct control (Sweeney et al., 2013). In our research, the framework served both as an organizational tool to structure interviews and as a guiding principle for their analysis. In practise, it helps shed light on barriers and support systems that energy-poor households encounter (Sweeney et al., 2013). Such knowledge is necessary for effective and efficient policy targeting (McKague et al., 2018).

The respondents were selected through purposive sampling. The goal was to include households considered to be energy vulnerable, as described in a study by Pye and Dobbins (2015), according to whom the term energy vulnerable customer includes not only those individuals and households at risk of energy poverty but also a broader group of consumers who may be at a disadvantage in the purchasing and use of energy in the electricity and gas retail markets (Pye & Dobbins, 2015). The interviews took place mainly in the autumn of 2022 and lasted between 30 and 95 minutes. They were conducted in person, over the phone, or online through the Whereby platform, depending on the respondent's preference (Babbie, 2016). All participants agreed with the recording. The interviews were then anonymized and transcribed by the interviewer in private. Both authors analysed the interviews inductively and independently by breaking the stories into thematic clusters and grouping them under the energy cultures framework's elements as superordinate deductively induced themes (Griffin & May, 2012) (Table 9.2).

Table 9.1 Elements of the energy cultures framework

Element	*Description*
Material factors	Physical aspects of energy use, such as the type of energy sources used and the physical characteristics of buildings
Norms	Social standards that influence how individuals and households think about and use energy
Practices	The daily routines and behaviours that shape how energy is used, such as the use of appliances, and heating and cooling practices
External influences	The contextual factors that shape energy behaviour, such as policies, regulations, social relations, and market forces that influence the availability and cost of different energy sources

Source: Stephenson et al. (2010, 2015).

Table 9.2 Summary of interviewees

Name	*Age*	*Status*	*Household composition*	*Type of housing*	*Housing size*	*Type of heating*	*Supplier bankruptcy*
Barbara Svatopluk	60-70	Married couple; pensioner and working woman	2	Own house	4+1 and 2+1 (unoccupied property)	Gas and solid fuels	Gas and electricity
Bianka	30-40	Single mother	1 and a child; flatmate	Rental housing	2+1	Gas	Gas and electricity
Bohdana Dušana	80-90 50-60	pensioner and working daughter	2; (2 and daughter at university)	Flat in a family house, easement	2+1 (3+1 daughter with family)	Gas	Electricity
Božena	80-90	Widow	1	Own house	3+1	Gas	Gas
Dalibor	50-60	One-person household	2; (2 and 1 in other flats)	Own house	3+1 (3+1 tenants and 3+1 mother)	Wood gasifier and gas	Gas electricity in unoccupied property
Jarmila (Denisa)	80-90	Couple of pensioners (Jarmila and husband)	2 (partly also son)	Own house	4+1 + unoccupied floor, only 2+1 is heated	Gas	Gas and electricity
Jaroslav	50-60	One-person household	1	Own house	3+1	Gas	Gas and electricity
Ludvík	60-70	Retired couple with a handicapped daughter	3; (and 2 pairs in other flats)	Own house	3+1 (3+1 tenants and 1+1 tenants)	Gas	Gas and electricity
Marie	40-50	Single mother	1 and 3 children (one at university)	Own house	4+1	Gas	Gas and electricity
Martin	20-30	Student with girlfriend	2	Rental housing	1+1	Gas	Gas and electricity
Mikuláš	30-40	One-person household	1	Own house	2+1 (unoccupied floor)	Electricity and solid fuels	Electricity
Radim	50-60	Married couple	2	Own house	5+1	Gas + solid fuels	Gas and electricity
Romana	50-60	One-person household	1	Own house	5+1	Solid fuels + gas	Gas and electricity
Tamara	30-40	Couple with a child	3	Own house	4+kitchen corner	Electricity	Electricity
Theodora	30-40	Couple with three children	5	Own house	4+kitchen corner	Electricity and solid fuels	Electricity

Source: Compiled by authors.

9.3 Lived experiences of affected households: prevailing factors and lacking practices

The upcoming sections utilize the elements of the energy cultures framework to depict the real-life experiences of households that have been impacted by the Bohemia Energy bankruptcy. Our analysis tracks the shifts in their material conditions, examines the norms and practises they share, and highlights the practises that they lack, which could impact their ability to deal with this unforeseen event. Moreover, we concentrate on external influences and scrutinize the barriers and support systems that were either present or absent for energy-deprived households (McKague et al., 2018).

Material culture played a role in the situation of virtually all affected households. Even before the energy supplier's bankruptcy and transfer to the SLI regime, most households had been experiencing heat leakages and difficulties maintaining an adequate temperature at affordable prices. They also faced problems with dampness and mould, aggravated by insufficiently heated spaces. Low-income households, mainly pensioners, paid a higher percentage of their income on energy due to inadequately insulated dwellings.

The SLI regime made the interconnectedness of inadequate material conditions (low energy efficiency) and low incomes/high energy expenses more visible. The event brought additional financial burdens, as indicated by the fact that 14 out of 15 households have fallen into energy poverty, as per the traditional 10% indicator (Fizaine & Kahouli, 2019). For example, Božena spent 60% of her income on energy bills, Jarmila and her husband over 50%, and Jaroslav around 35%. Only one household among our respondents did not exceed the threshold, though the more favourable situation was not enough to avoid concerns about future developments. Although during interviews households demonstrated awareness of the possibility of refurbishments or solar panel installations to reduce their energy bills, most were unable to afford such measures and thus did not consider them realistic options. Most households did not believe that the state could be of any help in this situation. They expressed doubts about the accessibility of state subsidy programmes in view of the needed co-financing, the various conditions that had to be fulfilled, and the fact that reimbursement came only after investment completion (see below). Similarly, students and young adults living in rented apartments faced inadequate housing conditions. However, their options for improving energy efficiency were limited by financial constraints and the need to secure the landlord's agreement.

The source of heating and hot water is undoubtedly another essential factor in the material culture of all households. In households that had more than one type of heating at their disposal, the event led to a shift towards higher usage of traditional heating sources (mainly fuelwood) (Stojilovska et al., 2023). The more well-off households began to pursue the acquisition of new energy technologies (such as heat pumps or solar panels with energy storage facilities). Overall, the event created a strong sense of insecurity and forced households to seek security by strengthening their own self-sufficiency, depending on the material conditions at their disposal.

Romana: *I have decided to use those subsidies to get a solar panel and backup battery storage. I want to be as independent as possible (...).*

Mikuláš: In principle, I thought about how energy-dependent we are. And if the price were to increase threefold now, there's nothing we can do against it. I can stop heating and walk around in warm clothes, but I cannot turn off the fridge. Okay, I can light a candle, but I just... I need to do laundry, heat water... So that's why I decided to be more self-sufficient and get that wood-fired boiler. I for sure want to cut off the electricity and the gas as much as possible.

Regarding household practises and norms, we first observed a strong role for emotions at play among our respondents. Suddenly losing their energy services supplier and finding themselves under the SLI regime, the respondents mentioned feelings of shock, fear, and anxiety (Longhurst & Hargreaves, 2019; Martiskainen & Sovacool, 2021; Petrova & Simcock, 2021). They mostly expressed concerns about future developments, shared an impression of unpredictability, and feared high or unmanageable energy prices, unexpected payments, and debts. New deposit payments were seen as a major burden, or at least as a source of negative impact on household well-being. No matter what the primary heating system was in the household, respondents also feared both the unaffordability and physical unavailability of fuelwood, gas, and/or electricity.

Božena: *It was terrible for the psyche. I was completely out of it. I cried a lot, but the kids told me not to cry and that I couldn't do anything about it (...).*

In a few instances the event had serious mental health effects (Liddell & Morris, 2010), contributing to vicious circles already present at the household level or creating new ones that further reduced the household's ability to cope with the situation and led to additional financial exigencies. Typically, such a household had to seek therapeutic treatment or face the negative consequences of an unfavourable energy contract concluded under pressure. Many households referred to the dubious practises of some energy market actors who tried to exploit households' uncertainty and coerce them into new contracts (see below).

Bianka: *I was regularly depressed. I was depressed for six months, then I got on medication that helped me. I had anxiety, I cried, and I didn't know how to deal with it. I was afraid I wouldn't be able to pay for it. When the bill came through, I didn't know where I would get the money. It hit me hard. I also had to go to therapy and spend a lot of money on different therapy sessions.*

Enhanced reliance on energy savings was another shared norm among our respondents. Especially those households already "trapped" in unsatisfactory material conditions and facing higher energy costs considered further savings as a solution to their situation. This belief was strongly reflected in households' everyday practises. Typically, households adopted new thermal standards and reduced indoor temperatures. In extreme cases, they even self-deprived for fear of additional energy expenses. We observed households trying to warm themselves, for example, by putting on additional layers of clothing and blankets, sipping hot drinks, being physically active, or working outside the house rather than raising the temperature in the room. We also witnessed so-called spatial shrink (Feenstra et al., 2021)—i.e., a reduction in the heating space—and a strong preference for cheaper heating and lighting sources. Some households learned to switch daily tasks to nighttime or changed their previous cooking habits and laundry practises (e.g., avoiding the dryer). Although some of these practises may not necessarily have negative impacts, many reduce the quality of life of the respondents and have negative health consequences.

Interviewer: *How do you cope with a low temperature at home?*
Bianka: A combination of everything. We heat a bit, sweaters, blankets. Sometimes we take clothes on, sometimes we heat up, it depends.
Svatopluk: For example, when I'm alone in the house, I don't heat, I do something all the time, there's always work around the house, I'm not in the house all day basically.

In the given situation of energy supplier bankruptcies, we believe it is also important to examine what was missing or lacking in practices. Firstly, we came across households that lacked knowledge on how to better maintain their appliances and heating systems to improve their performance and efficiency and thereby reduce energy spending. Others faced difficulties with metre readings or even could not access the metres for their (rented) apartments. In particular, senior households lacked the ability to use the Internet to find and choose an energy supplier and conclude an online contract. They had to rely on personal appointments at branches that were often crowded with long queues. Several households even reported difficulties reaching their prospective energy supplier (see below). Finally, households were mostly not used to asking for state social benefits and had difficulties navigating through the system (e.g., obtaining and filling in the needed paperwork). In these cases, feelings of shame often played a role as a barrier to asking for help.

Interviewer: *Haven't you considered asking for some social benefit?*
Jaroslav: Which one? No, I indeed haven't tried that. I don't have the right profile for that.
Interviewer: *Profile?*
Jaroslav: In which universe does a guy who lives alone get a social benefit? The most I can get is jail if I don't pay alimony for my daughter. But the state helping me? I've never asked the state for anything, and that won't change. I'm not some kind of a leech.

Relatives or friends who were willing and able to help substantially reduced the energy vulnerability of households lacking the abovementioned practises. We have observed both a norm (households' belief that they would get help from their relatives if needed) and a practise of addressing family members in difficult situations. Especially pensioners often asked their (grand)sons and (grand)daughters for assistance in securing a new supplier and/or social benefits. Internet-savvy households, on the other hand, were better equipped to find an adequate solution, i.e., to quickly switch to a new supplier, get out of the SLI regime, and/or familiarize themselves with state social benefits and subsidy programmes (if available), without the need for any assistance.

Božena: *I was left with only 1,600 CZK (approx. 67 €) for living expenses for the entire month. So, my kids had to ask for a housing allowance. But that brought its problems, with the paperwork I didn't understand. My daughter-in-law and daughter had to do it instead of me.*

This brings us to our last point—the role of external factors. It is already clear from the above that external influences were closely intertwined with both material factors and household norms/practices. Regarding material factors, interviews with households indicate state subsidy programmes for energy efficiency measures are inaccessible for many households. As one of the latest studies demonstrates, subsidies in their current form have not effectively reached energy-poor households in the Czech Republic. Instead, households that are more economically secure and have the means to invest in energy-efficient refurbishments (even without any assistance) have benefited (Frantál & Dvořák, 2022). Unfortunately, subsidy programmes designed and implemented this way have a negligible chance of positively influencing the situation of energy poverty in the country.

External factors also influenced households' norms and practises. For example, some of the respondents experienced unfair practises from prospective energy suppliers. They were contacted via phone by various companies, even though they had never shared their contact details. In other cases, the billing by the former energy supplier did not match the households' real energy consumption. Only if the household was prescient and took pictures of their energy metres immediately after their supplier ceased its activities would they able to seek a remedy. For example, despite the increased monthly payments for electricity, the SLI demanded an additional 12,000 CZK (approx. 500 €) from Dalibor's household. Since he had everything documented and was willing to defend himself in court, the SLI even had to refund him after a fair accounting. Unfortunately, not all households were vigilant against such practises by suppliers. One household even faced the dubious practises of energy hustlers. Another encountered a lack of transparency in billing from his landlord. This can undermine household confidence in market mechanisms.

Božena: *In September, a door-to-door salesman came by and wanted me to sign a contract amendment. I told him that it was in small print and that I couldn't read it and asked him to give me time to read it and sign it. He urged that he lived far away, had too many clients, and that it was not worth it. He insisted that it was nothing more than an amendment that they were my energy supplier. I signed it under duress, and by doing so, I made a huge mistake. Then, when I wanted to leave, I found out I couldn't and would have to pay tens of thousands to switch my supplier.*

Martin: Although we signed the contract with the energy company, the landlord did not want to show us the energy meters and charged us more every month. I demanded it several times, but he always said he would not show it to us and that we could move out if we had something against that.

Regarding more distant social relations, disappointment with the state apparatus was evident among most of our respondents. On one hand, the households believed the state underestimated the situation and failed to provide its institutions (i.e., Energy Regulatory Office—ERO) enough competencies to monitor and control the developments and provide early warnings to the customers of what might happen in the energy markets. On the other hand, already in the situation of energy supplier bankruptcy and under the SLI regime, the respondents considered the state's help insufficient or inaccessible, including absent or delayed information campaigns.

Radim: *The people who are elected don't care about ordinary people. That's the way I look at it. [...] If anyone cared, it couldn't happen that way. [...] The state has failed to set up the system.*

Marie: *I knew that the biggest supplier would take us over if the bankruptcy came. But no one ever told us what the prices would be. It never crossed my mind that when we have ERO, its jurisdiction is not to check if these companies are selling commodities they have secured. And they want me to do that? I'm a nurse. I didn't know what a spot price was.*

9.4 Discussion and conclusion

In conclusion, the empirical findings demonstrate the necessity for a broader understanding of energy vulnerability. The current approach used as a basis for future legislation sees energy vulnerability as a subset of energy poverty (see above). However, the collapse of energy service providers and subsequent customer transfers to the SLI regime emerged as a crucial external factor that increased households' susceptibility. Interviews with affected households highlighted the significant burden brought by the SLI pricing mechanism. The weak role of the state in carefully drafting legislation for the SLI regime and formulating instruments to help households cope better with unexpected developments worsened the situation.

The lack of communication from the state, especially to those without internet access, seriously undermined trust in the state and its institutions (Grossmann et al., 2021). The study also revealed a new norm of loss of confidence in smaller alternative suppliers among affected households, which could have serious implications for the functioning of liberalized energy markets if confirmed on a larger scale. The majority of our respondents firmly rejected the possibility of changing their energy supplier to a lesser-known company in the future. They mostly preferred large national energy providers, with whom they eventually entered into a contract. Among our respondents, only a few households kept weighing the credibility of the supplier against the price offered.

The energy cultures framework used in this study offers a comprehensive approach to understanding the internal and external factors that can lead to households slipping into energy poverty, as well as the conditions that can help them move out of it. Furthermore, it helps identify categories of energy-poor households that require specific policy interventions (McKague et al., 2018).

Specifically, there were differences in vulnerability factors (Middlemiss & Gillard, 2015) related to both material conditions and subjective norms and practises, with better socio-economic households being less affected. Many households lacked flexibility in the application of energy efficiency measures (Bouzarovski & Petrova, 2015). In terms of policy implications, the state must reconsider whether its subsidy programmes align with the EU's just transition objectives as energy-poor households currently lack the financial capital to avail themselves of the subsidy programmes aimed at improving energy efficiency and transitioning to low-carbon energy sources. Though some measures to help the most vulnerable are underway, such as the New Green Savings Light programme (The Ministry of the Environment, 2023), the state must continue to work on solutions to address this issue.

"Lacking practices" also played a role in household vulnerability. While many households attempted to address this issue by emphasizing their energy-conscious behaviour, avoiding the high and sometimes unaffordable deposit payments of the SLI regime required a swift exit from this system and enrolment in a more favourable energy supply contract. Unfortunately, our survey revealed significant disparities in the respondents' ability to navigate the offers of potential suppliers, use the internet, and enter into new contracts online. Those without this experience were at a disadvantage compared to those who possessed it. Moreover, in situations of energy poverty, we observed that households often lacked the practise of applying for state social benefits, primarily due to the fear of stigmatization and shame. However, more (internet) savvy family members were critical in helping overcome these challenges. Therefore, the state must undertake targeted information campaigns and public awareness initiatives to raise awareness about energy poverty and dispel the shame and fear that prevent households from seeking assistance.

Finally, the collapse of energy suppliers, the poorly defined SLI regime, and the unpreparedness of the state are not unique to the Czech Republic. We hope this study will contribute to a better understanding of energy poverty and vulnerability in ongoing efforts to define and address this issue not only in the Czech Republic but across the EU.

References

Allen, Katrina. 2003. "Vulnerability Reduction and the Community-Based Approach: A Philippines Study." In *Natural Disasters and Development in a Globalizing World*. London: Routledge. https://doi.org/10.4324/9780203402375

Babbie, Earl. 2016. *The Practice of Social Research*, Fourteenth Edition. *Teaching Sociology*. Boston: Cengage Learning.

Bouzarovski, Stefan, and Saska Petrova. 2015. "A Global Perspective on Domestic Energy Deprivation: Overcoming the Energy Poverty-Fuel Poverty Binary." *Energy Research and Social Science* 10. https://doi.org/10.1016/j.erss.2015.06.007

Bouzarovski, Stefan, and Harriet Thomson. 2018. "Energy Vulnerability in the Grain of the City: Toward Neighborhood Typologies of Material Deprivation." *Annals of the American Association of Geographers* 108 (3). https://doi.org/10.1080/24694452.2017.1373624

Bouzarovski, Stefan, Harriet Thomson, Marine Cornelis, Anaïs Varo, and Rachel Guyet, and European Commission. Directorate-General for Energy. 2020. *Towards an Inclusive Energy Transition in the European Union: Confronting Energy Poverty amidst a Global Crisis*. Luxembourg: Publications Office of the European Union.

ČTK, and Filip Horáček. 2009. "Moravia Energo Neplatí Dluhy, ČEZ Jí Proto Přestal Dodávat Elektřinu." *Idnes.cz*, February 19, 2009. https://www.idnes.cz/ekonomika/domaci/moravia-energo-neplati-dluhy-cez-ji-proto-prestal-dodavat-elektrinu.A090219_175123_ekonomika_fih

Feenstra, Mariëlle, Lucie Middlemiss, Marlies Hesselman, Koen Straver, and Sergio Tirado Herrero. 2021. "Humanising the Energy Transition: Towards a National Policy on Energy Poverty in the Netherlands." *Frontiers in Sustainable Cities* 3. https://doi.org/10.3389/frsc.2021.645624

Fizaine, Florian, and Sondès Kahouli. 2019. "On the Power of Indicators: How the Choice of Fuel Poverty Indicator Affects the Identification of the Target Population." *Applied Economics* 51 (11). https://doi.org/10.1080/00036846.2018.1524975

Frantál, Bohumil, and Petr Dvořák. 2022. "Reducing Energy Poverty in Deprived Regions or Supporting New Developments in Metropolitan Suburbs? Regional Differences in the Use of Subsidies for Home Energy Efficiency Renovations." *Energy Policy* 171 (December): 113250. https://doi.org/10.1016/j.enpol.2022.113250

Glaserová, Dominika. 2021. "Bohemia Energy Končí. O Klienty Se Postaráme, Slibují Dodavatelé Poslední Instance." *ČT24*. October 13, 2021. https://ct24.ceskatelevize.cz/domaci/3385150-dodavatel-elektriny-a-plynu-bohemia-energy-konci-skupina-ma-statisice-odberatelu

Griffin, Ann, and Vanessa May. 2012. "Narrative Analysis and Interpretative Phenomenological Analysis." In *Researching Society and Culture*. London: Sage Publications Ltd.

Grossmann, Katrin, George Jiglau, Ute Dubois, Anca Sinea, Fernando Martín-Consuegra, Malgorzata Dereniowska, and Robert Franke, et al. 2021. "The Critical Role of Trust in Experiencing and Coping with Energy Poverty: Evidence from across Europe." *Energy Research and Social Science* 76. https://doi.org/10.1016/j.erss.2021.102064

Kabrhelová, Lenka, Barbora Sochorová, and Zuzana Kubišová. 2021. "KAUZA BOHEMIA ENERGY: JSME SVĚDKY BEZRADNOSTI STÁTU, ŘÍKÁ REPORTÉRKA KLÍMOVÁ." *Irozhlas.cz*, November 15, 2021. https://www.irozhlas.cz/zpravy-domov/podcast-vinohradska-12-jana-klimova-bohemia-energy_2111150600_cen#transkripce

Kadlecová, Michaela. 2021. "Jak Postupovat Při Krachu Dodavatele Energií?" *Kalkulator.cz*, October 13, 2021. https://www.kalkulator.cz/clanky/244/jak-postupovat-pri-krachu-dodavatele-energii

Koďousková, Hedvika, and Lukáš Lehotský. 2021. "Energy Poverty in the Czech Republic: Individual Responsibility or Structural Issue?" *Energy Research and Social Science* 72. https://doi.org/10.1016/j.erss.2020.101877

Kubátová, Zuzana. 2021. "Poučení z Pádu Bohemia Energy: Dohled Nad Obchodníky Selhal." *Seznamzpravy.cz*, October 28, 2021. https://www.seznamzpravy.cz/clanek/pouceni-z-padu-bohemia-energy-dohled-nad-obchodniky-selhal-178899

Kulig, Michal. 2022. "Proč v Česku Krachují Dodavatelé Energií?" *Faei.cz*, April 25, 2022. https://faei.cz/proc-v-cesku-krachuji-dodavatele-energii/

Liddell, Christine, and Chris Morris. 2010. "Fuel Poverty and Human Health: A Review of Recent Evidence." *Energy Policy* 38 (6). https://doi.org/10.1016/j.enpol.2010.01.037

Longhurst, Noel, and Tom Hargreaves. 2019. "Emotions and Fuel Poverty: The Lived Experience of Social Housing Tenants in the United Kingdom." *Energy Research and Social Science* 56. https://doi.org/10.1016/j.erss.2019.05.017

Lopatka, Jan, and Jason Hovet. 2021. "Czech Power Company Bohemia the Latest Price Crisis Casualty." *Reuters.com*. October 13, 2021. https://www.reuters.com/world/europe/czech-firm-bohemia-energy-shuts-down-citing-surging-power-prices-2021-10-13/

Martiskainen, Mari, and Benjamin K. Sovacool. 2021. "Mixed Feelings: A Review and Research Agenda for Emotions in Sustainability Transitions." *Environmental Innovation and Societal Transitions* 40. https://doi.org/10.1016/j.eist.2021.10.023

McKague, Fatima, Rob Lawson, Michelle Scott, and Ben Wooliscroft. 2018. "Understanding Energy Poverty through the Energy Cultures Framework." In *Energy Poverty and Vulnerability*. London: Routledge. https://doi.org/10.4324/9781315231518-3

Middlemiss, Lucie, and Ross Gillard. 2015. "Fuel Poverty from the Bottom-up: Characterising Household Energy Vulnerability through the Lived Experience of the Fuel Poor." *Energy Research and Social Science* 6. https://doi.org/10.1016/j.erss.2015.02.001

Moravec, Václav. 2021. "Drtivý Dopad Krachu Energošmejdů." Ceskatelevize.cz. https://www.ceskatelevize.cz/porady/1126672097-otazky-vaclava-moravce/221411030511107/

OTE. 2023a. "Statistika – Počty OPM Dodavatelů." *Ote-cr.cz*. https://www.ote-cr.cz/cs/statistika/mesicni-zprava-plyn/pocty-opm-dodavatelu?date=2021-01-01

OTE. 2023b. "Statistika – Počty OPM Dodavatelů v CS OTE." *Ote-cr.cz*. https://www.ote-cr.cz/cs/statistika/mesicni-zprava-elektrina/pocty-opm-dodavatelu-v-cs-ote?date=2021-01-01

Petrova, Saska, and Neil Simcock. 2021. "Gender and Energy: Domestic Inequities Reconsidered." *Social and Cultural Geography* 22 (6). https://doi.org/10.1080/14649365.2019.1645200

Pye, Steven, and Audrey Dobbins. 2015. "Energy Poverty and Vulnerable Consumers in the Energy Sector across the EU: Analysis of Policies and Measures." Insight_e Policy Report, issue 2 Posted: 2015. https://energy.ec.europa.eu/system/files/2015-07/INSIGHT_E_Energy%2520Poverty-Main%2520Report_0.pdf

Robinson, Caitlin, Sarah Lindley, and Stefan Bouzarovski. 2019. "The Spatially Varying Components of Vulnerability to Energy Poverty." *Annals of the American Association of Geographers*. https://doi.org/10.1080/24694452.2018.1562872

Stephenson, Janet, Barry Barton, Gerry Carrington, Daniel Gnoth, Rob Lawson, and Paul Thorsnes. 2010. "Energy Cultures: A Framework for Understanding Energy Behaviours." *Energy Policy* 38 (10). https://doi.org/10.1016/j.enpol.2010.05.069

Stephenson, Janet, Barry Barton, Gerry Carrington, Adam Doering, Rebecca Ford, Debbie Hopkins, and Rob Lawson, et al. 2015. "The Energy Cultures Framework: Exploring the Role of Norms, Practices and Material Culture in Shaping Energy Behaviour in New Zealand." *Energy Research and Social Science* 7. https://doi.org/10.1016/j.erss.2015.03.005

Stojilovska, Ana, Dušana Dokupilová, João Pedro Gouveia, Anna Zsófia Bajomi, Sergio Tirado-Herrero, Nóra Feldmár, Ioanna Kyprianou, and Mariëlle Feenstra. 2023. "As Essential as Bread: Fuelwood Use as a Cultural Practice to Cope with Energy Poverty in Europe." *Energy Research and Social Science* 97. https://doi.org/10.1016/j.erss.2023.102987

Sweeney, Jillian C., Johannes Kresling, Dave Webb, Geoffrey N. Soutar, and Tim Mazzarol. 2013. "Energy Saving Behaviours: Development of a Practice-Based Model." *Energy Policy* 61. https://doi.org/10.1016/j.enpol.2013.06.121

The Ministry of Industry and Trade. 2019. "Vnitrostátní Plán České Republiky v Oblasti Energetiky a Klimatu." Prague. https://www.mpo.cz/assets/cz/energetika/strategicke-a-koncepcni-dokumenty/2019/11/Vnitrostani-plan-CR-v-oblasti-energetiky-a-klimatu.docx

The Ministry of the Environment. 2023. "Nová Zelená Úsporám Light." Prague. https://novazelenausporam.cz/nzu-light/

Veselovský, Martin. 2021. "Zákazníci Bohemia Energy Teď Zaplatí Násobně Víc, Starší Lidé Budou Trpět, Říká Šnobr." *DVTVcz*. https://video.aktualne.cz/dvtv/zakaznici-bohemia-energy-ted-zaplati-nasobne-vic-starsi-lide/r~89f347822c5d11ecbc3f0cc47ab5f122/

Vysoká škola ekonomická v Praze. 2021. "Zranitelný Zákazník a Energetická Chudoba." Prague. https://www.mpo.cz/assets/cz/energetika/vyzkum-a-vyvoj-v-energetice/resene-dokoncene-projekty-a-jejich-vystupy/projekty-podporene-v-ramci-1-verejne-souteze-programu-theta/2021/7/ECH_ZZ_studie.pdf

Walker, Ryan, Christine Liddell, Paul McKenzie, and Chris Morris. 2013. "Evaluating Fuel Poverty Policy in Northern Ireland Using a Geographic Approach." *Energy Policy* 63. https://doi.org/10.1016/j.enpol.2013.08.047

10 Insights on Energy Poverty from Street Vendors in Indonesia

Dinita Setyawati

10.1 Introduction

The Indonesian informal economy accounts for a significant proportion of total economic activity and employs between 61% and 70% of the total labour force (Alatas & Newhouse, 2010; Firdausy, 2000). It is estimated that about 22 million people are street vendors and market traders, an immediate result of both the 1997 Asian financial crisis and the population shift from rural to urban areas. In this study, street vendors are defined as micro-merchants who operate on the streets and in parks as informal traders. They are distinct from the kind of merchants who have more resources and operate in designated places such as food courts or shophouses. They play an integral part in people's lives, including providing affordable food and beverages in convenient locations. Many operate on the streets, in alleys, and in parks without legal registration. Jabodetabek (the local name for the Jakarta metropolitan area, comprising Bogor, Depok, Tangerang, and Bekasi) is one of the most highly populated areas on the island of Java, and most of its urban inhabitants can be considered energy vulnerable, or as facing a greater risk of energy poverty. This area also has the highest concentration of urban street vendors in Indonesia.

The street vendors face constant energy insecurity and challenges to their overall health and quality of life. There is an almost complete absence of policies or regulations to protect street vendors in Indonesia, leaving them with no social insurance or other state protection. According to research examining the management of street vendors in the Indonesian city of Makassar, the local government views street vendors as regulatory offenders (Mahsyar & Rijal, 2021). Although the research does not specifically refer to street vendors, people with lower incomes have been included among the groups at risk of energy and transport poverty (Martiskainen et al., 2021).

According to the International Energy Agency (IEA) (2017), energy poverty is defined as a lack of access to secure, affordable, and modern energy services, particularly electricity and clean cooking fuel. The European Union defines energy poverty as a situation in which households are unable to access essential energy services and products (Directorate-General for Energy, 2015). However, a broader conceptualisation of energy poverty encompasses meaning beyond a lack of access and insecurity of supply. It is also "the absence of sufficient choice in accessing adequate, affordable, reliable, high-quality, safe and environmentally benign energy services to support economic and human development" (Reddy & Reddy, 1994, p. 44). In addition, it includes non-income dimensions of poverty such as caloric intake, life expectancy, housing quality, and access to energy, among other factors (Barnes et al., 2011; Njiru & Letema, 2018).

Despite being considered a global challenge, the IEA (2017) shows that the energy poverty gap is greater in Asia and Africa. The nexus between energy and poverty in the Global South has received significant political and academic attention (Gunningham, 2013; Rafi et al., 2021;

DOI: 10.4324/9781003408536-13

Sagar, 2005; Soriano-Hernández et al., 2022), mainly due to the challenges around finding alternative energy sources to reduce the dependence of these households on traditional forms of energy. Traditional forms of energy include open fires burning low-quality fuels, particularly biomass, resulting in fumes and smoke that may impact areas such as health, education, labour productivity, and income (Bouzarovski, 2018). The literature also indicates that a lack of availability and affordability of clean energy services among poor households particularly affects female members and children, as they are responsible for the majority of household chores, such as cooking and collecting firewood (Kanagawa & Nakata, 2006).

The majority of the literature investigating energy poverty in developing countries focuses on electricity supply and reliability constraints (Batinge et al., 2019; Bouzarovski, 2018; Dagnachew et al., 2018). The emphasis is on the need to expand the electrical grid based on the experience of developed countries (Rahul Sharma & Chan, 2016) and to extend the coverage of the power grid into rural areas (Pereira et al., 2010). The literature also identifies limited access to funds as the leading impediment to energy access. Therefore, attracting private investors to invest in the electricity sector could mitigate the lack of adequate institutional infrastructure and financial capital (Batinge et al., 2019; Wilson et al., 2014).

Although most people consider rural areas to be the most affected by energy poverty, studies show that energy poverty remains a critical issue in urban areas (Barnes et al., 2004; Zhang et al., 2019). In large part, it is migrants and the urban poor who have trouble meeting their basic energy needs and have to pay higher prices for energy use, including electricity. These people also tend to reside in built environments with very poor living conditions and limited access to services such as water, sanitation, and hygiene. Energy poverty, therefore, deals with the inability of households to pay for satisfactory energy services and supplies of electricity or heat (Sovacool, 2012), regardless of whether they are located in urban or rural areas. It also reinforces the spatial marginalisation of the urban poor as a minority group (Sovacool & Furszyfer Del Rio, 2022).

To the best of my knowledge, there is limited research investigating energy poverty in urban areas of Indonesia, especially in terms of assessing energy access in relation to the economic activities of the urban poor (Brata, 2010). Discussion about Indonesian street vendors in the available literature mostly centres on the governance and institutional challenges facing the management of the urban informal sector (Mahsyar & Rijal, 2021; Permatasari et al., 2014; Wilson, 2010; Yatmo, 2009) and the economic vulnerability of street vendors (Fathy & Rachmawan, 2020; Timothy & Wall, 1997). In addition, the academic literature that deals with Indonesian policy on energy poverty analyses the impact of energy poverty on welfare in rural areas (Sambodo, 2008; Wirawan & Gultom, 2021). These studies find that the geographical characteristics of remote villages are often the main inhibiting factors for electricity supply and safe cooking methods. However, most of the urban poor in Indonesia face similar issues around access to electricity, reflecting the situation prevailing in rural areas. This is evidenced by shifts in energy use in urban household studies (Andadari et al., 2014; Barnes et al., 2004). Due to the unregistered nature of their businesses, street vendors do not have access to the electrical grid for their everyday operations. This study, therefore, documents the energy poverty of street vendors in Indonesia and its implications on their standard of living.

10.2 Research methodology and conceptual approach

This section explains the methodological and conceptual approach of this study. Based on the relevance of the context for this study, I focus explicitly on street vendors in Jabodetabek.

10.2.1 Case study approach

This study uses a case-study approach. Interviews were conducted with a set of actors, including street vendor community representatives and the vendors themselves. Participant observation was conducted in ten locations in Jakarta, Bogor, Depok, Tangerang, and Bekasi (i.e., Jabodetabek). These cities are home to the largest street vendor communities in Indonesia. They are attractive to rural-urban migrants and are home to eleven million people. They also have more developed infrastructure than other cities in Indonesia, creating a façade of good career prospects and attractive employment. In reality, rural people without good academic qualifications often find themselves working in non-standard forms of employment or working as street vendors to support themselves. Street vendors in Indonesia are classified as working in the informal economic sector and thus have no national insurance or pension cover.

These five Indonesian cities have similar geography, energy, and transport challenges. Bogor, Depok, Tangerang, and Bekasi are located immediately adjacent to Jakarta, the capital city of Indonesia. About 30 million people residing in Indonesian cities are deprived of access to energy and hygiene facilities (Asian Development Bank, 2022). While the Indonesian government has aimed to improve energy access, in practice it has only prioritised increasing access through the extension of the centralised energy grid, leaving behind people with no connection to the grid (Chelminski, 2016) in both urban and rural areas. In urban areas, power theft remains a serious problem, often disrupting the electricity supply to homes and affecting the performance of the electrical grid, as proven in other locations (Furszyfer Del Rio & Sovacool, 2023). This also holds for Jabodetabek. In addition, this area faces problems in terms of traffic congestion, high fuel consumption, and greenhouse emissions. Jakarta city itself is regarded as one of the most congested cities in the world, with one of the main causes of the commuting activities of people from Bogor, Depok, Tangerang, and Bekasi for activities such as work and schooling (Farda & al-Rasyid Lubis, 2018). The congested and unhygienic conditions in slum areas also put dwellers at risk from changing climatic conditions.

The empirical analysis draws on a dataset assembled over three years, from March 2019 to December 2022. Secondary sources include government policy reports, industry reports, and academic literature. Primary sources were obtained from semi-structured interviews with 28 street vendors and participant observations in ten locations in Jabodetabek. The semi-structured interviews took one hour on average and included questions on demographic characteristics, income, electricity expenses, daily activities, problems associated with the present energy service, and coping strategies. Participant observations were conducted for approximately four hours per day, between 15.00 hrs and 19.00 hrs, to observe the vendors' day-to-day activities. I selected interviews and participant observation because both methodologies can provide a first-hand perspective on the group and their activities, as well as the ability to experience events in the same way vendors experience them.

However, due to the nature of the methodology, the results and discussions arising from this study cannot be understood as representative of the highly complex Indonesian urban energy-poverty landscape; rather, they are a study of the implications of energy policy on livelihoods within a small community. Data collection occurred in three stages due to the situation caused by the ongoing pandemic. The first data collection period in June 2019 examined the basic economic drivers behind street vendors' operations and their barriers to electricity access. The second stage, occurring in August-September 2022, focused on interviewing street vendors operating in designated public places. The last stage, completed in December 2022, involved interviews to complete the previous stage and participant observations. Different vendors were interviewed during the second and third stages. Participant observations were undertaken in all three stages. The data coding process was ongoing and informed the subsequent stages. I employed two research assistants to help with the interviews and participant observations.

10.2.2 Conceptual approach: key aspects of energy poverty

This study builds on the concept of energy poverty introduced by Reddy and Reddy (1994). The concept emphasises the absence of choice in accessing energy services that are adequate, affordable, accessible, reliable, high-quality, safe, and environmentally friendly. These key aspects correlate with the physical, economic, and psychological dimensions of energy poverty, in which geographical locations reflect inequality and the inability to maintain decent living standards. Decent living standards are defined here as the ability to maintain comfortable temperatures at home, such as through the use of air conditioning or a fan, having lighting for night-time activities such as evening meals, watching television together, and school-age children doing their homework, and using electricity for electrical appliances such as blenders and electric stoves. The framework also includes coping strategies, which are psychological adaptations by which vendors exercise coping mechanisms as a response to their hardship and in the hope of recovering from energy poverty (Stojilovska et al., 2021) (Figure 10.1).

10.3 Findings and discussion

The results from interviews and participant observations show varying degrees of energy poverty, with respondents vending on public streets having the lowest income in comparison with respondents located in parks and markets. The research team noted that the first group of respondents is paying more for electricity services and is thus the most vulnerable. We also observed that all the street vendors experienced varying degrees of energy poverty in terms of their lack of choice of reliable, safe, and affordable electricity connections. When choosing between diesel-powered generators and illegal electricity connections, the latter option is considered more affordable. Illegal electricity connections involve vendors getting light for their stalls by connecting to nearby shophouses, where the latter tamper with a metre. This is an illegal practise that can cause fire hazards.

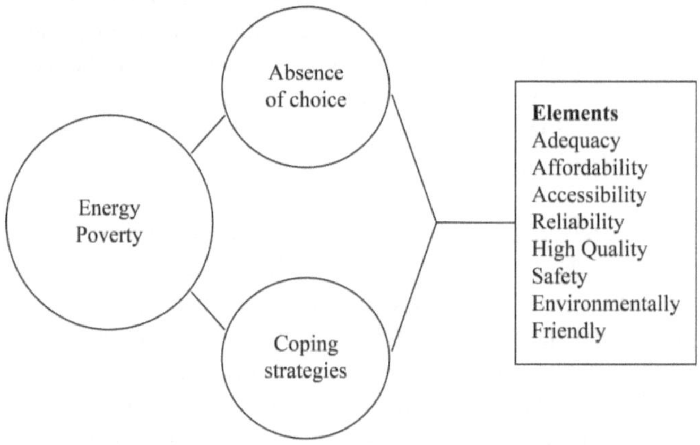

Figure 10.1 Key aspects of energy poverty.

Source: Adapted from Reddy and Reddy (1994).

10.3.1 Absence of choice

This study's findings suggest that the degree of difficulty in paying their accumulated energy costs is influenced by the absence of choice in selecting safe vending locations and adequate, affordable, accessible, reliable, safe, high-quality, and environmentally friendly energy services. The absence of choice can further exacerbate the vendors' physical, economic, and psychological vulnerabilities, exposing them to health threats, fire hazards, and a lack of future opportunities.

Participant observations show that vendors sell goods to the public without having premises or occupying walkways, roads, alleys, or other public spaces. Given that these people often operate in environments that can be dangerous or unpredictable, finding a safe and affordable vending location is perhaps the first and most significant issue this community faces. In this context, the negative consequences for the physical health of vending on the streets need to be emphasised. Vendors who spend most of their time outdoors are impacted by pollution from vehicles and industry, leading to health issues and non-communicable diseases. This also applies to the children of vendors, who often accompany them to their stalls. This could have long-term implications for educational attainment and employment opportunities, with consequences that endure into adulthood. From a wider societal perspective, the economic burden of non-communicable disease places a substantial burden on healthcare systems in terms of greater healthcare utilisation and costs.

Beyond impacts on health, energy-poor households face a choice between enduring uncomfortable temperatures, incurring debt, and cutting expenditure in other areas (Anderson et al., 2012; Gibbons & Singler, 2008; Hills, 2011; Mattioli et al., 2017), such as food consumption (Beatty et al., 2014). The research team found that all of our participants were living in slums and informal settlements with fans as cooling devices. The average temperature in Jakarta is between 26°C and 33°C. None of the respondents had air conditioning at home.

> I think air conditioning is a luxury. I do not have it in my home or my stall. I just use this small fan here and try to bear the heat. As long as my children can sleep at night, I am happy.
>
> (A09, food seller in Jakarta)

> I cannot be afraid of the pandemic. If I don't work, then who will feed my children? I do not know what I would do if I got sick.
>
> (A05, food seller in Bogor)

The slums are mostly located near active railway tracks or rivers, which lack hygiene and sanitation and utilise non-permanent materials for housing. Their houses are also located far from vending locations in the city centres. To reach their vending locations, the participants in our study had to walk between five and ten kilometres, dragging their carts across the city to maintain their livelihoods. There is an opportunity cost associated with the time used for travel, as they cannot be productive or do vending activities.

Presently, there is no legal electricity access provided by the government for street vendors operating on the street, for example, for lamps, blending juices, or cooking with electric stoves. These small businesses commonly use either diesel power generators that generate air pollution containing toxic air contaminants or illegal cable connections from nearby shophouses. Some respondents operate in front of shophouses where there are fixed vending spots with facilities for vending. These spots have canopies that provide shelter against the sun and rain and have (illegal) electricity points for stationary vendors. The absence of choice means that vendors are unable to purchase electricity legally, for example, by becoming registered customers of the State Electricity Company for their vending activities. It is also important to note that the only electricity connection available through the national grid belongs to the State Electricity Company. This means that customers do not have the option of subscribing to another electricity provider, creating little space for participation in energy decision-making.

> I pay about US $24 per month for my electricity bill. I think the price is affordable because my income is around US$350 per month. I use electricity for blending drinks, using a rice cooker and cooking. However, if there is a power cut, I will have less income because I do not have an alternative electricity supply.
>
> (A03, food and drink seller in Bogor)

The definition of "affordability" in energy poverty equates to spending no more than 10% of income on electricity services (Sovacool et al., 2019). Based on our data, most of our respondents ($n = 26$) spend less than 10% of their income on electricity services. However, the data does not necessarily confirm the premise that they are not energy poor. Their testimony states that electricity is not affordable, the electricity that they purchase is not adequate to maintain a comfortable temperature, and it is neither environmentally friendly (e.g., using polluting diesel generators), safe, nor reliable (if they depend on illegal electricity connections). More than 80% ($n = 22$) of the interview participants commented that they struggle to pay their energy and rent bills:

> I spend about US $8 (Indonesian rupiah Rp 100,000) on electricity bills. My income on average is about US $80 per month. For location rental, I spend about twice as much. Covering my monthly expenses is very difficult. I have to open my stall from 5 pm to 2 am.
>
> (A12, coffee seller in Jakarta)

Because energy expenses are high, they have little money to cover other services such as accommodation, healthcare, education, or transportation. This is the reason most respondents choose to travel to their vending location on foot.

> My income is between US $30 - US $50 per month. I pay around US$10-US$20 for electricity per month. It has been difficult due to the pandemic, I started selling here in 2020. I open my stall from 9 am to 12 midnight. I usually use electricity for the lamp, charging my phones, blending ice, and cooking. Electricity is mostly used at night. I need to find another location so I can get more customers.
>
> (A05, food and drink seller in Bogor)

Those who operate in front of private spaces have to pay rent to the stall or shophouse owner to sell in front of the building, in addition to their electricity bill.

> The electricity and rental prices are too high. But I do not have an option because I sell my products at night. The price of goods also keeps rising. Coffee and drink packages are more expensive now after the pandemic. I thought things would be better after the pandemic, but my shop still does not have a lot of customers.
>
> (A02, coffee seller in Bekasi)

In terms of adequacy and reliability, it seems that electricity connections do not satisfy energy requirements or comply with requirements on safety and quality of service. For example, one electricity terminal is used by numerous vendors.

> I choose to sell at night because more customers come here to eat after work. I have been selling here for three years. I get electricity from the building located nearby. Seven sellers use one electricity terminal. Sometimes there are conflicts if one of us uses more electricity than we agreed to. Then, the power is off and our customers are disturbed.
>
> (A03, food and drink seller in Bogor)

> It was quite easy to get a permit to sell here (in front of a small convenience store), I just asked permission from the local neighbourhood unit, who then submitted my request to the village administrative unit. I did not find it difficult to find a spot to open my stall. There are about six food and drink stalls in this area and we all purchase electricity from the coffee stall there.
>
> (A05, food and drink seller in Bogor)

10.3.2 Coping strategies

To survive, we found that the respondents employed coping strategies to access electricity, despite the apparent risks. For example, the local custom of using illegal electricity connections means that some respondents downplay the danger posed by this type of installation and believe that there is no need for standard installations. The chances of fire and electrical overloads are increased, which can cause serious damage to electrical appliances or the grid system.

> I know that using an electricity connection from another stall is dangerous. I often read and see news about fires in the stalls. We do not have proper fire safety guidelines. I think if there were cheap, designated places where we could vend for free, with reliable electricity connections, I would very much like to move there. I know that those places (referring to food courts) charge around USD$100 per month, which is almost half of my income.
>
> (A16, food seller in Jakarta)

> I did not know that purchasing electricity from a kiosk is illegal. I don't care about the fire hazards. I just need to do what I do to survive.
>
> (A28, food seller in Bekasi)

We also explored the choices that low-income vendors make when electricity competes with other essential needs, such as education and health.

> I used to sell during the daytime but I found that not many customers would buy my product (juice). Selling at night is much more expensive because I need to pay extra for the electricity. If I pay for electricity, I hardly have enough money left to pay for my son's education. I think about changing my product, but I have been doing this for the past seven years. I do receive cash transfers from the government sometimes, but it is not a regular income.
>
> (A15, juice seller in Jakarta)

> My son is going to graduate from high school soon. I worry about his future because I do not have the resources to fund his university studies. We have thought of a possible alternative, which is to send him to a training centre so he can work as a seafarer on foreign-owned vessels. Of course, I wish that electricity was cheaper so I could save more for his dream.
>
> (A21, food seller in Jakarta)

> If there is a power cut, I use an emergency lamp because diesel generators are not available in this area and they are expensive.
>
> (A16, food seller in Bogor)

> I think electricity is important because customers can see my stall from afar in the afternoon. I have regular customers but I want to attract new customers and grow my business.
>
> (A27, coffee seller in Bekasi)

There has been a suggestion to use solar panels as a cleaner electricity source for street vendors in India (Szakonyi & Urpelainen, 2015). This technology has the potential to be replicated in the Indonesian case. By using solar energy, there is potential for the vendors to label their products as "green" and increase their prices. The respondents in our study expressed an interest in using renewable energy to support environmental causes.

> Sometimes there are power cuts and I use candles to light my stalls. I did not know about renewable energy or that we could use the sun's power to generate electricity, but if we can use cleaner energy sources then I would be willing to use them. I saw on the news that people are protesting against coal and pollution.
>
> (A12, food seller in Jakarta)

> If I had the resources, I would prefer to use electricity from the sun.
>
> (A07, food seller in Jakarta)

> I think solar energy is very important. I know the trend now is toward sustainable living. Therefore, if I can use electricity sourced from renewable energy, I am contributing to the betterment of the environment. I think the government should build more solar-powered electric charging stations in convenient locations where we can operate our stalls.
>
> (A04, coffee seller in Tangerang)

There is research suggesting that we need to address energy poverty through a multidimensional lens (Furszyfer Del Rio & Sovacool, 2023). By observing the lived experience of energy-poor communities and identifying coping strategies, this study adds to the discussion that energy poverty should not be considered only in terms of the technical issues of affordability and accessibility but should also take into account the economic and psychological factors that create the problem. Frequently used coping strategies are bearing uncomfortable temperatures, restricting other needs, such as education, lowering the priority of transportation by walking to vending locations, and ignoring the impact on the health of being exposed to constant pollution.

We did not find this in our observations, but we found from the interviews, there are some cases of street vendors stealing electricity from public street lighting. This is very dangerous, as it can result in electrocution and disruption to the transportation system. Having streets left unlit due to vandalism increases the risk of road accidents.

It seems that the issue of energy poverty has become a systemic injustice. Vendors are left isolated in their individualised coping strategies rather than receiving help from the system (Bouzarovski & Simcock, 2017; Stojilovska et al., 2021). The absence of electricity choice stemming from Indonesia's monopolistic energy system leaves vulnerable communities with an inability to provide input into policymaking that should be targeting vulnerable citizens.

10.4 Conclusion

To examine the lived experience of energy poverty in Indonesia's urban informal economy, I interviewed 28 respondents and conducted site visits to ten locations in urban areas of Jabodetabek. This study shows that these street vendor communities experience energy poverty in terms of their lack of choice for reliable, safe, and affordable electricity connections. Moreover, this study reveals that energy poverty is exacerbated by geographical barriers and that coping strategies are used to access electricity. The street vendors experience varying degrees of energy poverty, depending on their income, vending location, and the availability of electricity. The poorest energy vendor has the lowest income, spends almost half of her income on electricity, and purchases electricity from a nearby kiosk, which connects to the electricity grid via an illegal cable connection. Some respondents expressed an interest in purchasing electricity sourced from renewable energy sources. Coping strategies to overcome energy poverty include choosing energy services over education and purchasing illegal electricity from a nearby kiosk.

From this study's exploration, I found that energy poverty is best understood through a multidimensional lens. It is not only a technical issue, but it also has social, cultural, and political implications. The street vendors, who spent a larger proportion of their income on energy, neglected other basic necessities such as health and education. There is also a loss in terms of the opportunity costs associated with the long travel time between homes and vending locations. Educational and health impacts related to expenditure on energy rather than schooling, as well as an increased incidence of sickness. For instance, time spent on the street may be related to acute respiratory infections. Most of the sellers also do not have sufficient health insurance, revealing the need for better social protection for these informal workers.

There is also an absence of choice; vendors have limited alternatives for accessing cleaner energy sources. They have no choice but to secure a connection to the electricity grid. Using solar energy has inherent advantages, but considering the cost of solar panels and the absence of a place to install the panels, this option is beyond the consideration of most street vendors. Despite this, the interest in using cleaner energy sources offers important opportunities for further action. For example, action by the government to provide access to solar energy could enable vendors who want to reduce their emissions to choose an electricity source that satisfies this desire.

Street vendors are one of the most politically and economically underrepresented communities in urban areas and are one of the most vulnerable to changes in the energy system. Their inclusion in the energy poverty bracket is systematically reinforced by the absence of a choice of adequate, reliable, affordable, and safe energy services in a centralised energy system. Going beyond the technical implications, this study also highlights the economic and psychological impacts in terms of more personal coping strategies that affect vendors' resilience against climate change, social development, and health consequences. Policies to combat energy poverty should address the ability of the vendors as individuals, and as economic actors, to prepare for, absorb, recover from and adapt to major shocks, for example, by providing free vending locations with safe, reliable, environmentally friendly electricity access as well as social safety nets accessible to all.

Acknowledgements

This research is part of the Toyota Foundation Research Grant Program 2019 deliverables, grant number D19-R-0042. I thank Mariam Kvaratskhelia for her comments on this chapter.

References

Alatas, V., & Newhouse, D. (2010). *Indonesia Jobs Report: Toward Better Jobs and Security for All: Ringkasan Eksekutif.* Washington, DC: World Bank Group. Retrieved from https://policycommons.net/artifacts/1453882/indonesia-jobs-report/2086360/. CID: 20.500.12592/j76081.

Andadari, R.K., Mulder, P., & Rietveld, P. (2014). Energy Poverty Reduction by Fuel Switching. Impact Evaluation of the LPG Conversion Program in Indonesia. *Energy Policy*, 66, 436–449.

Anderson, W., White, V., & Finney, A. (2012). 'Coping with Low Incomes and Cold Homes', *Energy Policy*, 49, pp. 40–52. https://doi.org/10.1016/j.enpol.2012.01.002

Barnes, D., Kerry, K., & William, H. (2004). *The Urban Household Energy Transition: Energy, Poverty, and the Environment in the Developing World.* Washington, DC: World Bank.

Barnes, D.F., Khandker, S.R., & Samad, H.A. (2011). Energy Poverty in Rural Bangladesh. *Energy Policy*, 39(2), 894–904.

Batinge, B., Musango, J.K., & Brent, A.C. (2019). Perpetuating Energy Poverty: Assessing Roadmaps for Universal Energy Access in Unmet African Electricity Markets. *Energy Research Social Science*, 55, 1–13.

Beatty, T.K.M., Blow, L., & Crossley, T.F. (2014). Is There a "Heat-or-Eat" Trade-Off in the UK? *Journal of the Royal Statistical Society Series A: Statistics in Society*, 177(1), 281–294. https://doi.org/10.1111/rssa.12013

Bouzarovski, S. (2018). *ENERGY POVERTY (Dis)Assembling Europe's Infrastructural Divide*. Basingstoke: Springer Nature.

Bouzarovski, S., & Simcock, N. (2017). 'Spatializing Energy Justice', *Energy Policy*, 107, pp. 640–648. https://doi.org/10.1016/j.enpol.2017.03.064

Brata, A.G. (2010). Vulnerability of Urban Informal Sector: Street Vendors in Yogyakarta, Indonesia. *Theoretical and Empirical Researches in Urban Management*, 5(5), 47–58.

Asian Development Bank (2022). *Building Resilience of the Urban Poor in Indonesia.* Metro Manila: Asian Development Bank.

Chelminski, K. (2016). The Political Economy of Energy Access and Sustainable Energy Transitions in Indonesia, *L'Europe en Formation*, 378(4), 146–165. https://doi.org/10.3917/eufor.378.0146

Dagnachew, A.G. et al. (2018). 'Trade-Offs and Synergies between Universal Electricity Access and Climate Change Mitigation in Sub-Saharan Africa', *Energy Policy*, 114, pp. 355–366. https://doi.org/10.1016/j.enpol.2017.12.023

Directorate-General for Energy (2015). *Energy Poverty and Vulnerable Consumers in the Energy Sector Across the EU: Analysis of Policies and Measures. Insight_E- European Commission.* Brussels. Available at: https://energy.ec.europa.eu/system/files/2015-07/INSIGHT_E_Energy%2520Poverty-Main%2520Report_0.pdf

Farda, M., & al-Rasyid Lubis, H. (2018). Transportation System Development and Challenge in Jakarta Metropolitan Area, Indonesia, *International Journal of Sustainable Transportation Technology* 1(2), 42–50.

Fathy, R., & Rachmawan, D. (2020). The Paradox of Informal Economy in Urban Area Indonesia: A Case Study Street Vendors in Jalan Salemba Raya, Jakarta, and Jalan Raya Sawangan, Depok, *MASYARAKAT: Jurnal Sosiologi*, 25(1), 83–106. https://doi.org/10.7454/mjs.v25i1.10871

Firdausy, C.M. (2000). 'The Social Impact of Economic Crisis on Employment in Indonesia', Report Center for Economic and Development Studies-Indonesian Institute of Sciences. Jakarta: LIPI.

Furszyfer Del Rio, D.D., & Sovacool, B.K. (2023). 'Of Cooks, Crooks and Slum-Dwellers: Exploring the Lived Experience of Energy and Mobility Poverty in Mexico's Informal Settlements', *World Development*, 161. https://doi.org/10.1016/j.worlddev.2022.106093

Gibbons, D., & Singler, R. (2008). *Cold Comfort: A Review of Coping Strategies Employed by Households in Fuel Poverty*. London: Centre for Economic and Social Inclusion.

Gunningham, N. (2013). 'Managing the Energy Trilemma: The Case of Indonesia', *Energy Policy*, 54, pp. 184–193. https://doi.org/10.1016/j.enpol.2012.11.018

Hills, J. (2011). *Fuel Poverty: The Problem and Its Measurement*. London: CASE report.

International Energy Agency (IEA) (2017). *Energy Access Outlook*. Paris: IEA.

Kanagawa, M., & Nakata, T. (2006). 'Socio-Economic Impacts of Energy Poverty Alleviation in Rural Areas of Developing Countries', in *Proceedings of the 26th USAEE/IAEE North American Conference*, Ann Arbor, MI, USA, pp. 24–27.

Mahsyar, A., & Rijal, R. (2021). Institutional Network Model in the Coordination of Street Vendor Empowerment in Makassar City, South Sulawesi, Indonesia. *Jurnal Ilmiah Ilmu Administrasi Publik*, 11(2), 583–590.

Martiskainen, M. et al. (2021). 'New Dimensions of Vulnerability to Energy and Transport Poverty', *Joule*, 5, pp. 3–7.

Mattioli, G., Lucas, K., & Marsden, G. (2017). 'Transport Poverty and Fuel Poverty in the UK: From Analogy to Comparison', *Transport Policy*, 59, pp. 93–105. https://doi.org/10.1016/j.tranpol.2017.07.007

Njiru, C.W., & Letema, S.C. (2018). 'Energy Poverty and Its Implication on Standard of Living in Kirinyaga, Kenya', *Journal of Energy*, 2018, pp. 1–12. https://doi.org/10.1155/2018/3196567

Pereira, M.G., Freitas, M.A. V., & da Silva, N.F. (2010). Rural Electrification and Energy Poverty: Empirical Evidences from Brazil, *Renewable and Sustainable Energy Reviews*, 14, 1229–1240.

Permatasari, A., Putro, U.S., & Nuraeni, S. (2014). 'Strategic Analysis Relocating Street Vendor Through 3D Negotiation Case Study: Street Vendor Surakarta, Indonesia', *Procedia – Social and Behavioral Sciences*, pp. 436–443. https://doi.org/10.1016/j.sbspro.2014.02.451

Rafi, M., Naseef, M., & Prasad, S. (2021). 'Multidimensional Energy Poverty and Human Capital Development: Empirical Evidence from India', *Energy Economics*, 101. https://doi.org/10.1016/j.eneco.2021.105427

Rahul Sharma, K., & Chan, G. (2016). Energy Poverty: Electrification and Well-Being. *Nature Energy*, 1, 16171.

Reddy, A.K.N., & Reddy, B.S (1994). 'Energy and Social Issues'. In *World Energy Assessment*, New York: United Nations Development Programme. p. 44.

Sagar, A.D. (2005). Alleviating Energy Poverty for the World's Poor. *Energy Policy*, 33, 1367–1372.

Sambodo, M.T. (2008). 'Energy Sector in Indonesia and Environment Impacts: From Fossil Fuel to Biofuel', *Jurnal Ekonomi dan Pembangunan LIPI*, 16(1).

Soriano-Hernández, P., Mejía-Montero, A., & van der Horst, D. (2022). 'Characterisation of Energy Poverty in Mexico Using Energy Justice and Econophysics', *Energy for Sustainable Development*, 71, pp. 200–211. https://doi.org/10.1016/j.esd.2022.09.005

Sovacool, B.K. (2012). 'The Political Economy of Energy Poverty: A Review of Key Challenges', *Energy for Sustainable Development*, 272–282. https://doi.org/10.1016/j.esd.2012.05.006

Sovacool, B.K., & Furszyfer Del Rio, D.D. (2022). '"We're Not Dead Yet!": Extreme Energy and Transport Poverty, Perpetual Peripheralization, and Spatial Justice Among Gypsies and Travellers in Northern Ireland', *Renewable and Sustainable Energy Reviews*, 160. https://doi.org/10.1016/j.rser.2022.112262

Sovacool, B.K., Lipson, M.M., & Chard, R. (2019). Temporality, Vulnerability, and Energy Justice in Household Low Carbon Innovations, *Energy Policy*, 128(December 2018), 495–504. https://doi.org/10.1016/j.enpol.2019.01.010

Stojilovska, A., Yoon, H., & Robert, C. (2021). 'Out of the Margins, into the Light: Exploring Energy Poverty and Household Coping Strategies in Austria, North Macedonia, France, and Spain', Energy *Research and Social Science*, 82. https://doi.org/10.1016/j.erss.2021.102279

Szakonyi, D., & Urpelainen, J. (2015). Energy Poverty Among Urban Street Vendors in India: Evidence from Patna, Bihar. *Energy for Sustainable Development*, 24, 44–49.

Timothy, D.J., & Wall, G. (1997). Selling to Tourists: Indonesian Street Vendors, *Annals of Tourism Research*, 24(2), 322–340. https://doi.org/10.1016/S0160-7383(97)80004-7

Wilson, I.D. (2010). 'The Streets belong to who?: "Governance" and the Urban Informal Sector in Jakarta, Indonesia', *Elephant in the Room: Politics and The Development Problem*, Perth: Asia Research Centre Policy Monograph, pp. 113–133.

Wilson, E., Rai, N., & Best, S. (2014). *Sharing the load: Public and private sector roles in financing pro-poor energy access*. www.iied.org.

Wirawan, H., & Gultom, Y.M. (2021). 'The Effects of Renewable Energy-Based Village Grid Electrification on Poverty Reduction in Remote Areas: The Case of Indonesia', *Energy for Sustainable Development*, 62, pp. 186–194.

Yatmo, Y.A. (2009). Perception of Street Vendors as "out of Place" Urban Elements at Day Time and Night Time, *Journal of Environmental Psychology*, 29(4), 467–476. https://doi.org/10.1016/j.jenvp.2009.08.001

Zhang, D., Li, J., & Han, P. (2019). A Multidimensional Measure of Energy Poverty in China and Its Impacts on Health: An Empirical Study Based on the China Family Panel Studies. *Energy Policy*, 131, 72–81.

11 Understanding Energy Poverty in South Africa

Nthabiseng Mohlakoana and Peta Wolpe

11.1 Introduction

For centuries, South Africa was plagued by colonialism and racial segregation. In 1948, formal Apartheid was established, cementing the separate and racist trajectory of the country and preventing the majority Black, Coloured, and Indian populations from having almost any human rights. These rights included access to education, basic services, and many other opportunities that contribute to a decent life. After years of struggle, in 1994 South Africa entered a new chapter in its history when the country underwent its first democratic election, which – also for the first time – allowed all citizens over the age of 18 to vote, regardless of race. This marked the official end of Apartheid and was followed by a new and democratic constitution, which pledged to:

> Heal the divisions of the past and establish a society based on democratic values, social justice and fundamental human rights
>
> (Republic of South Africa, 1996, p. 1)

Yet, almost 30 years later, despite having many pro-poor and development policies in place, the country is still beset by extremely high levels of poverty, inequality, unemployment, and energy poverty. In a world where sustainable and clean energy is increasingly a priority in response to the global climate crisis, the need to address energy poverty has also become more acute, despite being on South Africa's agenda since democracy. Not only is the country a high carbon emitter, built around a coal-driven energy-intensive industry sector, but because of apartheid, many poor households were deprived of access to basic services, including electricity. This is a double disadvantage for a country that is desperately trying to find relief from the triple challenge[1] of poverty, unemployment, and inequality (Parliament of South Africa, 2018).

Over the years in South Africa, the definition of household energy poverty has evolved from simply being understood as a lack of access to electricity to including a lack of access to basic energy services (ERC, 2002; Knox et al., 2018; Ledger, 2021; Mohlakoana & Wolpe, 2021; Thom et al., 2001). This consists of factors such as use, affordability, safety, health, gender, and many other elements. It is important to acknowledge that energy poverty is multi-layered, context-dependent and therefore requires context-based interventions.

This chapter takes a granular lens to understand the lived experiences of energy poverty in the South African context from both a municipal and community perspective. In doing so, it looks at the experiences of those living in energy poverty and those trying to find solutions.

DOI: 10.4324/9781003408536-14

11.2 Lived experiences of energy poverty

What is striking about these photographs is that they capture the plight of people living in an unelectrified informal settlement next to one of the 12 coal-fired power stations and the coal-to-liquid fuel plant in Mpumalanga. Coal mining and coal led industries have had an enormously negative impact on the environment and on people's lives and health. There is clear evidence of premature deaths, asthma, and other lung diseases from air pollution. The land and air are coated in coal dust, which can be seen when one drives through the area. Rivers are affected by acid mine drainage, which leads to illness and affects land for agriculture. Many are without work and do not pass health tests in order to work in the mines because of respiratory illness, further compounding unemployment (Annecke et al., 2022; Bega, 2021; groundWork, 2019; Nowicki, 2020) (Figures 11.1 and 11.2).

This highlights the many contradictions present in South African society. In 2016, approximately one in seven households lived in informal dwellings, and in high density urban centres, the numbers were higher. According to a recent report by the Socio-Economic Rights Institute of South Africa (SERI), 79% of households lived in formal dwellings, 13.9% in informal dwellings, and 5.9% in traditional homes (SERI, 2018). In a country where housing and electrification are seen as having high political priority and given significant inroads into resolving the challenges, this is concerning.

A survey of 41 low-income electrified households in KwaGuqa, a small township in Mpumalanga province in the heart of South Africa's coal region, brings to light the lived experiences of

Figure 11.1 Betty Zulu, a resident of the Big House settlement, with her laundry and a generator. This settlement is situated next to Komati power station in Mpumalanga.

Source: Photograph by Daylin Paul.

Figure 11.2 Klaas Mtsweli sits outside his home in Big House settlement, near Komati power station in Mpumalanga.

Source: Photograph by Daylin Paul.

energy-poor people 29 years into democracy. For the purposes of this chapter, quotations have been extracted from the interviews conducted to illustrate the lived experiences. Participants have been anonymised, and each quote is assigned a number based on the household number of the survey. In this case, "HH" represents household, and the number refers to the number given to the questionnaire from 1 to 41.

Whilst the country has a policy of Free Basic Electricity (FBE) for poor or indigent households (discussed in more detail below), only nine of the 41 households in the sample knew that they were receiving this subsidy. This may be because if and when they purchase additional units of electricity, the receipt does not show the initial free units they are entitled to. The households explained to the researchers that they had applied for indigent status but did not know if this was granted as they had not received confirmation from the municipality. This led to their being confused about their indigent status and eligibility for receiving FBE. Registering as indigent is not easy, and often people either do not realise that they were granted the subsidy or simply do not get it because they do not fulfil the criteria. All the households made it clear that even with the FBE allowance, they did not have enough electricity to meet all their energy needs, and most were resorting to "cheaper" alternative energy sources such as coal, kerosene, and wood. Of the sample, 27% said that they opt for multiple sources of alternative energy based on affordability, and 73% relied on a single alternative source. Many indicated that they mainly use electricity for lighting and boiling water. People are only able to buy small amounts of fuel, such as a one-litre bottle size of paraffin or a few candles. Ultimately this is not only unsafe but a more expensive option as a result of lack of disposable income, which leaves most households with little choice (Mohlakoana & Wolpe, 2023).

HH1: "Electricity is expensive because we have to use all our money on energy and food."
HH10: "It is expensive and sometimes it's difficult to get paraffin (kerosene) as it is not always available in the area. This means you can't cook, and you end up eating bread only."
HH7: "High electricity costs leave us no option but to use the energy that is harmful for our lives."

Most households indicated that whilst these alternative energy sources are also expensive, they are more affordable than electricity. The sample showed that households were spending between 18% and 37% of their income on energy costs, and many were using food money to service their energy needs. An impossible choice to have to make.

As with many informal settlements across the country, such as Imizamo Yethu in Cape Town, every winter season, there are fires caused by unsafe types of energy sources. Given the high density of informal dwellings present, an outbreak of fire in one shack leads to a domino effect with many others also going up in flames. In September 2022, up to 500 people were displaced in Imizano Yethu due to the fire. These fires not only lead to loss of irreplaceable belongings, they also lead to losses of life and the knock-on associated consequences for families and the local community (Flores Quiroz et al., 2021; Viljoen, 2022).

The majority of respondents in the KwaGuqa sample were clear that lower electricity prices would make their lives much easier. They could afford to use more electricity and reduce the use of unsafe fuels, and most importantly, they would have more money for food.

HH10: "It will be better because I will be able to save my pension money and cook as I wish."
HH9: "It will help me to pay the rent and be able to cook healthy food as I am suffering from high blood pressure."
HH11: "It will give us a chance to afford it and maybe it will last a month."
HH38: "I would be able to save and bake cookies instead of buying them."
HH3: "It will help a lot because there will be no need for alternative fuels? maybe I will be able to heat my house and cook whatever I want."
HH5: "It will help a lot because I can be able to use my electric stove to do what I cannot do right now."

From these statements, it is obvious that households would benefit in many ways from reduced electricity costs and/or higher subsidies and the ability to prepare healthier food without worrying about the high costs of energy. This would significantly improve their livelihoods.

All the sampled households were clear that the current electricity prices are too expensive, which impacts how much electricity they can realistically afford and results in them buying other less safe energy fuels such as wood, coal, and liquid petroleum gas (LPG). This corresponds with the literature on energy poverty, which indicates that electricity connections do not necessarily lead to the use of the supply to satisfy all household energy needs (Ledger, 2021; Thom et al., 2001; Wolpe & Reddy, 2014). The free basic allowance for those who receive it is inadequate for all their requirements, and buying electricity is too expensive for these households.

The following statements are some of the responses from households that were asked how they would be impacted by higher electricity prices.

HH3: "As for me, I will go back to the coal stove even though it is hard to get that coal. It is expensive, 50kg of coal is R200 (≈USD 11) and cannot last for a month, it will be hard."
HH12: "I will not be able to use electricity, I will have to look for other alternatives."

HH22: "It is going to kill us, and I will not be able to buy it. I will be done."
HH15: "It will be too expensive. Things will go from better to worse."

This small snapshot of life for people living in energy poverty illustrates how hard their lives are and how they are making choices between energy and food. These impossible choices continue to face poor households despite the numerous political efforts and policies that have been developed over the years to directly tackle this problem.

11.3 South African policies to alleviate energy poverty

In 1994, the newly elected government was faced with the enormous need and obligation to redress the many negative legacies of the apartheid regime. This meant giving priority to development; by providing housing, health care, education, and other basic services such as electricity to the majority of the population who had previously been denied these rights. Almost all policy development in the democratic era has concentrated on addressing these challenges whilst at the same time trying to grow the economy, reduce inequality, and increase employment.

The first of such policies was the Reconstruction and Development Programme (RDP) White Paper (White Paper on Reconstruction and Development, 1994). It was heralded as a key vehicle for the transformation process away from apartheid. The policy outlined a national electrification programme, the building of homes, job creation, and land redistribution, as well as other reforms. It was successful in providing basic services to many poor people (Baker & Phillips, 2019; Baker, 2017; Wolpe & Reddy, 2014). Today, more than 85% of the population now has access to electricity, and over 3 million low-cost formal houses have been built. Even though many were built on the periphery of cities far from work and opportunities and without ceilings, they did provide low-cost formal housing to those who previously lived in shacks.

The 1996 Constitution of South Africa set out the functions of the three-tier system of government for the country. Local government (municipal level) was given a clear role and mandate to provide the delivery of basic services to the communities in their areas of jurisdiction in "a sustainable and equitable manner" (Republic of South Africa, 1996; SALGA, 2014; SEA, 2017). Whilst it is the developmental arm of the system of government and does receive grants from the national government, it must also generate revenue to deliver on its mandate. It does this largely through property rates and the sale of electricity to households and businesses, amongst other mechanisms.

The South African White Paper on Energy Policy (1998) also had several important objectives, which included goals of increased energy access and security, improved governance, and economic growth (DME, 1998). Of critical importance, the aim of this policy was to increase access to affordable energy services for as many households as possible, especially those that were previously disadvantaged by apartheid government policies.

Access to basic energy services is the first step in overcoming energy poverty, but access on its own is not sufficient as it does not translate to the full use of that service. This is because poor households generally do not have the means to afford the full range of services on offer (Pye et al., 2015), making affordability a huge issue. Further, decisions on which energy source and what appliances to use are often influenced by household dynamics such as power and gender relations and remain an understudied area (Abbas et al., 2020; Beshilas, 2019; Johnson et al., 2020). Measuring access to affordable and clean energy as per the Sustainable Development Goal (UN, 2015) is achieved through the number of households with an electricity connection. These figures can be misleading, as they do not indicate the extent of energy being used by those connections. The case of South Africa illustrates this point well. As noted above, the statistics indicate high household electricity connections at 85% in 2018 (StatsSA, 2018) compared to

36% in 1994 (Wolpe & Reddy, 2014). But this still leaves more than a million households without grid connected electricity (Taylor, 2023). South Africa may be seen as being on track to achieving the SDG 7 goal, but in reality, the rate of energy poverty was shown to be at 58% in 2015 (Ye & Koch, 2021). Given the rising unemployment (32.9% in the third quarter of 2022), poverty, and inequality rates in the country, it is without doubt that household energy poverty numbers have increased (StatsSA, 2022). It is therefore clear that electricity connections do not translate to electricity use or a reduction in energy poverty.

One of the main challenges faced by the government was that even with the rapid electrification of low-income households, many could not afford to use this electricity supply. In the late 1990s, the electricity utility Eskom undertook research to determine the minimum amount of electricity required by low-income households and deemed that 50 kWh per month (the equivalent of R125 or USD 9) was sufficient. This resulted in the government's energy department publishing and initiating the implementation of Free Basic Electricity (FBE) to alleviate the challenges of electricity affordability in 2003 (DME, 2003). Most municipalities provide an electricity subsidy of 50 kWh monthly – although some give slightly higher amounts depending on their budgets and subsidy allocations for basic services. As much as this subsidy brings relief to many low-income households in the country, it is not enough to cover all their household energy needs. In fact, recent research (Ledger, 2021; Ledger & Rampedi, 2022) contends that this amount is grossly insufficient to meet even the basic energy needs of most households, and this view is supported by many municipalities and civil society groups. The subsidy is rolled out by local municipalities, and they are expected to provide the subsidy even in Eskom-distributed areas, which has created some challenges. Eskom shares the distribution of electricity with 165 licenced municipalities (Calitz & Wright, 2021; Mohlakoana & Wolpe, 2023).

Generally, low-income households supplement this electricity subsidy by buying extra electricity units and/or using a mix of energy sources such as wood, coal, kerosene, LPG, and other unsafe fuels as described in the fieldwork study above. The FBE allocation comes from the Equitable Share Grant and is for all basic services. However, to receive the benefit, households must be registered as indigent, and the registration process is onerous and often means that those who should be getting the benefit are not. In addition, because local governments are struggling to meet all their mandated operations and must contend with financial constraints, the grant – which is not ringfenced – does not reach all beneficiaries and is sometimes used for other municipal expenses. Further, many households are using the money and grants they receive for food to supplement or simply provide for their energy needs (Ledger & Rampedi, 2022).

11.4 Deepening energy poverty and informal electricity connections

In South Africa, many municipalities and Eskom lose revenue due to informal or illegal electricity connections. These are connections made by households either directly from the electricity supply source, such as the distribution sub-station, or from another household that has a formal electricity connection. These informal connections are regarded as illegal by Eskom and the municipalities because they do not comply with the electricity connection regulations. These illegal connections lead to their inability to recoup revenue for the cost of supplying electricity and pose serious safety risks. This not only results in technical losses for the municipality or Eskom, but it also overloads the electricity distribution grid due to having to supply more households than it is capable of.

There are several reasons why households end up with informal or illegal electricity connections (Mbandlwa & Geyevu, 2022). The most obvious is the lack of affordability, which leads to the inability to use electricity. Most of the households with informal connections are in low-income areas where unemployment and poverty levels are high. Their inability to pay for the formal electricity supply leads to households using alternative sources of energy. These alternatives also come at a cost that is still too high for some. Another reason for informal connections is the income generation aspect, whereby those with formal connections "sell" electricity informally and at a higher cost. Such informal connections are usually to people with informal homes living across those with formally connected housing.

Over the years, as people flock to cities from rural areas, there has been a growth in informal settlements. This situation worsened during the COVID-19 pandemic, which led some people to give up their formal dwellings for informal settlements due to loss of income and unaffordable rental rates. This informal land occupation occurs on land not deemed suitable for the provision of basic services such as water, electricity, and sanitation by local municipalities. Therefore, households that are living in informal dwellings end up using dangerous fuels such as kerosene because they are not able to access and use modern energy services, such as formal electricity connections.

Reliance on biomass and unhealthy fuels such as kerosene due to both a lack of access and the ability to afford modern energy services such as electricity also contributes towards energy poverty. It is a known fact (Simcock et al., 2018; Thom et al., 2001) that households experiencing poverty lack sufficient income to cater for their household needs, such as food, energy, and suitable housing, and are often classified as living below the poverty line. The most common benchmark to determine the affordability of energy for households is if they spend 10% or more of their income on their energy needs (Bouzarovski et al., 2021; Wolpe & Reddy, 2014). This percentage is considered unaffordable, especially when compared to mid- and high-income households, which spend 2-3% of their income on energy. For low-income households, spending such high amounts of their income on energy often means less available income for other household priorities, such as food.

The informal connections are indicative of deepening poverty and energy poverty levels in the country (Mbandlwa & Geyevu, 2022). With high unemployment, many households cannot afford basic household needs, which include access to energy, including electricity. Similar to other countries, the South African inflation rate has risen due to several factors, including the COVID-19 pandemic and the loss of employment for many as the economy dipped. The ongoing Russian invasion of Ukraine has also led to an increase in food and energy prices. For South African low-income households, high inflation has made it much more difficult to afford basic services such as food, housing, medical care, and now increasingly energy.

11.5 Delivery of energy services

The cost of supplying a household with electricity is high and difficult for a distributing municipality to recoup, particularly if the customer is not using sufficient electricity for them to generate income (Creamer, 2022; Eskom, 2022). This has to be managed by electricity departments, which are under constant pressure to work within a budget whilst providing the service. The sale of electricity is one of the main mechanisms for municipalities to generate revenue as they can add a levy to their tariffs. Electricity tariffs are not uniform throughout the country and the implementation of electricity subsidies is also dependent on whether the distributor is Eskom or the municipality. There is a lack of awareness on the part of customers in terms of how the system operates and why there are different tariffs. This lack of awareness leads to a lack of understanding, which heightens the strain on distributors and customers (Eberhard, 2006). For instance, in conversations with households and municipal officials, communities want to know why their neighbours pay less than they do. Officials know that communities do not understand the tariffs in place or how the subsidies work, but there are not enough staff and interventions to remedy this situation. They recognise that there is not adequate communication to raise awareness, and yet they do not have the resources to change this situation.

Local government officials and civil society groups are calling for an increase in the FBE grant. It is well documented that 50 kWh of free electricity per month is simply not enough to keep poor households out of poverty. Recent research suggests a minimum of 250 kWh is more appropriate, which is equivalent to R625 or USD 34 (City of Cape Town, 2022; Ledger, 2021). However, the national grant they receive to roll out FBE will not enable them to implement such an increase. More recently, the national government has developed a Just Energy Transition Investment Plan (JET-IP) in response to the USD 8.5 billion pledged at the 2021 Glasgow Conference of Parties (COP) 26. The funds were pledged by the European Union, the United Kingdom, France, Germany, and the United States as the Climate Investment Fund to accelerate the coal transition for developing countries (Presidential Climate Commission, 2022). The plan includes a focus on local government and a small increase in the FBE allowance to 100 kWh (Republic of South Africa, 2022).

The City of Cape Town municipality has been striving to address energy poverty for many years. They have a 98% electrification rate. However, despite this high rate, they continue to experience electricity theft and energy poverty. The municipality has rolled out a ceiling retrofit programme as many of the RDP houses built in the early phase (national programme) were without ceilings to keep costs down and keep numbers high (City of Cape Town, 2022). However, a home with no ceiling and no insulation makes no sense from an energy perspective (Phillips et al., 2011). Even though the programme was regarded as helpful for low-income households, it could not go further due to a lack of further funding (Halsey & Wolpe, 2020). The municipality has a long-standing low-income energy services programme to reduce energy poverty, but it has not been able to move sufficiently beyond research and implement at scale. The City of Cape Town municipality recognised that they are understaffed and need to drive a more collaborative approach with other departments. The City of Cape Town was one of the first municipalities to electrify an informal settlement using what is known as the "maypole method" whereby several households are connected to an electricity supply via a transformer that is placed on a pole. An area that this municipality is focusing on is increasing information and communication about energy use and efficiency, not only across departments but also with communities. Communication channels do not always work well, and this is another area of work that needs to be developed. Conversations with municipal officials reveal that a lack of understanding of the issues and challenges both within a municipality and with communities hinders the work that is happening on the ground.

There is no clear constitutional mandate on energy poverty, and the fiscal framework under which municipalities operate further hinders what they can do. As one official has said, they are hamstrung not by a lack of will but by a lack of money and a lack of legislation to enable them to do what needs to be done. The municipal official went on to say that they have to deal with this emotional trauma and have no power to resolve the problem of illegal connections and growing informal settlements. No sooner have they disconnected an illegal connection and it is reconnected. Many municipalities have teams in place to deal with illegal connections and non-technical losses often caused by metre tampering.

A survey of a formal settlement in Joe Slovo, Cape Town, which has had a number of sustainable interventions such as solar water heaters, ceilings, roof overhangs, LED lights, and street lighting installed in houses, demonstrates how a holistic approach has helped the community (City of Cape Town, 2013). Key results from respondents included how living in an energy-efficient home as opposed to a shack was so much better and safer and that life is more dignified.

The solar water heaters were great in the summer but had no electrical backup for the winter and so were less effective. People liked having hot water on tap rather than having to use a kettle or stove. However, over time, it has become clear that maintenance of solar water heaters was not included in the budget, and many are no longer functioning. The addition of ceilings made a huge difference to people in terms of less dust, less noise, and better ambient temperatures.

11.6 Conclusion

This chapter has given a brief overview of what energy poverty means in the South African context, both for communities and for municipalities. Managing this enormous challenge requires understanding context and multi-dimensional interventions. This is not always easy where there are both financial and capacity constraints and people are fighting to simply survive.

Why has the national government failed to shift the energy poverty trajectory despite numerous policies and political intent? The answer to this question is highly complex and this chapter has only been able to touch the surface. In part, it is because municipalities must both generate revenue and operate in a developmental manner without sufficient resources to support such a pathway and, therefore, poverty reduction. The economic policies in place since democracy that should have supported and upheld the development agenda have in fact slowed down or even blocked a redistributive transformation of the energy system. The very nature of driving a free market approach, particularly at the local government level, has put a strain on their ability to deliver on their service delivery mandates. It is also evident that, at the very least, the FBE allowance should be increased to a minimum of 250 kWh per month with greater checks to ensure that those entitled to this benefit actually receive it. This chapter has highlighted the importance and need to make a distinction between (energy) access and (energy) use as these do not necessarily lead to alleviation of energy poverty. In South Africa, as the levels of poverty and unemployment rise, so do other societal inequalities, which make it more difficult for low-income households to comfortably meet their household energy needs without compromising other needs such as food. There is some hope that the newly launched Just Energy Investment Plan (JET-IP) by national government of how the USD 8.5 billion should be allocated which includes a review of local government's fiscal framework and an increase in the FBE allowance, could help strengthen and support a reduction in energy poverty in the country.

Acknowledgements

Daylin Paul, Photographer and Visual Journalist for two photographs from his book *Broken Land*, published by Jacana Media Ltd., 2019.

Note

1 The term "triple challenge" is widely used by the South African government after the South African High Level Parliamentary Committee assessing key legislation and acceleration of fundamental change, refers to the high levels of unemployment, poverty and inequality as the triple challenge currently faced by the country (Parliament of South Africa: 2018).

References

Abbas, K., Li, S., Xu, D., Baz, K., & Rakhmetova, A. (2020). Do Socioeconomic Factors Determine Household Multidimensional Energy Poverty? Empirical Evidence from South Asia. *Energy Policy*, 146. https://doi.org/10.1016/j.enpol.2020.111754

Annecke, W., Wolpe, P., David, A., & Godin, A. (2022). *What role for social policies in the framework of the just transition in South Africa?* Editions ADF. February 2022, No. 230. Paris.

Baker, Lucy. (2017). Post-Apartheid Electricity Policy and the Emergence of South Africa's Renewable Energy Sector. In Arent Douglas, Arndt Channing, Miller Mackay, Tarp Finn, & Zinaman Owen (Eds.), *The Political Economy of Clean Energy Transitions*. Oxford University Press. Oxford. https://doi.org/10.1093/oso/9780198802242.001.0001

Baker, L., & Phillips, J. (2019). *Tensions in the Transition: The Politics of Electricity Distribution in South Africa. 37*(1), 177–196. https://doi.org/10.1177/2399654418778590

Bega, S. (2021, February 13). *'We're fighting to breathe clean air.'* Mail and Guardian. https://mg.co.za/environment/2021-02-13-were-fighting-to-breathe-clean-air/

Beshilas, L. (2019, December 10). *Tackling Energy Poverty with Gender Mainstreaming*. https://www.nrel.gov/state-local-tribal/blog/posts/tackling-energy-poverty-with-gender-mainstreaming.html

Bouzarovski, S., Thomson, H., & Cornelis, M. (2021). Confronting Energy Poverty in Europe: A Research and Policy Agenda. *Energies*, *14*(4). https://doi.org/10.3390/en14040858

Calitz, J. R., & Wright, J. G. (2021). *Statistics of utility-scale power generation in South Africa in 2020*. http://hdl.handle.net/10204/11865

City of Cape Town. (2013, February). N2 Gateway Community Newsletter. *N2 Gateway Community Newsletter*.

City of Cape Town. (2022). *The City of Cape Town's pathway to alleviate energy poverty*. https://resource.capetown.gov.za/documentcentre/Documents/Graphicsandeducationalmaterial/CCT_Alleviation_of_energy_poverty_booklet.pdf

Creamer, T. (2022, September 21). *Affordability of Eskom's 32% hike comes under intense scrutiny on last day of hearings*. Engineering News. https://www.engineeringnews.co.za/article/affordability-of-eskoms-32-hike-comes-under-intense-scrutiny-on-last-day-of-hearings-2022-09-21/rep_id:4136

DME. (1998). *White Paper on the Energy Policy of the Republic of South Africa*. Department of Minerals and Energy. Pretoria.

DME. (2003). *Electricity Basic Services Support Tariff: Free Basic Electricity Policy*. Department of Minerals and Energy. Pretoria.

Eberhard, A. (2006). The Political Economy of Power Sector Reform in South Africa. In Victor & Heller (Eds.), *The Political Economy of Power Sector Reform* (pp. 215–253). Cambridge University Press.

ERC. (2002). *Options for Basic Electricity Support Tariff: Analysis, issues and recommendations*. University of Cape Town. Cape Town.

Eskom. (2022). *Summary of the Retail Tariff Plan*. Eskom Website. https://www.eskom.co.za/distribution/retail-tariff/

Flores Quiroz, N., Walls, R., & Cicione, A. (2021). Towards Understanding Fire Causes in Informal Settlements Based on Inhabitant Risk Perception, *Fire*, *4*(3), 39. https://doi.org/10.3390/fire4030039

groundWork. (2019). *Down to Zero*. www.groundwork.org.za

Halsey, R., & Wolpe, P. (2020). *Energising our future – Together: Tackling energy poverty and improving community engagement in the Western Cape*. www.90by2030.org.za

Johnson, O. W., Han, J. Y. C., Knight, A. L., Mortensen, S., Aung, M. T., Boyland, M., & Resurrección, B. P. (2020). Intersectionality and Energy Transitions: A Review of Gender, Social Equity and Low-Carbon Energy. In *Energy Research and Social Science* (Vol. 70). Elsevier Ltd. https://doi.org/10.1016/j.erss.2020.101774

Knox, A. J., De Groot, J. R., & Mohlakoana, N. (2018). *Post-apartheid spatial inequalities and the built environment Drivers of energy vulnerability for the urban poor in South Africa*. https://www.ebsco.com/terms-of-use

Ledger, T. (2021). *Broken Promises*. https://rekordeast.co.za/318517/claiming-free-electricity-promise-north-residents/

Ledger, T., & Rampedi, M. (2022). *Hungry for Electricity*. Public Affairs Research Institute.

Mbandlwa, Z., & Geyevu, M. (2022). Economic Conditions That Leads to Illegal Electricity Connections at Quarry Road Informal Settlement in South Africa. *International Journal of Special Education*, *37*(3). https://surveyheart.com/form/5e5d59a62e417841d9710612

Mohlakoana, N., & Wolpe, P. (2021). *A Just Energy Transition to Facilitate Household Energy Access and Alleviate Energy Poverty*. www.tips.org.za

Mohlakoana, N., & Wolpe, P. (2023). *Unbundling Eskom: How would a new distribution system impact on energy poverty*. Editions AFD. February 2023, No. 269. France.

Nowicki, L. (2020, January 14). Court orders dissolution of Makana municipality. *Ground Up*. https://www.groundup.org.za/article/court-orders-dissolution-makana-municipality/

Parliament of South Africa. (2018). *Overcoming the Triple Challenge: A Report to the High Level Parliamentary Committee from the Centre for Development and Enterprise*. Government Printing Works. Pretoria.

Phillips, C., Silver, J., & Rowswell, P. (2011). *Mamre Ceiling Insulation Evaluation Energy retrofitting in Low Income Communities*.

Presidential Climate Commission. (2022). *A PRESIDENTIAL CLIMATE COMMISSION REPORT A Framework for a Just Transition in South Africa*. Pretoria

Pye, S., Brajković, J., Dobbins, A., & De Miglio, R. (2015). *Energy poverty and vulnerable consumers in the energy sector across the EU: analysis of policies and measures*. Insight_E, Policy Report 2. European Commission. Brussels.

Republic of South Africa. (1996). *The Constitution of the Republic of South Africa* (108 of 1996). South African Government. Government Printing Works. Pretoria.

Republic of South Africa. (2022). *South Africa's Just Energy Transition Investment Plan (JET IP): for the initial period 2023–2027*. Government Printing Works. Pretoria.

SALGA. (2014). *The role of municipalities as service authorities for electricity provision*. https://www.scribd.com/document/324899964/The-Role-of-Municipalities-as-Electricity-Service-Authorities-Draft2-SALGA-Aug-2014-docx#

SEA. (2017). *Sustainable Energy Solutions for South African Local Government: A Practical Guide*. Sustainable Energy Africa.

SERI. (2018). *Informal Settlements and Human Rights in South Africa: Submission to the United Nations Special Rapporteur on adequate housing as a component of the right to an adequate standard of living*. https://www.blacksash.org.za/

Simcock, N., Thomson, H., Petrova, S., & Bouzarovski, S. (2018). *Energy Poverty and Vulnerability: A Global Perspective* (N. Simcock, H. Thomson, S. Petrova, & S. Bouzarovski, eds.). Routledge Taylor and Francis Group.

StatsSA. (2018, June). *Energy and the Poor: a Municipal Breakdown | Statistics South Africa*. Statistics South Africa Department. http://www.statssa.gov.za/?p=11181

StatsSA. (2022). *Quarterly Labour Force Survey (QLFS) Q3:2022*. https://www.statssa.gov.za/publications/P0211/Presentation QLFS Q3 2022.pdf

Taylor, J. (2023, March 7). Just an energy transition – a gendered analysis of energy transition in South Africa. *ARUA Climate Change and Inequalities Symposium*. University of Cape Town. Cape Town.

Thom, C., Mohlakoana, N., Dekenah, M., & Heunis, S. (2001). *Case studies on the impact of electrification in rural areas*. Eskom. Johannesburg.

UN. (2015). *The Millennium Development Goals Report 2015*. United Nations. New York.

Viljoen, N. (2022, September 20). *500 people displaced as yet another blaze rips through Imizamo Yethu*. NEWS24. https://www.news24.com/news24/community-newspaper/peoples-post/500-people-displaced-as-yet-another-blaze-rips-through-imizamo-yethu-20220920

White Paper on Reconstruction and Development (1994). (testimony of RSA). https://www.gov.za/sites/default/files/governmentgazetteid16085.pdf

Wolpe, P., & Reddy, Y. (2014). *Tackling Urban Energy Poverty in South Africa*. Sustainable Energy Africa. Cape Town.

Ye, Y., & Koch, S. F. (2021). Measuring Energy Poverty in South Africa Based on Household Required Energy Consumption. *Energy Economics Journal*. 103(*November 2021*), 105553. https://doi.org/10.1016/j.eneco.2021.105553

12 Resources and Skills of Low-Income Households to Tackle Energy Poverty in France

Coralie Robert

12.1 Introduction

Low-income households, due to socio-economic conditions or housing characteristics, are particularly sensitive to energy environment tensions such as sudden price increases. Moreover, they face difficulties adapting to and coping with them (Bouzarovski et al., 2014). Inequalities in terms of access to resources affect their ability (i.e., their room to manoeuvre) to adjust their practices and cope with adversity.

Energy poverty has been recognised as a systemic issue in many European countries, including France, since the early 2000s, both in academic literature and public policy (Brunner et al., 2010). Several surveys have highlighted the complex interrelationships, such as building quality, and repercussions, such as social well-being and health impact, of energy poverty in individuals' daily lives (Devalière, 2008; Dubois, 2012). Energy poverty is considered a multifaceted problem involving dynamic and interconnected factors (Bouzarovski et al., 2014; Baker, Mould, and Restrick, 2018; Duvoux, 2009; Middlemiss et al., 2018). It encompasses situations in which access to sufficient and "appropriate" levels of energy for heating, lighting, and cooking is not ensured. Energy poverty often arises from a complex interplay of technical and social factors. Some authors (Bouzarovski et al., 2014; Thomson et al., 2016; Walker, 2014) have explored the relationship between building thermal quality, energy consumption, and energy poverty, highlighting the ways in which poor thermal quality can exacerbate difficulties in accessing affordable energy. This chapter builds upon the research of Bartiaux et al. (2018), which proposes a systemic approach to analysing energy poverty as a restriction of multiple capabilities in the context of Belgium. In a similar vein, the present chapter focuses on the French context and aims to identify the various capabilities that individuals may possess or lack in order to cope with fuel poverty. In France, the link between energy and inequality has indeed crystallised around the notion of energy poverty (Villalba & Semal, 2018), which encompasses the barriers to energy consumption and all associated services, making it an exclusionary phenomenon.

In this context, a lived experience of energy poverty approach (Misset & Siblot, 2019) allows us to go beyond concepts of energy poverty in terms of deficiencies by also highlighting the resources and skills available to the most precarious when coping with such a phenomenon. Indeed, several solutions to tackle energy poverty are deployed in everyday life by those who are directly affected. These solutions include energy-saving measures, such as closing doors and windows to prevent heat loss; community-based solutions, such as using mutual support for online administrative procedures; sharing resources, such as using a relative's washing machine; and reducing comfort measures, such as staying in a heated public place for several hours.

The aim of the conducted study is to gain insight into the diverse practices of individuals in dealing with energy poverty by examining the perspectives and practices of those directly

DOI: 10.4324/9781003408536-15

affected. The inability to access energy services is not solely an economic issue, but also impacts fundamental aspects of human well-being, including "deprivations of freedom" (Duvoux, 2009). Such deprivation can undermine human dignity by challenging core capacities such as physical integrity or health (Oris, 2009). Capabilities are central to understanding energy poverty. Energy is not an end in itself (Duvoux, 2009), but a crucial means to enable the mobilisation of capabilities. Capabilities refer to the question, "what is this person able to do and be?" (Oris, 2009, p. 39), which is best understood as the ability to convert resources into capabilities. Among the range of potential responses to this question, the focus is on the "ability to develop strategies to overcome constraints and cope with difficulties" (Hitchings & Day, 2011), commonly known as coping strategies.

Therefore, this chapter contributes to the existing literature on energy poverty through the prism of the lived experience of individuals facing energy poverty and the emotional dimension associated with it. While recent literature has explored the concept of capabilities in the context of energy poverty, little attention has been given to valuing the existing capabilities of energy-vulnerable citizens. This study focuses on the resources and skills already present in the households surveyed, with the aim of rethinking public policies based on these findings. While the energy transition has the potential to improve living conditions for all, it also poses risks for the most vulnerable. Therefore, it is important to consider the current experiences of those in energy poverty when thinking about a fair and just transition[1], for all and developing solutions for energy justice (Middlemiss et al., 2019).

This chapter is based on a case study conducted in France among individuals directly affected by energy poverty. Qualitative research was carried out to identify the strategies employed by these individuals to cope with energy poverty. Through in-depth interviews and observations, I was able to grasp the diversity of resources and capabilities available to vulnerable populations.

12.2 Views from the literature

A lived-experience understanding of energy poverty allows us to go beyond viewing it solely as deficiencies and sheds light on the resources and skills available to those who are most vulnerable in coping with it.

In discussing the phenomenon of energy poverty, I adopt a bottom-up approach by providing a precise description of the lived experience. This allows us to capture the discourse of individuals and understand the strategies, emotions, and repercussions of energy poverty in various aspects of daily life such as work, family, education, and health. My work builds on previous research on the lived experience of energy poverty by highlighting the multiple dimensions of resources for dealing with this phenomenon (Chatman, 1996; Giuliani, 2009; Hobfoll, 2001; Middlemiss et al., 2018; Stojilovska et al., 2021).

The phenomenon of energy poverty is commonly understood in terms of three aspects: limited financial resources, energy prices, and the poor energy performance of buildings. However, some authors (e.g., Middlemiss et al., 2019) provide a more systemic perspective by studying the relationships between different factors that influence the likelihood of experiencing energy poverty. These factors include not only people's life circumstances but also social and political dimensions (Middlemiss & Gillard, 2015).

Recent research highlights the dynamic nature of energy poverty, which is not a fixed determinant but an expression of instability among various elements such as low income and inadequate housing quality, as well as externalities like rising energy prices (Brunner et al., 2010). It is the fluctuating interplay of these elements that allows us to fully grasp the phenomenon. The shifting nature of energy poverty is largely influenced by the instability of individuals' situations, including economic privilege, family circumstances, health, residential environment, and other factors (Middlemiss et al., 2018).

I place capabilities at the heart of understanding energy poverty, drawing on Day et al.'s (2016, p. 260) definition of energy poverty as "an inability to realise essential capabilities as a direct or indirect result of insufficient access to affordable, reliable and safe energy services, and taking into account available reasonable alternative means of realising these capabilities". Consuming energy provides access to capabilities and services that are essential for individuals to fully participate in society. This perspective adds to the previous discussion by highlighting the importance of capabilities in assessing the impacts of energy poverty beyond just the material aspects of affordability and availability of energy services.

The conceptual framework of capabilities captures the complex relationship between the experience of energy poverty and the capabilities of individuals and how they mutually influence each other (Middlemiss & Gillard, 2015). Understanding the interplay between energy poverty and capabilities is crucial for developing effective policies and interventions that address the multi-dimensional nature of energy poverty and its impacts on individuals' well-being and opportunities. By adopting a capabilities perspective, we can better grasp the nuanced and dynamic nature of energy poverty (Middlemiss, 2022) and develop strategies that promote holistic and sustainable solutions.

Unlike previous scholarly work that has primarily focused on the negative repercussions of energy poverty, which often reduce or disempower individuals (Bartiaux et al., 2018; Duvoux, 2009; Retière, 2003), the current research takes a different approach. In addition to examining the challenges faced by individuals experiencing energy poverty, this study also highlights the capabilities and resources that are available to them through qualitative research (discourse analysis and observation). I, therefore, mobilise the conceptual framework of capabilities while reinscribing their mobilisation in a specific context: temporal and spatial scale, socio-technical framework (energy market, the energy efficiency of the housing stock, etc.) (Lees, 2014). By adopting a more nuanced perspective, this research sheds light on the agency and resilience of individuals facing energy poverty and underscores the importance of recognising and leveraging the strengths and capabilities of these individuals. This approach can inform the development of more empowering and sustainable strategies to address energy poverty and promote social and environmental justice.

I view energy poverty as a dynamic phenomenon (Middlemiss, 2022). Like Jouffe et al. (2015), I adopt a capabilities approach that recognises the significance of social relationships and collective action in the acquisition and utilisation of capabilities rather than relying solely on an individualistic perspective, such as low income or a lack of education. Social relations play a crucial role in shaping access to resources, opportunities, and capabilities, and these factors are crucial for understanding the experience of poverty. In the context of energy poverty, Ibrahim' and Evans' works suggest that it is not enough to focus solely on individual-level factors such as income or energy efficiency in the home. Rather, it is important to consider how social relations and power structures may shape people's access to affordable and reliable energy services and how these factors may interact with other dimensions of poverty and inequality. By taking a more relational approach to energy poverty, we can gain a more nuanced and comprehensive understanding of this complex phenomenon.

It is crucial to recognise that resources and skills vary greatly among individuals, shaped by their unique situations and life trajectories. Not everyone is equally equipped to face the challenges associated with energy poverty. While it is important to acknowledge that individuals are capable and competent, it is also important to acknowledge that these competencies may not always be readily mobilised in the face of all challenges, and they may not be equally distributed across societies.

12.3 Context

This chapter is based on the results of a doctoral research project, provides an analysis of the lived experience of energy poverty in France, and therefore situates itself in the context of the Global North. It is important to consider the socio-technical framework and the energy market in France, including the price of energy and the various public action measures that exist, such as financial aid for bill payments and energy renovation. In addition, in France, there are energy-saving awareness programmes targeted towards the most vulnerable population with the objective of "greening the poor" (Malier, 2019; Robert, 2021).

12.3.1 Study area

The European Commission has insisted on the need for a "universal energy service", that is, a service that "must be provided in all the member states at a reasonable price for all users and guarantee non-discriminatory access conditions" (Day et al., 2016). In Europe, poverty is typically measured using a poverty rate5[2], which is a threshold set at 60% of the median living standard. In Europe, the indicators for identifying the phenomenon are based on ratios measured in terms of energy bill expenditure to income.

Energy poverty in France is primarily identified by the indicator of households whose income per unit of consumption falls below the 3rd income decile, which corresponds to the poorest 30%. In 2022, 10.5% of French households were identified as experiencing energy insecurity, spending more than 8% of their income on energy bills (Tableau de bord de la précarité énergétique – Edition 2022, 2022). Moreover, in France, a subjective indicator, the feeling of cold, is also considered in understanding energy poverty. About 40% of households reporting discomfort from the cold attribute it to poor thermal insulation of their housing, while 36% attribute it to financial limitations. Indeed, 60% of French people have declared reducing heating at home to avoid high bills, as per the National Energy Mediator (*ibid*).

The statistics of the National Observatory of Energy Poverty highlight the proportion of French people affected by the phenomenon through different variables. These include the socio-economic characteristics of households, housing characteristics, and energy practices. According to indicators in France, those most affected by energy poverty are typically individuals who are unemployed, disabled, or living in rented housing, both in social and private markets (constituting 52.6% of households in energy poverty). Other vulnerable groups include single-parent families and foreign households. Energy insecurity in France is predominantly characterised (by the government) by challenges in paying energy bills, with 19.1% of the low-income population reporting unpaid bills in the last 12 months of 2021 or facing difficulties in accessing energy services.

The structural characteristics of buildings and heating conditions play a crucial role in energy consumption. Indeed, electricity is the primary source of Energy, used by a third of the French population, and its cost, along with that of gas more recently, has been steadily increasing for over a decade[3].

12.3.2 Research approach

This research was carried out in France between 2016 and 2018, with a sample of low-income households from the first three deciles. The households were diverse in terms of living arrangements, social components, family composition, and professional status.

The methodology employed for this research utilised a mixed qualitative approach, which involved several components: (1) observation and participation in awareness-raising workshops;

(2) longer interviews with professionals and with residents who were facing various challenges related to energy poverty.

This analysis was conducted at two locations in France. The first location was in La Courneuve, a suburb of Paris that has been identified by the local authorities as being affected by energy poverty due to its old and energy-inefficient housing stock. In La Courneuve, the survey was conducted through door-to-door visits with the residents.

The second study area was in Nantes, one of the largest cities in France. Access to the field activities of interest was obtained through the city's social action department (CCAS, Community Centre for Social Action). One activity was the teaching flat, or "energy cafe". Additionally, participation in these collective workshops on energy saving provides an opportunity to meet and interact with individuals who were later interviewed individually to gain a better understanding of their reception and perception of energy advice.

A categorisation of resources and skills was established based on discourse analysis and observation of household practices in energy poverty, as shown in Figure 12.1. This wheel depicts the various barriers and opportunities for the deployment of coping strategies. It maps the different types of resources that were mobilised and observed during the study, illustrating the multidimensional approach to coping strategies. For example, to be able to request financial assistance, multiple types of resources, such as relational, administrative, or territorial resources, could be mobilised. This diagram helps identify the types of resources that are necessary to carry out specific activities related to coping with energy poverty.

12.4 Findings

Individuals facing precarious circumstances deploy a set of strategies in their daily lives to constantly adjust to unstable living conditions, which may include health issues, financial challenges, and residential instability. To provide a concise overview of the diverse resources available to these individuals, Figure 12.1 presents a summary diagram based on examples from the field survey.

It should be noted that a single practice may simultaneously involve the activation of several resources at the same time. For instance, the use of energy meters necessitates physical resources (such as access to an energy meter, as well as socio-linguistical capabilities to understand the data.

Energy poverty is not solely determined by economic and material disparities but also by the social resources that individuals possess. For instance, support from friends and family in tasks like cutting firewood, moving gas bottles, or assisting with administrative procedures can play a significant role in coping with energy poverty. These everyday supports not only challenge the perception of individuals as lacking resources but also highlight the coping strategies employed by the most economically disadvantaged to mitigate the deprivation of access to essential resources.

Possessing social skills has a positive impact on various aspects of an individual's life, including engagement in social groups, the development of emotional bonds with others, and self-confidence. According to Riggio et al. (2020), social skills are crucial in building self-assurance and ease in interaction with others. These skills encompass both social control skills and social expressiveness and are acquired through experiences with peers, influencing interpersonal abilities (Siblot, 2006, p. 339). The authors highlight that the acquisition and deployment of social skills are lifelong. These skills play a significant role in social situations where "role-playing" skills are essential for maintaining social dignity, in other words, "saving face" (Groves et al., 2019). Possessing such skills allows individuals to make a positive first impression and receive favourable judgements from professional counterparts, particularly in evaluative contexts.

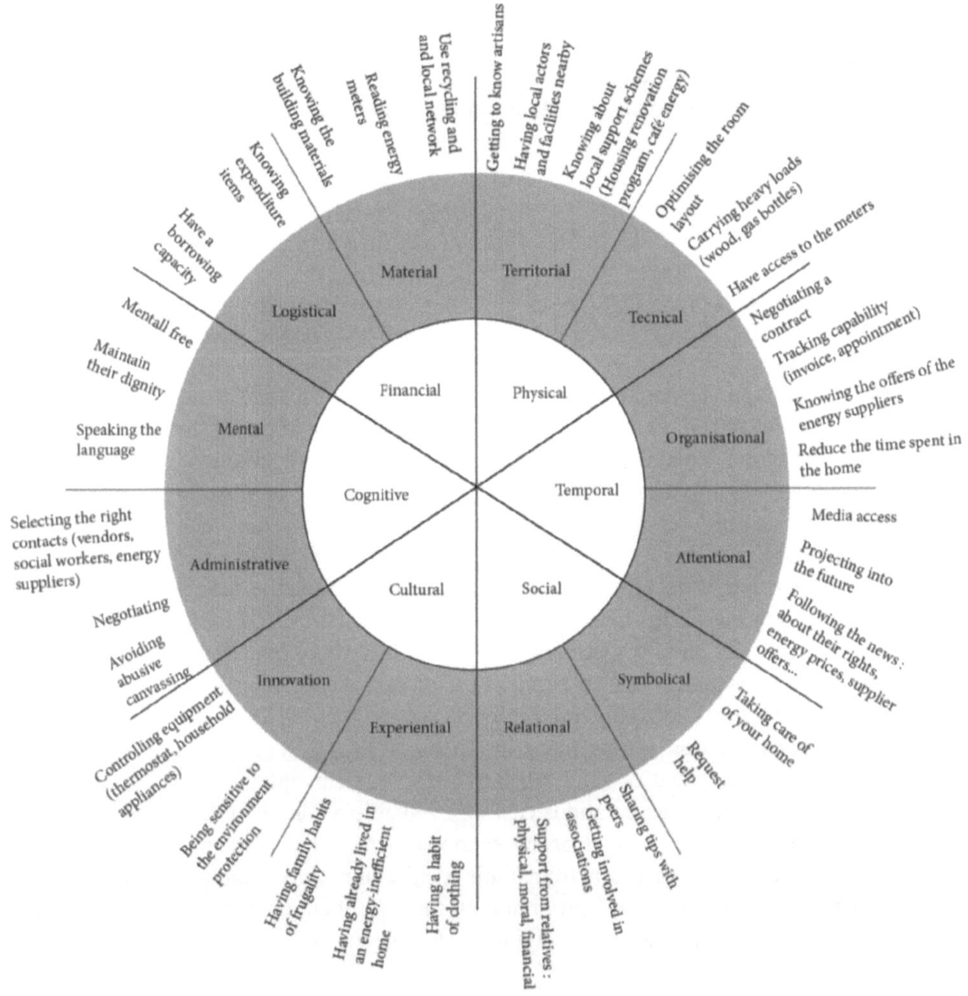

Figure 12.1 Types of resources for dealing with energy poverty.

The use of donated objects is also prevalent among the respondents analysed. However, this can lead to an "energy boomerang" effect (Chard & Walker, 2016), where the satisfaction of receiving donated equipment, such as additional heating, leads to a sense of belonging to consumer society while energy efficiency takes a backseat. Moreover, these individuals may not be able to choose items that best meet their needs, such as those with lower energy consumption.

Resources are also mobilised in accessing rights, namely, the ability to defend oneself or seek help in doing so, which impacts citizenship (Longhurst & Hargreaves, 2019). Non-take-up of social rights is a significant issue in poverty alleviation policies, including those addressing energy poverty. The determinants of non-take-up can be at both the policy and individual levels, where the process may seem complex or time-consuming, for example. At the same time, low-income households may claim their rights when they access public services, such as the right to housing[4] (Sifres, 2018). For example, one respondent mentioned the support of an associative actor to set up an insalubrity file on her former home and to carry out requests for social housing. This resource person "*made her aware of everything that she was entitled to*".

Solidarity networks play a crucial role in times of hardship, as highlighted by Stojilovska et al. (2021). The survey reveals the significance of associative involvement in accessing resources, such as knowledge of aid mechanisms, sharing of tips, moral support, and more. These associative networks serve as important sources of assistance for individuals facing challenging circumstances.

> *"I do not have a big income, but I am quite independent in my red tape (with welfare service), in all of the following: phone calls, going to see people, meeting people, I am quite sociable so that does not bother me."*
>
> Sophie, <1st income decile, 43 years old, looking for work, tenant of an apartment in the private sector, lives in a shared apartment in Nantes

Access to information is a deciding factor, as emphasised by Chauffaut and Dauphin (2012) and Boubaker Nobilet (2017) in the context of social groups, underlining the influence of belonging to a social group. I concur with this perspective, viewing it in terms of "informational poverty" or wealth, which relates to the ability to access, process, and use information. Some respondents in the survey keep a variety of leaflets or guides to retain relevant information, particularly for future use. This requires the ability to sort through diverse information sources, prioritise documents, and project future needs addressed by these leaflets. Such resources prove valuable in navigating the energy market (Fawcett & Boardman, 2009, p. 234), especially in negotiating contracts with energy suppliers, where administrative resources are mobilised for comparing offers. For instance, one respondent mentioned conducting their "own market research" on regulated tariffs and commercial offers.

Commitment to association and participation in social life, as highlighted by Riggio et al. (2020), fosters the development of communicative resources, including the ability to express oneself in public. Investing in associations can be a valuable resource in the context of energy insecurity, contributing to the "capital of autochthony" (*ibid*). For instance, two respondents shared how they received mutual aid from their local community. One participant reported learning about specific financial aid programmes through their neighbourhood association, while another respondent mentioned receiving assistance from their neighbour to complete administrative procedures online.

Local anchoring, despite the injunction to mobility for employment and social inclusion (Lee et al., 2021; Vignal, 2016), can offset such pressures. Furthermore, it is not solely the length of time lived in a particular location that determines local integration. Local embeddedness can occur quickly, without any residential seniority. The resources derived from residential seniority are more related to individuals' investment in mobilising a local network. However, it's important to note that people's inner circle can help to improve daily life without necessarily leading to an effective exit from poverty. Local anchorages can serve as a support system for resources, including symbolic ones, for the most vulnerable through donations, the exchange of services, and other means.

The local context also plays a significant role in shaping people's capabilities. Local anchoring provides resources, particularly symbolic ones, for the most vulnerable. The availability of services and amenities near residential areas is a crucial factor in determining people's ability to cope with daily challenges. Access to services and amenities, such as second-hand purchasing networks, is facilitated by the presence of accessible service offerings. Living far from services results in time and budget costs to access them, as highlighted by Guirdarlal (Hamilton & Catterall, 2008, p. 39), who underscores the impact of discount supermarkets located close to areas with high levels of precariousness on reducing consumption inequalities.

The proximity, especially in terms of walkability, to essential services such as the post office and customer service agencies of energy suppliers has an impact on the capacity of individuals. The gradual disappearance of local services and physical meeting places contributes to the vulnerability of the most disadvantaged.

In fact, the implementation of dematerialised procedures requires electronic equipment (computer, smartphone) and internet access, resources that are unequally shared.

> *"In fact, at that time, we still had the possibility to go to EDF, to move. If we had a question to ask, we could ask someone! And not to a wall (...) Now, we can't do that anymore!"*
>
> Valerie, < 1st income decile, 45 years old, inactive, tenant of a social housing apartment, lives alone in La Courneuve

The importance of proximity to a service area is a dimension that was emphasised several times during the interviews. This contributes to a feeling of trust and consideration of not being left out. Geographical distance from institutions and services has a negative impact not only on access to services but also on close relationships with professionals. Valerie is disabled and unable to walk. To apply for energy bill payment assistance (municipal aid), she must make an in-person application at her local council. Since Valerie cannot walk, she has to pay for transport in a medical vehicle.

Understanding energy poverty requires an appreciation of time as a critical resource (Groves et al., 2019). Over time, changes in both individuals' circumstances and their perception of those circumstances can impact their ability to take action (Middlemiss et al., 2018). The households surveyed have experienced many challenges in their life trajectories, all of which have led to moments of vulnerability that have sometimes undermined their confidence in the future. Consequently, the ability to plan ahead or deal with unexpected events becomes a resource that is unequally distributed among individuals, based on their personal experiences.

For some respondents, as for inactive individuals with limited financial resources, the amount of available time may serve as a variable for adjustment. This time can be used to opt for less costly but slower modes of transportation or to schedule the use of household equipment during off-peak hours. Such adjustments can help tailor energy usage to the household's routine, such as heating only when the children return from school. But the time resources are also more solicited for the precarious people who have to accomplish a diversity of administrative steps, for example.

Postponing certain purchases or waiting to turn on the heating until a more cost-effective time can be effective ways to save money. Taking the time to look for a "good deal" and accumulate savings for new equipment can also be beneficial. However, the most vulnerable individuals may find themselves waiting due to various factors, such as lengthy administrative procedures, slower modes of transportation, and so on.

Being attentive to one's budget, consumption, and comfort levels is essential for managing energy usage. However, this level of energy and attention is not evenly distributed across society, with the most vulnerable individuals often bearing a greater mental burden in order to ensure that they are not overspending or wasting energy unnecessarily. These individuals may have limited financial resources and may need to make difficult decisions about how to allocate their limited funds. They may also be dealing with substandard living conditions or a lack of access to basic energy services, which can further exacerbate their mental burden.

> *"I got up at night to do the accounts huh I'm not kidding!"*
>
> *Nadia, 3rd income decile, 70, retired, homeowner, lives in Nantes with her husband*

Attention to others (caring) also shapes the way individuals manage their energy consumption (Groves et al., 2019, p. 17). Indeed, individuals with caregiving responsibilities for children or frail individuals adjust their consumption practices by prioritising certain expenditure items and focusing on the comfort of other household members.

This is evident during the Christmas period, where individuals use illuminations to please children, whereas during the rest of the year, they pay more attention to the lights (Chard & Walker, 2016). This underscores the importance of recontextualising resources and skills in relation to specific contexts (seasonality, religion).

The maintenance of dignity is observed in personal narratives as well as in the presentation of oneself and one's home to others. This management of appearance is a strategic process (Groves et al., 2019; Siblot, 2006, p. 352) that involves avoiding negative judgements, such as being perceived as cheap, a bad house guest, or a bad parent (Chevallier, 2012; Goffman, 1973). The institutional biographical injunction[5] (Fawcett & Boardman, 2009) also calls for the activation of resources on the part of the individuals concerned to put their experiences into words. Those who are more familiar with social action tend to be more comfortable with the exercise and are able to take advantage of existing mechanisms to either mask or highlight certain living conditions, depending on the audience.

"Being active" is a way to avoid being perceived as "assisted" and to meet the expectation of being "proactive" (Nussbaum & Sen, 1993, p. 80). This enables individuals to differentiate themselves from those who are perceived as disengaged or passive. According to Misset and Siblot, volunteering is particularly important for working-class individuals, but its relevance depends on the context. Volunteer work can serve both civic and social purposes and lead to the development of specific social practices. For individuals who are experiencing precariousness, the act of "giving of one's time" can serve as a source of support (2019, 88).

Low-income households often rely on coping skills at a moral and psychological level to maintain their dignity and self-worth (Groves et al., 2019). Individuals adapt their practices based on what they deem acceptable, often using the "least worst" logic. This resilience also involves embracing frugality and reducing their needs to gain greater control over their material and financial environment one must negotiate. Participating households demonstrate skills in "knowing how to live" through the daily management of their home, using do-it-yourself or sewing skills to improve their domestic comfort. They show ingenuity and resourcefulness in repurposing household objects to cope with sensations of cold and drafts, using them in new and unconventional ways, such as making draught excluder of old clothes.

12.5 Discussion and conclusion

Coping with energy poverty is an ordeal in which individuals mobilise resources based on their lived experience and the previous challenges they have faced. In this sense, it does not contribute to the empowerment of individuals. On the contrary, the constrained framework in which the practices of energy consumption or budget management are carried out hinders the mobilisation of their resources and skills. In a constrained situation, individuals deploy their resilience and adaptation capacities by mobilising available psychological and emotional resources. Resources and skills both play a crucial role in shaping energy consumption practices, and their value can vary depending on the society and culture in which they are situated (Ibrahim, 2013). This means that access to resources and the ability to acquire certain skills are highly dependent on social and cultural factors. For instance, individuals from disadvantaged backgrounds may have limited access to financial resources or the education necessary to develop certain skills, while those from more privileged backgrounds may have easier access to such resources and skills.

This understanding of the social and cultural situatedness of resources and skills is important in the context of fuel poverty, as it can help explain why certain groups may be more vulnerable to energy poverty than others.

This chapter highlights the dual reading of capabilities, in that they can be both a support for emancipation and help fight against energy poverty thanks to their mobilisation. They can also be reduced and constrained during limited access to energy services. As Middlemiss and Gillard (2015, p. 227) point out concerning the link between social relations and energy poverty, these resources can "both enable access to energy services, and be a product of such access".

Recognising that individuals are capable and competent does not mean that these skills are mobilised in the face of all challenges, nor does it mean that skills are equally distributed among individuals. Resources are unevenly distributed with respect to social class, gender, age, or socio-professional category (Gillard et al., 2017; Pellicer-Sifres et al., 2021). As such, not everyone has the same room for manoeuvre to deal with challenges and situations. Individuals are also constrained by a material and technical environment. The environment conditions the "possibility of", in addition to the "competence to". Being competent is not enough.

It is also necessary to deconstruct the projection made by institutions and professionals, as policymakers but also street-level bureaucrats, of the resources and skills of the most precarious. Indeed, if institutional discourses participate in the promotion of a daily ecology largely based on "small gestures", it is advisable to recall that these are already practices anchored in the lifestyles of the most precarious. These individuals already mobilise resources and skills to save energy. By deconstructing the resources deployed to deal with energy poverty, it is then possible to identify potential blockages or, conversely, points of support using the logic of empowerment.

The lived experience approach has allowed us to highlight the multidimensional nature of the resources involved. It also underlines the diversity of domains where positive repercussions can be observed during the deployment of energy poverty measures, including the fight against social isolation, access to rights, and symbolic valorisation of applied knowledge. Our results underline the importance of acting on different resources in order to improve public action instruments. These include the improvement of mobility options such as the public transport network, the maintenance of physical resource spaces in local areas, and action against social isolation. Public action must focus on providing conditions conducive to the deployment of individuals' resources, in addition to improving material and technical conditions. To create effective public policies that promote social and energy justice (Bartiaux et al., 2021; Misset & Siblot, 2019), it is important to have a deep understanding of people's everyday lives. This includes understanding their daily routines, their access to resources and services, and the challenges they face. Only by understanding people's lived experiences can policymakers create policies that will truly address their needs and promote fairness and energy justice.

Notes

1 By "just transition" I mean considering the impact on different social groups so as not to impact them negatively in the various changes in the energy transition. The goal is to interweave social justice and environmental transition objectives. See Chapter 1 for more details on this issue.

2 In France, the standard of living of individuals corresponds to household income divided by consumption units (1 unit for the first adult, 0.5 for the second adult and 0.3 for children under 14).

3 This rise in electricity prices is the result of a combination of various factors related to economic and political circumstances (promotion of renewable energies), investments in infrastructure (almost half of the French nuclear fleet being non-operational during the winter of 2022–2023), as well as changes in the energy market (price fluctuations due to supply and demand, production costs, energy policies, and geopolitical developments).

4 In France, "right to housing" is included in the law since 2007, understood as the obligation of the courts to protect this right and to find a place for families to stay in an adequate home.
5 In other words, a form of narration of one's history, one's path, and the difficulties encountered, adjusted for an agent of the institution. This autobiography is translated into "the internalization of the norms promoted by the institution" (Duvoux, 2009).

References

Åberg, A., Höffken, J., & Lidström, S., 2018. Looking for Perspectives! EU Energy Policy in Context. *Advancing Energy Policy: Lessons on the Integration of Social Sciences and Humanities* 47–59. https://doi.org/10.1007/978-3-319-99097-2_4

Baker, K. J., Mould, R., & Restrick, S. 2018. Rethink fuel poverty as a complex problem. *Nature Energy*, 3(8), Article 8. https://doi.org/10.1038/s41560-018-0204-2

Bartiaux, F., Day, R., & Lahaye, W., 2021. Energy Poverty as a Restriction of Multiple Capabilities: A Systemic Approach for Belgium. *Journal of Human Development and Capabilities* 1–22. https://doi.org/10.1080/19452829.2021.1887107

Bartiaux, F., Vandeschrick, C., Moezzi, M., & Frogneux, N., 2018. Energy Justice, Unequal Access to Affordable Warmth, and Capability Deprivation: A Quantitative Analysis for Belgium. *Applied Energy* 225, 1219–1233. https://doi.org/10.1016/j.apenergy.2018.04.113

Boubaker Nobilet, P., 2017. *Les capacités d'agir des personnes défavorisées en matière d'information* (thesis). http://www.theses.fr. Paris 8.

Bouzarovski, S., & Petrova, S., 2015. A Global Perspective on Domestic Energy Deprivation: Overcoming the Energy Poverty–Fuel Poverty Binary. *Energy Research & Social Science* 10, 31–40. https://doi.org/10.1016/j.erss.2015.06.007

Bouzarovski, S., Petrova, S., & Tirado-Herrero, S., 2014. From Fuel Poverty to Energy Vulnerability: The Importance of Services, Needs and Practices. Working paper (No. 2014-25). SPRU-Science Policy Research Unit, University of Sussex Business School. http://dx.doi.org/10.2139/ssrn.2743143

Brunner, K-M., Christanell, A., & Spitzer, M. (2010). Energiekonsum, Armut, Nachhaltigkeit. In Institut für Elektrizitätswirtschaft und Energieinnovation (Ed.), Tagungsbeitrag des 11. Symposiums Energieinnovation: Alte Ziele, neue Wege (pp. 1–10).

Chard, R., & Walker, G., 2016. Living with Fuel Poverty in Older Age: Coping Strategies and Their Problematic Implications. *Energy Research & Social Science, Energy Demand for Mobility and Domestic Life: New Insights from Energy Justice* 18, 62–70. https://doi.org/10.1016/j.erss.2016.03.004

Chatman, E.A., 1996. The Impoverished Life-World of Outsiders. *Journal of the American Society for Information Science* 47, 193–206. https://doi.org/10.1002/(SICI)1097-4571(199603)47:3<193::AID-ASI3>3.0.CO;2-T

Chauffaut, D., & Dauphin, S., 2012. Normes de parentalités: production et réception [Revue de littérature]. *Revue des politiques sociales et familiales* 108, 108–115. https://doi.org/10.3406/caf.2012.2694

Chevallier, J., 2012. *Le statut, in: Que sais-je?* Presses Universitaires de France, Paris cedex 14, pp. 77–100.

Day, R., Walker, G., & Simcock, N., 2016. Conceptualising Energy Use and Energy Poverty Using a Capabilities Framework. *Energy Policy* 93, 255–264. https://doi.org/10.1016/j.enpol.2016.03.019

Devalière, I. 2008. Au-delà des impayés d'énergie, comment appréhender la précarité énergétique ? Espace populations sociétés [Space Populations Societies], 2008/1, 191–201.

Dubois, U. 2012. From targeting to implementation: The role of identification of fuel poor households. *Energy Policy*, 49, 107–115. https://doi.org/10.1016/j.enpol.2011.11.087

Duvoux, N., 2009. L'injonction biographique dans les politiques sociales. *Informations sociales* n° 156, 114–122.

Fawcett, T., & Boardman, B., 2009. Housing market transformation, in: ECEEE Summer Study Proceedings. Presented at the Act! Innovate! Deliver! Reducing energy demand sustainably, pp. 225–235.

Gaudet, S., & Turcotte, M., 2013. Sommes-nous égaux devant l'« injonction » à participer? : Analyse des ressources et des opportunités au cours de la vie. *socsoc* 45, 117–145. https://doi.org/10.7202/1016398ar

Gillard, R., Snell, C., & Bevan, M., 2017. Advancing an Energy Justice Perspective of Fuel Poverty: Household Vulnerability and Domestic Retrofit Policy in the United Kingdom. *Energy Research & Social Science* 29, 53–61. https://doi.org/10.1016/j.erss.2017.05.012

Giuliani, F., 2009. Éduquer les parents ? les pratiques de soutien à la parentalité auprès des familles socialement disqualifiées. *Revue française de pédagogie. Recherches en éducation* 83–92. https://doi.org/10.4000/rfp.1769

Goffman, E. (1973). *Présentation de soi. la mise en scène de la vie quotidienne*. Paris, France: Minuit.

Groves, C., Shirani, F., Henwood, K., Cherry, C., Thomas, G., & Pidgeon, N., 2019. 'We're struggling now': energy vulnerability, adaptive preferences, and values forms of life 29.

Guirdarlal, V. (2016). *Vulnérabilité et changement des comportements de consommation. : recherche sur les consommateurs d'énergie en France et au Royaume-Uni*. Doctoral dissertation. Pau, France. https://www.theses.fr/2016PAUU2003/document

Hamilton, K., & Catterall, M., 2008. "I Can Do It!" Consumer Coping and Poverty. *Advances in Consumer Research*, 35, 551–555. https://strathprints.strath.ac.uk/id/eprint/9208

Hitchings, R., & Day, R., 2011. How Older People Relate to the Private Winter Warmth Practices of Their Peers and Why We Should Be Interested: *Environment and Planning A*. https://doi.org/10.1068/a44107

Hobfoll, S.E., 2001. The Influence of Culture, Community, and the Nested-Self in the Stress Process: Advancing Conservation of Resources Theory. *Applied Psychology* 50, 337–421. https://doi.org/10.1111/1464-0597.00062

Ibrahim, S. (2013). Collective Capabilities: What Are They and Why Are They Important? *Maiyteree* 22, 4–8.

Jouffe, Y., Caubel, D., Fol, S., & Motte-Baumvol, B., 2015. Faire face aux inégalités de mobilité. *Cybergeo: European Journal of Geography*. https://doi.org/10.4000/cybergeo.26697

Lee, J., Kim, H., & Byrne, J. (2021). Operationalising Capability Thinking in the Assessment of Energy Poverty Relief Policies: Moving from Compensation-Based to Empowerment-Focused Policy Strategies. *Journal of Human Development and Capabilities* 22, 292–315.

Lees, J., 2014. Ethnographier la précarité énergétique: au-delà de l'action publique, des mises à l'épreuve de l'habiter (Thèse de doctorat). École doctorale de l'École des hautes études en sciences sociales, France.

Longhurst, N., & Hargreaves, T., 2019. Emotions and Fuel Poverty: The Lived Experience of Social Housing Tenants in the United Kingdom. *Energy Research & Social Science* 56, 101207. https://doi.org/10.1016/j.erss.2019.05.017

Malier, H. 2019. Greening the poor: The trap of moralization. *The British Journal of Sociology*, 70(5), 1661–1680. https://doi.org/10.1111/1468-4446.12672

Middlemiss, L., 2022. Who Is Vulnerable to Energy Poverty in the Global North, and What Is Their Experience? *WIREs Energy and Environment* n/a, e455. https://doi.org/10.1002/wene.455

Middlemiss, L., Ambrosio-Albalá, P., Emmel, N., Gillard, R., Gilbertson, J., Hargreaves, T., Mullen, C., Ryan, T., Snell, C., & Tod, A., 2019. Energy Poverty and Social Relations: A Capabilities Approach. *Energy Research & Social Science* 55, 227–235. https://doi.org/10.1016/j.erss.2019.05.002

Middlemiss, L., & Gillard, R., 2015. Fuel Poverty from the Bottom-up: Characterising Household Energy Vulnerability Through the Lived Experience of the Fuel Poor. *Energy Research & Social Science* 6, 146–154. https://doi.org/10.1016/j.erss.2015.02.001

Middlemiss, L., Gillard, R., Pellicer, V., & Straver, K., 2018. Plugging the Gap Between Energy Policy and the Lived Experience of Energy Poverty: Five Principles for a Multidisciplinary Approach, in: Foulds, C., Robison, R. (Eds.), *Advancing Energy Policy: Lessons on the Integration of Social Sciences and Humanities*. Springer International Publishing, Cham, pp. 15–29. https://doi.org/10.1007/978-3-319-99097-2_2

Misset, S., & Siblot, Y. (2019). Donner de son temps » pour ne pas être des « assistés. *Sociologie* 10, 73–89.

Nussbaum, M., & Sen, A. (1993). *The Quality of Life, Wider Studies in Development Economics*. USA: Oxford University Press.

Oris, M. (2009). *Transitions dans les parcours de vie et construction des inégalités*. PPUR presses polytechniques.

Pellicer-Sifres, V., Simcock, N., & Boni, A., 2021. Understanding the Multiple harms of Energy Poverty Through the Nussbaum's Theory of Central Capabilities. *Local Environment* 1025–1041.

Retière, J.-N., 2003. Autour de l'autochtonie. Réflexions sur la notion de capital social populaire, Abstract. *Politix* 121–143. https://doi.org/10.3406/polix.2003.1295

Robert, C. 2021. *Faire face à l'expression d'un impératif de sobriété énergétique: Du conseil à l'injonction, les stratégies de ménages précaires en France* [Thèse de doctorat, Université Paris Nanterre]. http://www.theses.fr/2021PA100100/document

Riggio, R.E., Eaton, L.G., & Funder, D.C., 2020. Skill in Social Situations: The Essence of Savoir-Faire, in: Sternberg, R.J., Kostić, A. (Eds.), *Social Intelligence and Nonverbal Communication*. Springer International Publishing, Cham, pp. 333–357. https://doi.org/10.1007/978-3-030-34964-6_12

Siblot, Y. (2006). *Faire valoir ses droits au quotidien*. Presses de Sciences Po.

Sifres, V.P. (2018). Ampliando la comprensión de la pobreza energética desde el enfoque de capacidades: hacia una mirada construida desde las personas afectadas. *Revista Iberoamericana de Estudios de desarrollo= Iberoamerican Journal of development Studies* 7, 138–163.

Stojilovska, A., Yoon, H., & Robert, C., 2021. Out of the Margins, into the Light: Exploring Energy Poverty and Household Coping Strategies in Austria, North Macedonia, France, and Spain. *Energy Research & Social Science* 82, 102279. https://doi.org/10.1016/j.erss.2021.102279

Tableau de bord de la précarité énergétique – Edition 2022 (Chiffres-clés), 2022. Observatoire National de la Précarité Energétique.

Thomson, H., Snell, C. J., & Liddell, C. 2016. Fuel poverty in the European Union: A concept in need of definition? *People, Place & Policy Online*, 1(10), 5–24. https://doi.org/10.3351/ppp.0010.0001.0002

Vignal, C., 2016. L'ancrage local, une ressource pour les classes populaires des territoires désindustrialisés. *Mobilités résidentielles, territoires et politiques publiques* 197–210. https://doi.org/10.4000/books.septentrion.3203

Villalba, B., & Semal, L. (2018). *Sobriété énergétique : Contraintes matérielles, équité sociale et perspectives institutionnelles*. Editions Quae, Versailles, France. ISBN: 9782759228843.

13 Too Cool or Too Hot

Thermal Comfort in Low-Income Homes and Lived Experiences of Energy Poverty in India and Austria

Tania Berger

13.1 Introduction

A substantial body of literature already explores the planning and design of housing for the urban poor in the Global South. However, this scholarly discourse primarily focuses on questions of affordability (Manna, 2018; Mukhija, 2001; Nastar, 2020). So far, little work has been done on heat stress and adaptation to the heat in these socioeconomic groups. In the Global North, particularly in Europe, research on energy poverty has concentrated chiefly on heating requirements[1] (Brunner et al., 2012; Dubois, 2012; Hernández & Bird, 2010). The need to adapt to extreme summer heat is only now appearing as a concern to reckon with, given that heat-related deaths are rising even in more temperate climates.

13.1.1 Energy poverty and thermal comfort[2]: in the Global North (mostly) concerned with heating

Energy poverty is generally understood as an interplay of high energy prices, low household incomes, and low energy efficiency homes (Boardman, 2010, p. 32). When looking at energy efficiency, it has to be noted that low-income households need to reside in comparably cheap accommodations to be able to cover all their living costs, including housing costs and energy bills. In the case of Austria, cheap accommodation is predominantly found in rentals[3] built before 1960. Low-income households are, on average, more frequently found to live in privately rented apartments in this era (Statistik Austria, 2022). These apartments are more cheaply available than recently constructed rentals (Global Alliance for Buildings and Construction, 2018). With the first legal requirement on the insulation of buildings enacted in Austria only in 1974[4], most buildings in this age bracket display relatively poor energy efficiency standards. Buildings built before 1960 therefore have lower energy efficiency values than more recently constructed ones.

Households residing in poorly insulated buildings will need to consume more energy to keep their home at a specific threshold temperature (e.g., 20–21°C as recommended by the World Health Organization; ; WHO, 1987) than households in well-insulated buildings with an efficient heating system. This low energy efficiency of the buildings concerned means tenants must consume more energy to satisfy their basic needs in terms of keeping warm during cold winters. This higher energy consumption rate will generally tend to effectuate higher energy bills for the household concerned. Households with low budgets at their disposal are at risk of spending a considerable proportion of this budget on energy if they live in such buildings (which, as shown above, they tend to do above average). This proportion thentends to be well above the average household expenditure on energy. Households affected by this mismatch between disposable income and energy bills are considered energy poor.

DOI: 10.4324/9781003408536-16

Costs of real estate – either in sale or rent – do not necessarily reflect a property's energy efficiency in the first place; features such as location, infrastructure provision, and equipment may play a more decisive role. In a bid to tip real estate markets towards valorizing more energy efficient buildings, the European Parliament first enacted the Energy Performance of Buildings Directive (EPBD) in 2002. This directive required the issuing of a standardized energy certificate in the event of the sale or rent of a building or apartment to inform consumers about the energy efficiency of the building they are about to inhabit. Notably, such certificates cannot predict customers' future energy bills in the specified unit, as these bills are determined by residents' actual energy consumption rather than the standardized demand forming the basis of the certificates' calculation. These certificates, however, render comparable different properties on offer in regard to their respective energy efficiency. Current research, however, tends to indicate that the introduction of EPBD's energy certificates has not yet succeeded in raising consumers' awareness of energy efficiency and thereby increasing the value of more efficient buildings (Ayala et al., 2016; Mutter & Sterk, 2016+02:00).

Energy poor households residing in homes with low energy efficiency standards struggle to afford thermal comfort – besides covering other basic energy needs such as cooking warm food regularly, bathing, etc. Lack of thermal comfort is therefore one prominent consequence of energy poverty. However, maintaining healthy lifestyles below certain heating, cooking, and bathing thresholds is impossible. If low-income households restrict their energy consumption for heating below certain thresholds to avoid high energy bills, they risk jeopardizing their long-term health. Thermal comfort is thus closely linked to health (; Santillán Cabeza, 2010).

13.1.2 ... and with heat (mostly) in the Global South

India comprises a significant proportion of the global population that is energy poor. The country houses 700 million people primarily using solid fuels for cooking (Tripathi et al., 2015) and 244 million without access to electricity who use lower-rung fuels such as kerosene for lighting purposes (International Energy Agency, 2016). There is also an unaccounted portion of households that do have electricity connections but have unreliable and sporadic access to electricity (Kumar et al., 2019). Finally, there are those who do have stable access to electricity, mostly in urban areas, yet are struggling to afford it, especially when it comes to high energy-consuming appliances such as air conditioners to safeguard thermal comfort in times of high temperatures (Gupta et al., 2020).

Due to climate change, prolonged heat waves with elevated temperatures are expected for most of the globe in the decades to come. In its latest report, the International Panel on Climate Change states that extreme weather events, including heat waves, have intensified in cities and globally (IPCC, 2022). Population exposure to heat waves will continue to increase with additional warming. The IPCC links the increased occurrence of extreme heat to impacts on human health, mentioning that extreme heat events already result in human mortality and morbidity today as observed results of climate change. It acknowledges significant geographical differences in heat-related mortality without additional adaptation. Overall, outdoor temperatures will increase in both the Global South as well as the Global North. This heat stress will further amplify these cities' existing urban heat islands (UHIs). Therefore, cities will be most affected by heat events, and their populations will generally be most exposed to heat. Adaptation to heat is therefore a task of global scale (ibid).

UHIs significantly impact buildings in urban areas (Akbari et al., 1992) because increased urban temperatures exacerbate the energy demand for cooling buildings. Urban heat islands are one of the best-documented examples of anthropogenic climate change (Santamouris, 2007).

Put simply, these phenomena consist of the fact that air temperatures are often higher in cities than in their surrounding countryside (Asimakopoulus & Santamouris, 2011; Oke, 1982; Oke, 1988). At the same time, high urban temperatures considerably decrease the cooling potential of natural and night ventilation (Santamouris, 2007). If the surrounding outdoor areas in cities are hardly cooler than the indoor spaces in buildings, opening windows at night will only slightly help to cool these spaces.

While UHIs are the result of a combination of several physical phenomena[5], not per se specific to neither the North nor the South, more research has been conducted on UHIs in Northern cities (Oke, 1982). Already, heat-related concerns are fast gaining attention in the more temperate climates of the North in light of rising average temperatures due to global climate change. Also, energy demands related to cooling are foreseen to rise at the global level (Davis & Gertler, 2015). With the additional impact of UHIs, urban areas even in temperate climates will have to adapt to heat. Those groups in society who are already struggling to afford energy for heating will most probably face difficulties covering their cooling demands too (ibid).

In 2022, India was the highest ranked among 54 high-impact countries, in which the rural and urban poor are at high heat-related risk because they lack access to cooling; roughly 190 million urban residents were assumed to fall within this group nationwide (Sustainable Energy For All, 2022). Ownership of air conditioning appliances in 2016 amounted to 30 million units in India, corresponding to just 4% of all households (as compared to 570 million in China). The International Energy Agency points out that AC ownership rises very steeply with income and predicts that sales of AC units in India will continue to grow rapidly (International Energy Agency, 2018), with a current annual increase of 8% in electricity demand for cooling in the residential sector (International Energy Agency, 2016: p. 92).

There are well-documented links between excessive heat and increased morbidity and mortality globally (Vicedo-Cabrera et al., 2021). To counteract heat stress, electrically powered air conditioning systems are being installed to provide respite. Low-income households, however, often lack the financial resources to afford electricity for such devices to keep themselves cool (if, at all, they have the resources to buy the devices). For these groups, again, lack of thermal comfort is a prominent consequence of energy poverty. And as in the case of cold homes, prolonged periods of thermal discomfort in hot environments likewise carry health risks (Vicedo-Cabrera et al., 2018).

This chapter presents the findings of two research projects that examined the diverse lived experiences of energy poverty resulting in frequent lack of thermal comfort in Global South and North contexts. It found similarities despite the apparent discrepancies in overall framework conditions such as climate, living standards, welfare regimes, or housing provisions.

13.2 Methods

Two research projects conducted independently and at different points in time form the basis of this chapter. Their approaches, methods, and results are presented hereafter separately at first before drawing joint conclusions.

13.2.1 Methods of Study I on lack of heating in Austria

A study on energy poverty (hereafter Study I), concentrating on households experiencing difficulties keeping their homes warm in Austria, was conducted in 2014 in the small town of Krems (approximately 25,000 inhabitants), about 80 km to the west of the capital city of Vienna. Based on group discussions with experts from a local social housing provider and a local NGO, six buildings in Krems were selected for further investigation with *Wachaustrasse*, the case study area later selected, being among them right from the beginning.

A standardized questionnaire was developed to investigate residents' housing and living conditions. It contained 38 questions in 7 blocks relating to the building/apartment, heating and hot water generation, rental and energy costs, household income, health, and demographic data. Thus, the survey not only investigated households' housing and living conditions but also gave a clear indication of whether or not the particular household could be regarded as energy poor. This identification was especially relevant in the case of the residential building block *Wachaustrasse,* which became the focus for developing renovation concepts. The block comprises three floors each, accessed via eight staircases, and houses 85 apartments. The net habitable areas of these apartments vary between 37 and 65 m^2. Ten semi-structured telephone interviews, 25 questionnaires manually filled in and sent by mail, and nine interviews done on the spot were conducted in six buildings between January and June 2014 in Krems. Out of these, two telephone interviews and eight in-person interviews were conducted in *Wachaustrasse* alone. Draft reports were presented to respondents during three residents' meetings to determine whether the raw data was correctly interpreted. During the third occasion, municipal and local social service representatives were present and invited to give feedback to residents and researchers alike.

13.2.2 Methods of Study II to assess heat-related thermal comfort in India and Austria

A second study (hereafter Study II) was conducted in 2019 in the Indian cities of Vijayawada and Jodhpur (with a population of approximately 1 million inhabitants) and the Austrian capital Vienna (1,8 million inhabitants). It looked at how heat affects the daily lives of low-income households. The dwellings of low-income respondents investigated in Vijayawada are characterized by poor thermal standards of buildings and poor ventilation. The government-built rehabilitation colony[6], in which respondents live, was selected for field measurements of indoor temperatures and qualitative interviews. Most interviewees in Jodhpur reside in self-built, one-storey units while all Viennese respondents are tenants in multi-storey apartment blocks that were built before the advent of building by-laws on energy efficiency.

For this research, interview questions were developed for semi-structured qualitative questionnaires. Fourteen semi-structured qualitative interviews were conducted in low-income households not only in the colony in Vijayawada (six interviews) but also in Jodhpur (five interviews) in India and Vienna (three interviews) in Austria, Europe. It is worth pointing out that, contrary to Study I, this study explored one aspect of thermal comfort – namely heat – in both South and North contexts. Not least due to its limited sample size, Study II did not seek to outrightly compare the very different situations in India vs Austria. Rather, it aimed at exploring potential common concerns regarding thermal comfort in the interviewed, low-income households in both countries.

Interviews in India were translated and transcribed from the local Telugu language. Local translators translated Hindi/Marwari into English, and interviews in Vienna were done in German. In this chapter, the results obtained from qualitative interviews are presented.

13.3 Results

13.3.1 Too cold … (Study I)

Survey results revealed considerable levels of poverty and, as a consequence, self-restrictions in energy consumption for many of the interviewed households in *Wachaustrasse.* The household income for 45% of the interviewees was less than 901 EUR per month, while the income of a further 27% of the interviewees ranged between 901 and 1,500 EUR monthly. Residents' average income ranges far below the national average of around 3,400 EUR per month and the median income for Austrian households (Statistik Austria, 2016). The living and heating costs of 45% of the interviewed households amount to between 5 and 7 EUR/m^2 per month. These expenses were below the Austrian average of 8 to 9 EUR per month (Höltl et al., 2015) at the time of the study. This demonstrates that the interviewees are restricting their expenditure on energy. This reduced expenditure is reflected in reduced thermal comfort in their homes.

Seven out of ten interviewed households reported that they could not afford to sufficiently heat their home. They all had lived experiences of diminished thermal comfort while residing in cold homes. Out of the seven affected households, six reported mould growth in their apartment[7], with five of these mould-affected households indicating that at least one household member was suffering from chronic health constraints. In contrast, all three households in the building that indicated they were able to keep their home warms were free of mould. Several reasons were mentioned for being unable to heat some rooms. These included leaky windows and doors, insufficient or lacking insulation, old and inefficient heating systems, and the technical impossibility of heating some rooms (e.g., due to a lack of connection to the chimney), in addition to the economic reasons that applied to all affected households. Some households reported multiple reasons for not being able to heat all rooms.

Given respondents' limited energy consumption, installing thermal insulation in their building does not hold the potential to further decrease their consumption. However, it could significantly increase thermal comfort for them. Nevertheless, several respondents voiced considerable concern over and fear of a future where such refurbishment is implemented as they expect associated rental costs to increase as a result. Their building was constructed and owned by a limited-profit-housing-association[8] obliged to offer affordable rents. Rents are therefore legally prohibited from being increased beyond the national? inflation rate. The associations, however, can still increase those components of rent that are dedicated to savings for maintenance and future refurbishments. Several respondents stated that they would have to look for alternative housing should these rent components increase after thermal insulation has been applied to the building.

13.3.2 … or too hot (Study II)

Most male respondents interviewed in the Indian cases of Study II are engaged in manual labour, which they carry out outdoors. Therefore, men are more exposed bodily to direct sunlight and heat during the day due to their work. Working less or not at all due to heat-induced strains on their bodies during the day or outright heat-related illnesses represents a significant problem for daily workers and self-employed alike. Being unable to sleep sufficiently due to a lack of thermal comfort in hot homes over prolonged periods of time was cited as contributing to such health concerns. If, as a consequence, they are unable to work, then heat has a negative impact on the incomes of poor households engaged in the informal economy. Not working at all or working less diminishes their earnings during days of extreme heat.

In eight of the 11 interviewed households in India, women try to stay inside their homes during hot days and take care of their children. For most of the day, they regard these homes as a means of protection against hot, direct sunlight. Staying inside is perceived as advantageous during the morning and midday hours. The picture starts to change during and after the day's hottest hours. Indian interviewees indicated heat to be most prevalent between 11 a.m. and 4 p.m. Four households in Vijayawada – both in single-family units and apartment blocks– reported that they would start sitting outside in the shade (of trees or buildings) around this time.

They partly do this because their houses and flats become overheated, turning uncomfortably hot in the afternoon. However, respondents stressed that they mostly sit outside to chat with neighbours.

Buildings cannot fully satisfy the protective properties ascribed to and expected of them by residents: temperature recordings in interviewees' homes in Vijayawada revealed that indoor temperatures were higher than outdoors for the entire recording period, including nights and mornings. Due to their limited resources, most households—especially those in the resettlement colony in Vijayawada—had no choice in selecting their homes, whether these were row houses or units in multi-storey buildings. They also had no say in the design of their homes to cater for expected climatic conditions, for example, in terms of shutters, shades, and openings to enable ventilation. Seemingly unrelated issues likewise impact the thermal comfort of residents. For example, with drains clogged by uncollected garbage, water gets stagnant close to the homes, mosquitoes breed, and consequently residents need to block windows to keep out mosquitoes even during the night when the admission of cooler outside air could potentially improve indoor thermal comfort.

Even though many interviewees conceded that they perceive heat to be intense at times, they mostly put it into perspective with other difficulties they face in their day-to-day lives (such as water and electricity shortages or the lack of proper sanitation). They stated that, compared with these, the heat was one of their lesser concerns. In a similar vein, Austrian interviewees were more concerned with being able to afford their daily necessities (covering bills for energy in general, rent, medicine, food supplies, expenses for children's schooling, etc.) than with heat and felt that heat was something they simply had to endure. Prioritizing their daily problems also contributed to paying less attention to the lack of thermal comfort in their homes. Heat tolerance can thus be understood as a low-cost coping strategy applied by many of the interviewees. However, this strategy bears the risk of not paying attention to health-related heat risks.

13.3.3 Comparative heat stress assessment in India and Austria

Study II mainly conducted interviews in Austria as a control group mechanism. There are, of course, apparent differences between these two countries. The hot, arid climate of Jodhpur and the hot and humid one of Vijayawada are distinctively different from Vienna's more temperate conditions, not least in terms of absolute daily temperatures. Furthermore, housing structures are different. All interviewed Viennese households live in multi-storey buildings, and all of them are tenants (as are nearly 80% of the city's total population). By contrast, homeownership was more prevalent in both Indian samples[9]. However, interviewees in both India and Austria belong to low-income groups. Due to their limited resources, the interviewed households had little or no choice in selecting their homes, whether single, mostly owned houses, or rented apartments in multi-storey buildings. Consequently, they hardly had any agency over the design of their homes with regard to climatic conditions, especially when it comes to ventilation, shutters, and shades to keep out solar radiation or thermal mass to dampen temperature peaks.

Examples of this lack of agency include – besides the above-mentioned issue of mosquito breeding – residents in India being unable to sleep on cool rooftops due to a lack of ladders to get there. In Austrian households, they often cannot – in the absence of shutters – shield their rooms from hot sunlight during the daytime or open windows for ventilation during the night because of the noise of heavy traffic on nearby roads. For the case of winter cold, respondents in Study I had pointed out that some of their rooms did not even have access to chimneys (so even if they would be able to afford both a heater and fuel, they still could not heat these rooms).

This lack of agency influences how residents can handle thermal discomfort in their homes. While low incomes force them to live in poorly planned and built buildings that constitute sources of thermal discomfort rather than comfort, this lack of resources keeps them from affording any but the most basic remedies A lack of thermal comfort can even lead to a loss of income for Indian informal workers, which makes it even harder for them to change anything about their situation.

In contrast to the heat-related situation found in the Austrian sample of Study II, heat in India threatens to impact interviewees' household incomes more negatively. Those working outside – primarily men – may see themselves incapacitated by the heat, so they will not be able to work for several days or weeks. In the context of primarily informal working relationships, they cannot fall back on sick leave payments in such cases and consequently earn less than during cooler months. While these findings may not be surprising, they are a stark reminder that people's socioeconomic limitations can play out similarly even under very distinct geographical circumstances.

13.4 Discussion and conclusion

The two studies presented here showcase how low-income households in both India and Austria struggle to afford thermal comfort in their homes. Interviewees restricted their energy consumption to be able to pay their energy bills. Consequently, they cannot either sufficiently heat their homes or not mechanically cool them at all and therefore suffer from thermal discomfort.

Mould growth is the most apparent consequence of a lack of heating in the Austrian cases. While in the Indian cases, buildings constitute the single most important and effective means of protection against heat stress for the interviewees. However, design restrictions and appliances of cheap building materials limit the effectiveness of this protection, especially during the evening and night. Building design and materials are difficult to change or improve after their original construction. Therefore, people must remain with what was provided by local authorities. Buildings thus play an ambivalent role: they can constitute a source of thermal discomfort or comfort – the former in case they are poorly planned and built, the latter in case they are constructed with sufficient thermal mass, equipped with shading, and provide suitable openings for ventilation.

The most low-profile measures for alleviating heat stress in existing settlements in Indian cases can only indirectly target related problems. Safeguarding regular and sufficiently frequent garbage removal, mostly from open drains near homes, represents such an indirect approach to increasing household thermal comfort. In Austria, by contrast, low-profile measures could include funding schemes for tenants to install external shading on windows.

The research findings presented here highlight deficiencies in thermal comfort in the studied homes. Thermal discomfort, as a prominent consequence of energy poverty, is the common denominator linking the lived experiences of energy-vulnerable households in both India and Austria (which, as is exemplary demonstrated, links, at least partly, to health and sometimes even to livelihood issues). On this common denominator, the dialogue over energy poverty can hinge in these distinctively different countries.

Lessons learned from these findings should not only be applied to the planning of future housing, especially in the Indian context, but dedicated efforts are also necessary to ameliorate people's sufferings in existing structures despite their constrained resources (Mahadevia et al., 2020). More research is required to assess the effectiveness of small-scale and low-cost interventions in these buildings. Future extreme heat events induced by global climate change will impact low-income residents' lives and livelihoods, especially in densely built urban areas such as resettlement and rehabilitation colonies in India. Residents' health and livelihood in times of heat waves hinge on the heat-protective qualities of their homes. These interdependencies would have remained unclear when only looking at health, housing, or livelihood separately. By contrast, an integrated approach to the topic reveals how improvements in housing can foster resilience to heat regarding both health and livelihood. Considering the well-established link between poverty and ill health (Vaid & Evans, 2017), the critical role played by housing in promoting health and well-being under climate change becomes apparent.

Retrofitting the homes of low-income households increases residents' thermal comfort and well-being in both the Global South and North. The scholarly discourse in the South has yet to shift from focusing on more sustainable, new construction projects towards improving and invigorating existing building structures. In the North, retrofitting and refurbishment are already on the agenda. However, socioeconomic consequences, such as the displacement of poor residents, due to unaffordable rent increases after refurbishment, are still partly overlooked. Additionally, heat in housing is only now starting to become an issue of concern, even in more temperate climates, despite the fact that climate predictions for the coming decades clearly indicate an increased frequency of prolonged heat waves in these regions. In conclusion, awareness of the unfavourable thermal conditions in low-income households must be raised to initiate positive change in both contexts presented here.

Notes

1 Notwithstanding the fact that the broad term "Global South" encompasses many regions in which heating is also a necessity for parts of the year.

2 Thermal comfort is defined by the American Society of Heating, Refrigerating and Air-Conditioning Engineers (ASHRAE 55-2020) and the Austrian Standards Institute (ÖNORM EN ISO, 7730:2005) Klicken oder tippen Sie hier, um Text einzugeben.) as "that condition of mind which expresses satisfaction with the thermal environment" (Benton et al., 1990). Thermal comfort herein is understood as a desirable state according to social norms. People's practices linked to thermal comfort have to be seen as activities that they undertake to heat and cool their bodies and homes. These are not tendencies of luxury (Madsen & Gram-Hanssen, 2017). Habits, meanings, and knowledge are the main components of collectively shared practice. Comfort practices, then, are habitual actions that vary with different ideas of identity and social histories (Gram-Hanssen, 2010). Thermal comfort is thus understood as a socio-technical issue, as a negotiable socio-cultural construct that is both an idea and a material reality (Chappells & Shove, 2005).

3 In urban housing markets with a prominent rental sector – such as in Germany and Austria – renting will mostly be low income households' choice (Reinprecht, 2014; Statista, 2022; Whitehead & Scanlon, 2014) as they may not be able to afford the upfront costs for buying a home. To put this into perspective with India: here, the share of rentals in the overall housing market varies greatly between different union states, ranging from roughly 15% to 40%. Of all rented households, 80% are located in urban areas (Knight Frank India PVT LTD & Khaitan & Co, 2019).

4 ÖNORM B 8110/Beiblatt "Erläuterungen zu Abschnitt 6 der ÖNPRM B 8110 Hochbau-Wärmeschutz", Dezember 1974.

5 The following factors are contributing to UHI formation (Santamouris et al., 2001):

- Complex geometry of the built environment decreases heat loss from within streets.
- Thermal properties of construction materials increase the storage of heat in the fabric of the city.
- Anthropogenic heat is released from the combustion of fuels (e.g., cars).
- Removal of heat by wind from within streets is reduced.

6 The term "rehabilitation colony" in the Indian context broadly refers to residential estates of considerable size (often housing several hundred families) built to accommodate previously informal settlers affected by eviction from and removal of their makeshift homes.

7 Mould can cause many health effects, such as a stuffy nose, sore throat, coughing or wheezing, burning eyes, or a skin rash. People with asthma or who are allergic to mould may have severe reactions. Immune-compromised people and people with chronic lung disease may get infections in their lungs from mould (Mudarri & Fisk, 2007).

8 Limit-profit housing associations in Austria are legal entities under a dedicated Act. Their main goal is to construct and run affordable housing for which purpose they receive public subsidies (e.g., favourable loans). Any profit generated on a particular residential building has to be reinvested in the same.

9 Further gross differences, of course, exist regarding, for example, standard of living and welfare systems, which are, however, not explored in more detail here.

References

Akbari, H., Davis, S., Dorsano, S., Huang, J., & Winnett, S. (Eds.) (1992). Lawrence Berkeley Laboratory report: LBL-31587. *Cooling Our Communities: A Guidebook on Tree Planting and Light-Colored Surfacing*. Washington, DC: U.S. Environmental Protection Agency Office of Policy Analysis Climate Change Division.

ASHRAE 55-2020 (American Society of Heating, Refrigerating and Air-Conditioning Engineers) Thermal Environmental Conditions for Human Occupancy. Includes ANSI/ASHRAE addenda listed in Appendix N.

Asimakopoulus, D. N., & Santamouris, M. (2011). Energy and Climate in the Urban Built Environment (1st edition). *BEST (Buildings Energy and Solar Technology)*. London, New York: Routledge.

Ayala, A. de, Galarraga, I., & Spadaro, J. V. (2016). The Price of Energy Efficiency in the Spanish Housing Market. *Energy Policy*, 94, 16–24.

Benton, C. C., Bauman, F. S., & Fountain, M. E. (1990). A Field Measurement System for the Study of Thermal Comfort. *ASHRAE Transactions*, 96, 623–633. Retrieved April 23, 2022, from https://escholarship.org/uc/item/9cb6t2wz

Boardman, B. (2010). *Fixing Fuel Poverty: Challenges and Solutions*. London: Earthscan.

Brunner, K.-M., Spitzer, M., & Christanell, A. (2012). Experiencing Fuel Poverty. Coping Strategies of Low-Income Households in Vienna/Austria. *Energy Policy*, 49, 53–59.

IPCC (2022). Summary for Policymakers. In *Climate Change 2022. Impacts, Adaptation and Vulnerability. Contribution of Working Group II to the Sixth Assessment Report of the Intergovernmental Panel* (pp. 3–33). Cambridge: Cambridge University Press.

Chappells, H., & Shove, E. (2005). Debating the Future of Comfort: Environmental Sustainability, Energy Consumption and the Indoor Environment. *Building Research & Information*, 33(1), 32–40.

Davis, L. W., & Gertler, P. J. (2015). Contribution of Air Conditioning Adoption to Future Energy Use Under Global Warming. *Proceedings of the National Academy of Sciences of the United States of America*, 112(19), 5962–5967.

Dubois, U. (2012). From Targeting to Implementation: The Role of Identification of Fuel Poor Households. *Energy Policy*, 49, 107–115.

Global Alliance for Buildings and Construction. (2018). *2018 Global Status Report: Towards a Zero-Emission, Efficient and Resilient Buildings and Construction Sector*. Nairobi: United Nations Environment Programme.

Gram-Hanssen, K. (2010). Residential Heat Comfort Practices: Understanding Users. *Building Research & Information*, 38(2), 175–186.

Gupta, S., Gupta, E., & Sarangi, G. K. (2020). Household Energy Poverty Index for India: An Analysis of Inter-State Differences. *Energy Policy*, 144, 111592.

Hernández, D., & Bird, S. (2010). Energy Burden and the Need for Integrated Low-Income Housing and Energy Policy. *Poverty & Public Policy*, 2(4), 5–25.

Höltl, Andrea, Berger, Tania, & Faustmann, Anna (2015). Sustainable Housing and Energy Poverty. In: *Proceedings of STS Conference Graz 2015*.

International Energy Agency (2016). *World Energy Outlook 2016 (World Energy Outlook)*. Paris. Retrieved March 10, 2023, from https://iea.blob.core.windows.net/assets/680c05c8-1d6e-42ae-b953-68e0420d46d5/WEO2016.pdf

International Energy Agency (2018). *The Future of Cooling: Opportunities for Energy Efficient Air Conditioning*. Paris.

Knight Frank India PVT LTD, & Khaitan & Co (2019). *Institutionalising The Rental Housing Market In India – 2019*. Mumbai. Retrieved March 09, 2023, from https://content.knightfrank.com/research/1004/documents/en/india-topical-reports-institutionalising-the-rental-housing-market-in-india-2019-6718.pdf

Kumar, P., Rao, S., & Yadama, G. N. (2019). Energy Poverty in India. In: Kumar, P., Rao, S. & Yadama, G. N. (eds.) *Encyclopedia of Social Work*. NASW Press: Washington DC, USA & Oxford University Press: Oxford, UK.

Madsen, L. V., & Gram-Hanssen, K. (2017). Understanding Comfort and Senses in Social Practice Theory: Insights from a Danish Field Study. *Energy Research & Social Science*, 29, 86–94.

Mahadevia, D., Pathak, M., Bhatia, N., & Patel, S. (2020). Climate Change, Heat Waves and Thermal Comfort—Reflections on Housing Policy in India. *Environment and Urbanization ASIA*, 11(1), 29–50.

Manna, P. (2018). Measuring Functionality of the Low-Cost Housing for the Urban Poor: Unheard Voices of Asansol City, West Bengal. *Space and Culture, India*, *6*(1), 91–106.

Mudarri, D., & Fisk, W. J. (2007). Public Health and Economic Impact of Dampness and Mold. *Indoor Air*, 17(3), 226–235.

Mukhija, V. (2001). Upgrading Housing Settlements in Developing Countries. *Cities*, 18(4), 213–222.

Mutter, A., & Sterk, L. (2016+02:00). Increasing Consumer Awareness in Energy Efficient Housing. In P. Aguilera, S. Beckham Hooff, N. Cheong, H. M. Gutiérrez, A. Leshchynskyy, R. Mathew, et al. (Eds.), *Strategic Energy Solutions: A Case Study of the Öresund Region* (pp. 33–38). Lund: International Institute for Industrial Environmental Economics, Lund University.

Nastar, M. (2020). Message Sent, Now What? A Critical Analysis of the Heat Action Plan in Ahmedabad, India. *Urban Science*, 4(4), 53.

Oke, T. R. (1982). The Energetic Basis of the Urban Heat Island. *Quarterly Journal of the Royal Meteorological Society*, 108(455), 1–24.

Oke, T. R. (1988). Street Design and Urban Canopy Layer Climate. *Energy and Buildings*, 11(1–3), 103–113.

ÖNORM EN ISO, 7730:2005 (2006 05 01): Ergonomics of the thermal environment – Analytical determination and interpretation of thermal comfort using calculation of the PMV and PPD indices and local thermal comfort criteria. Vienna.

Reinprecht, C. (2014). Social Housing in Austria. In K. Scanlon, C. M. E. Whitehead, & M. Fernández Arrigoitia (Eds.), *Real Estate Issues. Social Housing in Europe* (pp. 61–74). Chichester, West Sussex: Wiley Blackwell.

Santamouris, M. (2007). Heat Island Research in Europe: The State of the Art. *Advances in Building Energy Research*, 1(1), 123–150.

Santamouris, M., Papanikolaou, N., Livada, I., Koronakis, I., Georgakis, C., Argiriou, A., & Assimakopoulos, D. (2001). On the Impact of Urban Climate on the Energy Consumption of Buildings. *Solar Energy*, 70(3), 201–216.

Santillán Cabeza, S. (Ed.) (2010). *Opinion of the European Economic and Social Committee on 'Energy Poverty in the Context of Liberalisation and the Economic Crisis' (Exploratory Opinion: (Exploratory Opinion)* (2011/C 44/09). Brussels: European Economic and Social Committee.

Statista (2022). Residential real estate in Europe: Industries & Markets. Retrieved March 10, 2023, from file:///C:/Users/TBerger/Downloads/study_id69412_residential-real-estate-in-europe.pdf

Statistik Austria (2016). Tabellenband EU-Silc 2015: Einkommen, Armut und Lebensbedingungen. 2016. Retrieved February 03, 2023, from https://media.salzburger-armutskonferenz.at/42~42/2015-tabellenband-eu-silc.pdf

Statistik Austria (2022). Dimensionen Der Energiearmut In Österreich: Hohe Energiekosten bzw. Nicht-Leistbarkeit von Energie für Wohnen. Wien.

Sustainable Energy For All (2022). *Chilling Prospects 2022: Tracking Sustainable Cooling for All (Chilling Prospects)*. Vienna: Sustainable Energy For All. Retrieved March 09, 2023, from https://www.seforall.org/system/files/2022-07/seforall-chilling-prospects-2022.pdf

Tripathi, A., Sagar, A. D., & Smith, K. R. (2015). Promoting Clean and Affordable Cooking: Smarter Subsidies for LPG. *Economic and Political Weekly*, 50(48), 81–84. Retrieved March 10, 2023, from http://www.jstor.org/stable/44002904.

Vaid, U., & Evans, G. W. (2017). Housing Quality and Health: An Evaluation of Slum Rehabilitation in India. *Environment and Behavior*, 49(7), 771–790.

Vicedo-Cabrera, A. M., Guo, Y., Sera, F., Huber, V., Schleussner, C.-F., & Mitchell, D., et al. (2018). Temperature-Related Mortality Impacts Under and Beyond Paris Agreement Climate Change Scenarios. *Climatic Change*, 150(3–4), 391–402.

Vicedo-Cabrera, A. M., Scovronick, N., Sera, F., Royé, D., Schneider, R., & Tobias, A., et al. (2021). The Burden of Heat-Related Mortality Attributable to Recent Human-Induced Climate Change. *Nature Climate Change*, 11(6), 492–500.

Whitehead, C., & Scanlon, K. (2014). Social Housing in Europe. In K. Scanlon, C. M. E. Whitehead, & M. Fernández Arrigoitia (Eds.), *Real Estate Issues. Social Housing in Europe* (pp. 8–34). Chichester, West Sussex: Wiley Blackwell.

WHO (1987). Health Impact of Low Indoor Temperatures: Report on a WHO (World Health Organization) Meeting Held at Copenhagen on November 11–14, 1985. *Environmental Health Series (Environmental Health Series).* Copenhagen: World Health Organisation.

14 Heating Deprivation in the Southern Cone

Sensitivities and Resilience Shaping the Vulnerability Experience

Paz Araya, Tamara Oyarzún, and Betina Cardoso

14.1 Introduction

Energy poverty is common in the Southern Cone and regions with cold climates, such as Argentina and Chilean Patagonia. Energy deprivation and air pollution are evidence of this condition in these regions. Several studies concern air pollution from firewood use (Calvo et al., 2022; Schueftan et al., 2016). Still, heating deprivation remains a field that requires further exploration (Navarro-Espinosa & Thomas-Galán, 2023). Heating deprivation is a fundamental aspect of people's lived experience, as it shapes the conditions that increase vulnerability and precariousness (Reyes et al., 2019).

The definition of energy poverty presented by RedPE (2019) allows for a situated approach to the phenomenon. According to this definition, a household is in energy poverty when it does not have equitable access to quality energy services. That is, adequate, reliable, non-polluting, and safe services to cover fundamental needs to sustain its inhabitants' human and economic development. In this sense, energy needs and services are defined in a specific temporal and spatial context and under specific socio-cultural conditions (Cortés & Amigo, 2022). In turn, we will refer to heating deprivation as the condition in which a household cannot sustain a minimum tolerable thermal comfort temperature. Therefore, heating deprivation is the result of two states. One is the material inability of the building to maintain a minimum tolerable temperature due to its poor thermal isolation. The second is the physical and economic inability of the household to produce the heat necessary to sustain a minimum tolerable temperature because of budgetary constraints and a lack of physical access to energy resources and heating technologies.

The conditions and manifestations of energy poverty and heating deprivation vary according to households' geographical, socio-cultural, and economic conditions. It is relevant to observe the shared characteristics of the central-southern areas of Chile and Argentina, the southernmost countries of the continent with socio-historical and climatic similarities. The present ecological and climatic crises are similarly affecting the general conditions of life in the region (IPCC, 2022). Moreover, these conditions are experienced differently from other countries in Latin America, which may share common social-political characteristics but differ in their climatic conditions. Furthermore, while finding similar climatic conditions in countries from the global north, their livelihoods fundamentally differ, particularly in the extent of inequality and multidimensional poverty levels (IPCC, 2022).

In this chapter, we explored the concept of vulnerability conditioned by heating deprivation in the central-southern regions of Chile and Argentina. We interpreted it as a significant component in shaping the lived experience of energy poverty (Middlemiss & Gillard, 2015). The central-southern areas of Chile and Argentina feature a long winter season (ranging from four to eight months), which constrains the population from maintaining healthy minimum

DOI: 10.4324/9781003408536-17

temperatures inside the home during this period (Cardoso & González, 2019; Schueftan & González, 2015). Despite having similar environmental constraints, state responses differ between the two countries. While Chile has had a long history of radical neoliberal policies resulting in a weak state in implementing effective social policies, Argentina has been more effective in introducing energy subsidies for residential natural gas. This primary policy has been absent in Chile. However, the similarity between the two countries lies in the infrastructure, where thermal insulation is not standard in the housing stock. Households tend to resolve issues around comfort with coping strategies and significant overconsumption of cheap fuel sources (Cardoso & González, 2019).

Deprivation constitutes a specific experience of permanent negotiation in the face of heating or budget restrictions (Anderson et al., 2012). Therefore, people may perceive situations of vulnerability differently in their everyday lives (Adger, 2006). Observations from a vulnerability approach (Thomas et al., 2019) aid in understanding this condition as a "state of being" that extends beyond the indicators that allow us to quantify it (Middlemiss & Gillard, 2015). This viewpoint accounts for the lived experience of energy poverty in this way (Hargreaves & Longhurst, 2018). Our objective is to understand the factors that shape the energy vulnerabilities of populations exposed to heating deprivation. We will examine heat care and regulation practices to better understand how they shape sensitivities and response capacities to living in a low-temperature house. We carried out a bottom-up and qualitative analysis of the practices, which considered their complexities from the understanding of the people (Eisfeld & Seebauer, 2022; Middlemiss & Gillard, 2015).

14.1.1 Energy poverty: a three-dimensional approach

The phenomenon of energy poverty is multidimensional, encompassing issues of access, equity, and quality of energy (Urquiza et al., 2019). Observing the deprivation of heating in these three dimensions allows us to distinguish the complexity of its expressions in socio-cultural, socio-technical, and socio-ecological spaces (Cardoso et al., 2022). This approach to energy poverty, developed by RedPE (2022), makes it possible to link these three dimensions, as shown in Figure 14.1.

This framework recognises that a household reaches energy poverty when it falls below minimum thresholds in three dimensions, as shown in Figure 14.1. These thresholds are expressed materially in a physical, technological, and economical manner (González-Eguino, 2015) and subjectively as thresholds of tolerance and acceptance (RedPE, 2019; Urquiza et al., 2019). The quality dimension concerns the thresholds of acceptance and tolerance of the household members, which determine their energy needs, adequacy, safety, and health conditions. The economic dimension includes the economic thresholds of the minimum expenditures that allow each household to meet its energy needs. Finally, the access dimension relates to the physical and technological thresholds above which it is possible to meet the energy needs of each household.

When applying this approach to observe the conditions of heating deprivation, the following aspects are determinants in each dimension:

- Quality dimension (tolerance and acceptance thresholds): What minimum temperatures are tolerated and accepted by the household members? We will use the term a*ccepted temperature* to refer to the minimum condition of thermal comfort for the inhabitants of a household, which varies according to cultural and individual conditions. We will use the term *tolerated temperature* to refer to the temperature ranges outside of which there are higher disease risks.

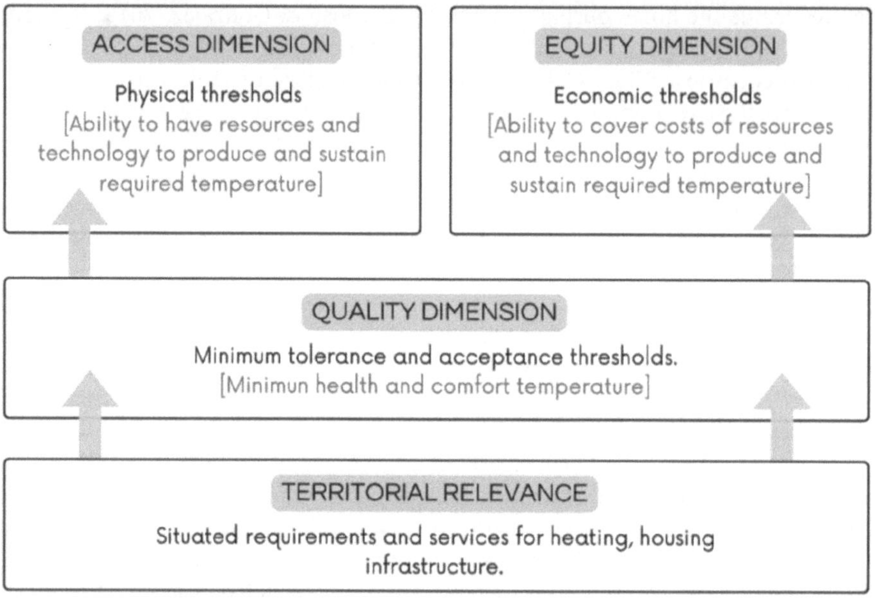

Figure 14.1 Shows the three-dimensional analysis framework of energy poverty and its application in the case of deprivation of heating.

Source: Own elaboration based on RedPE (2019).

According to the World Health Organization, living spaces' minimum healthy conditions are 18°C to 24°C (World Health Organization, 2018). The heat demand is, then, a function of the temperature needs and the thermal performance of the dwelling, given by its material capacity to sustain a temperature over time.

- Equity dimension (economic thresholds): What minimum economic conditions guarantee a household's ability to have a heating service according to its needs? This dimension considers a household's income in relation to covering energy resources and technology costs to meet its minimum heating needs. In this regard, costs include the minimum expenditure on energy sources for heat supply and the investment in heating technologies or thermal insulation.
- Access dimension (physical and technological thresholds): What minimum physical and technological conditions guarantee a household's ability to have a heating service according to its needs? The minimum heat demand of a household will be satisfied to the extent that there is the capacity to have sufficient energy and technological resources adequate for producing the required heat, thus sustaining the minimum tolerated and accepted temperatures in safe conditions.

14.1.2 Vulnerability to heating deprivation

We will consider the concept of vulnerability as the propensity of someone to suffer the adverse effects of a hazard, whether physical, social, economic, or environmental (Centro de Ciencia del Clima y la Resiliencia (CR)2, 2018; ONU, 2007). While there are various formulations of vulnerability, several scholars agree that vulnerability is modulated by sensitivity and resilience, with the latter composed of adaptive capacity and response capacity (Adger, 2006;

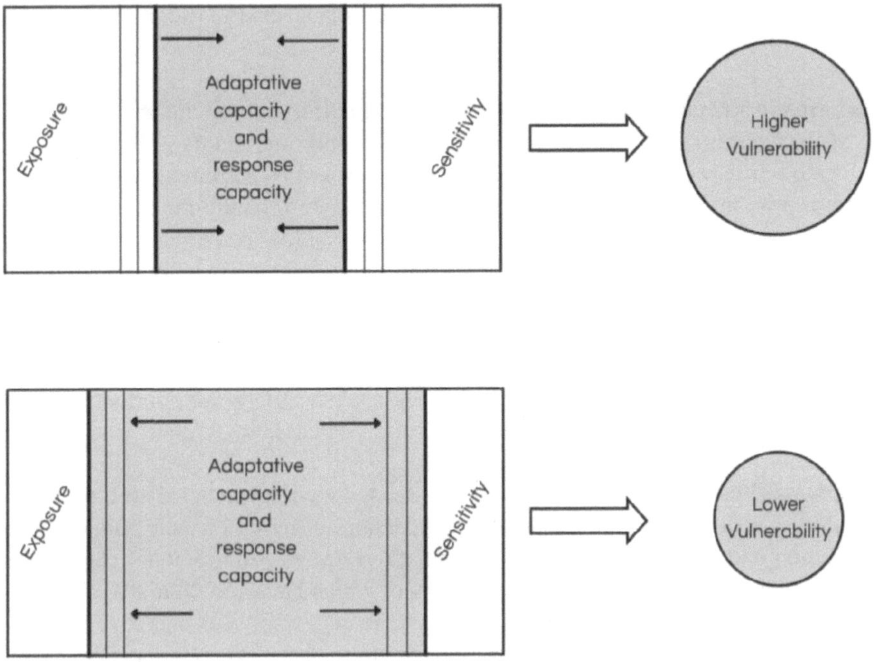

Figure 14.2 When facing an exposure, vulnerability is modulated by sensitivity and the capacity to respond and adapt.

Source: Based on Thomas et al. (2019, p. 2).

Middlemiss & Gillard, 2015). Limited adaptive and response capacity combined with high sensitivity and exposure contribute to higher vulnerability. In comparison, higher adaptive and response capacities combined with lower sensitivity and exposure would contribute to lower vulnerability (Figure 14.2).

Sensitivity is often described as the ability to feel and manifest emotional and physical responses to stimuli (Mertens et al., 2020). Intrinsic factors making a territory or population susceptible to being impacted by a threat can affect sensitivity (Urquiza & Billi, 2020). Coping and adaptive capacities allow people to resist or repair the effects of an external disturbance, so it is possible to maintain or recover the equilibrium of pre-exposure conditions (Estévez et al., 2011). Finally, response capacity enables the ability to respond proactively to a shock, and anticipatory adaptive capacity allows one to act proactively and accommodate to mitigate the impact that will arrive (Adger, 2006; Urquiza & Billi, 2020).

14.2 Ethnography on living with heating deprivation in three sites

We analyse the data from three cases collected during different periods with similar objectives and common elements. We identified these elements for qualitatively interpreting the factors affecting vulnerability to heating deprivation in the three populations. We have conducted visits to each of the sites and the interviewees' homes. The ethnographic methodology consisted of contextualised dialogue with each of the families visited. We carried out semi-structured interviews and participant observation. In some cases, we added video tours, photographs, and house mapping to understand daily life dynamics further. We based on an interview schedule, and the researchers spent time with community members in an everyday context for participant observation.

i **Rural area in Argentinean Patagonia:** In the community of Laguna Blanca, in the province of Ro Negro (40° 43′ S and 69° 50′ W), with 180 inhabitants (35 families in total) and located in an environment of shrub-steppe, limited by an uncultivated region. The climate is arid and cold, with rain and snow concentrated in autumn and winter (Bustos, 2006). All the families in the village have incomes below the poverty level compared to cities in the region. Therefore, we can consider that they live on a subsistence level. The main activity is sheep and goat raising for wool, hair, and the self-consumption of meat. The houses are made of baked bricks, and the roofs and windows lack insulating materials, so they do not have good thermal quality beyond the heat provided by the fuel used (mainly firewood and gas) (Cardoso & González, 2019). We conducted interviews in Laguna Blanca during 2019 in January, April, August, and October, where we visited 20 families.

ii **Medium-sized cities in southern Chile:** The second case covers small and medium-sized cities in the Araucanía region of Chile that maintain, in terms of their public policies on air pollution and habitability, a functional relationship with the regional capital Temuco. The Araucanía region has 35 municipalities, 14 of which have 15,000 to 50,000 inhabitants (BCN, 2021). These medium-sized cities have some public services that cover the most relevant needs of the population, while for more sophisticated services, the people must travel to the regional capital. These cities present a cold season lasting more than six months of the year and rainy weather, which implies a high demand for heating in the residential sector, mainly covered by firewood consumption. The dwellings in these locations are predominantly semi-detached and detached houses made of lightweight materials (CDT, 2019) and without reinforced insulation. We conducted the ethnography mainly in the city of Victoria with 18 families between April and June of 2022.

iii **Elderly population in a big urban city:** The third case focuses on households inhabited by older people in the municipality of Quinta Normal, located in Santiago, Metropolitan Region, Chile. This district is entirely urban (INE, 2021) and presents the Mediterranean climate of Santiago (Pérez-Fargallo et al., 2020). The region also has a higher multidimensional poverty rate than the Metropolitan Region and scores higher in the Ageing Index (BCN, 2021). Among the homes visited were houses and flats with various construction materials, such as cement, brick, adobe, and wooden partitions, and roofs of metal sheets, slate (fibre cement), and tiles or shingles. On the other hand, the flats have brick walls and partition walls, and their roofs consist of the floor of the upper apartment. We conducted ethnographic interviews with ten families between December 2021 and December 2022.

We conducted ethnographies for each case according to the guidelines developed by Corona (2016). During this shared time, the researchers interacted with the interviewee and recorded the activities carried out in the so-called dialogue of knowledge (Gúber, 2004). From previous research, some families were already engaged, which allowed us to get to know new families in a snowball-type sampling. Aiming to assess the present results, we covered different subjects during the visits. In terms of in-house dynamics, the interviews covered heating strategies, personal and interpersonal dynamics of temperature regulation, heating technology use, and the dynamics of fuel substitution. Also, we covered subjective narratives such as perceptions of thermal comfort in housing and work/study places, stories and memories around heating practices, emotions and knowledge about heating and thermal comfort. Finally, budgeting dynamics were also covered, as were decision-making and the impacts of heating costs, as well as willingness and capacities to invest in technologies and thermal insulation.

Using the previously described frameworks, we proceed through three stages of categorisation: open coding, axial coding, and selective coding (Strauss & Corbin, 1998). As a result, the segments and descriptions in Chapter 3 follow. In this way, the interpretation and analysis of the interviews illustrate the range of the biographical-narrative approach in connection to individual and group experiences (Gúber, 2004).

14.3 How does heating deprivation affect the vulnerability of the household members?

In our undertaking, we connect the analytical framework of the three-dimensional analysis of energy poverty with the thresholds and constituents of the vulnerability framework. The articulation of the frameworks is shown in the scheme set out in Figure 14.3. The frameworks' articulation allows us to identify the most critical factors influencing sensitivity to, and capacity to respond to, exposure under heating deprivation conditions. These factors will be determinants of the members of a household's vulnerability to heating deprivation.

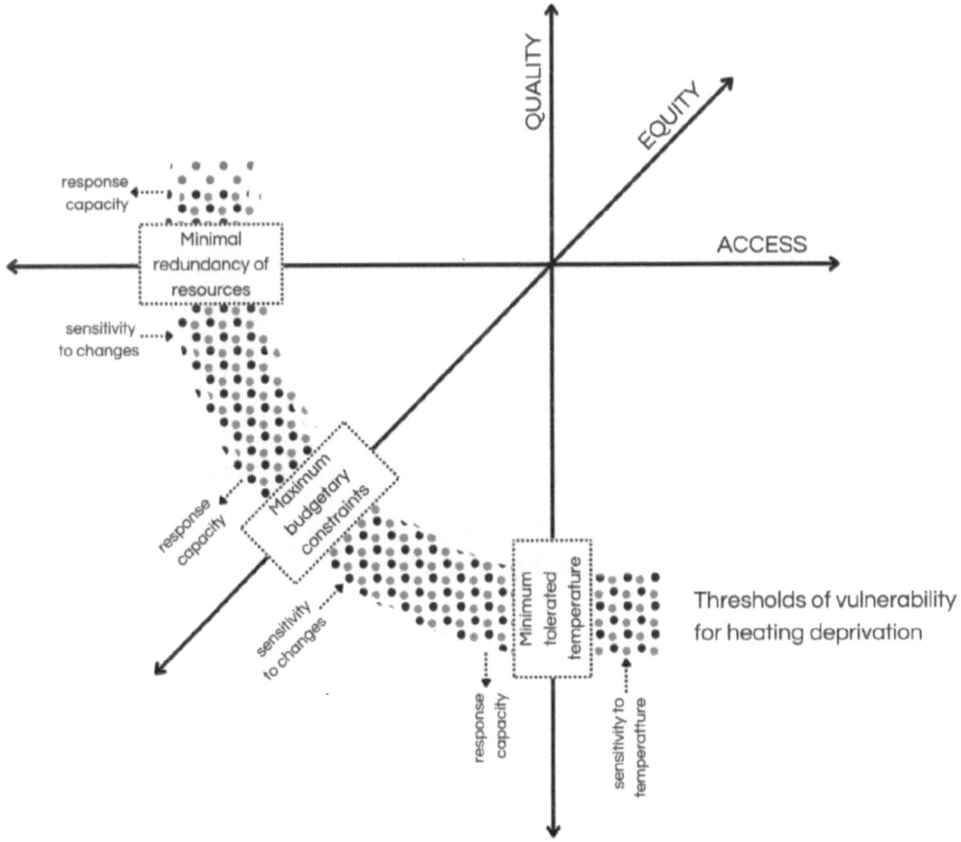

Figure 14.3 Thresholds of vulnerability for the three dimensions of heating deprivation.

Source: Own elaboration.

In the quality dimension, vulnerability is modulated by temperature sensitivity, which affects health and comfort, and by adaptative and response capacity when exposed to temperatures below these minimum thresholds. In this sense, high sensitivity to low temperatures and a limited response capacity to such exposure increase individual and shared vulnerability. Therefore, heating deprivation is the inability to produce and sustain the minimum temperatures accepted or tolerated by the household members inside the dwelling.

The equity dimension is determined by the constraints on the household budget, which enable it to cover the costs of heat production and sustain the minimum tolerated and accepted temperature. Sensitivity to changes in budget and the capacity to respond to these changes set the vulnerability thresholds. High sensitivity to budget changes coupled with a low capacity to respond to budget constraints increases vulnerability. Heating deprivation results from the inability to adequately afford the resources and technologies necessary for heat production and maintenance, allowing household members to sustain the minimum temperature required. High vulnerability is primarily related to resignation processes in the face of budgetary constraints.

In the access dimension, the critical element will be the redundancy of alternative energy sources and technologies for heat production and sustaining temperatures above the minimum tolerated and accepted thresholds. Then, vulnerability is shaped by sensitivity to changes in the redundancy of energy and technological resources and the capacity to respond to a scarcity of accessible alternatives. A high sensitivity to changes in heating alternatives, coupled with a low capacity to respond to shortages, increases vulnerability. The experience of heating deprivation will result from scarcity, referring to energy sources or technologies for heat production or thermal insulation. In the face of vulnerability resulting from heating deprivation, modulated by the access dimension, household members rely on inadequate solutions to keep warm, putting their health and well-being at risk.

14.3.1 *Quality dimension: low-temperature acceptance and tolerance thresholds*

14.3.1.1 Practices affecting sensitivity

Sensitivity to low temperatures is a subjective issue that is co-dependent with everyday dynamics. Although the lack of adequate temperatures was transversal, inhabitants are not manifesting their sensitivity similarly. Some people reported not feeling comfortable, and others, while noting that they are satisfied with the temperature in their homes, also mentioned they need to change their dynamics to make conditions in their homes tolerable.

Interviewees often define sensitivity to low temperatures using a comparative benchmark. This benchmark results from comparing thermal conditions between houses and workplaces, between rooms in the same place, or between memories of other homes. Housekeepers and retired older adults who spend more time in the house consider themselves more aware of indoor temperatures than workers and students who spend their day elsewhere. Also, many reported that insulation conditions and temperatures were poorer in their childhood homes than they are today. Hence, they felt that the temperatures in their current homes were acceptable compared to their life history.

For carers, the needs of others shape temperature acceptance thresholds, as they attach significant importance to maintaining a comfortable temperature for people under their care. We also observed this situation in the case of caring for pets, especially in the case of older adults who consider the company of pets fundamental. Moreover, care work affects people's mental burdens. In many cases, the stressful conditions arising from the concern of being unable to provide a "healthy home" for the household member/s under care meant an increased intolerance to the perceived cold inside the home. These cases have an essential gender component, as they are more common in women than men.

People's intrinsic conditions, whether physical, psychological, or cognitive, also shape sensitivity. Older adults and children below three years of age tend to have a high sensitivity to low temperatures, which is also expressed in more severe illnesses when exposed to low temperatures. People with chronic diseases report the same effects; being unable to reach comfortable temperatures means spending most of the day in bed during the cold season. These conditions can deepen vulnerability in a sense that goes beyond the energy poverty phenomenon towards multidimensional poverty, precarity, and disempowerment.

14.3.1.2 Practices affecting adaptive and responsive capacity

Coping strategies to deal with low temperatures are many and varied. They can range from concentrating the use of the house in a single room; wearing warm clothing; blanket choices; spending more time in bed; consuming specific foods and beverages (like the traditional drink "yerba mate"[1] in Argentina); and using complementary heating devices (like *Guateros*[2]). They are fundamental coping strategies in cold climate zones; interviewees commonly perceived them as a skill to overcome the cold.

Since poorly insulated buildings are one of the foremost causes of heat deprivation, housing interventions are essential to the government's strategies for responding to heat deprivation. These interventions depend on technical, administrative, and economic abilities. In this regard, it is common to adopt low-cost and easily accessible strategies, such as door and window interventions, to reduce air infiltration. In some cases, families buy weatherstripping for doors and windows. However, it is a widespread practice that instead of weatherstripping, they use homemade or temporary solutions from locally available materials, such as fabric, cardboard, or plastic, often reusing objects. Home retrofitting is less common, as it is limited to those who own their house, those who can make a significant investment, or those who know about and can apply for government grant schemes.

There are also intrinsic psychological, cognitive, social, cultural, and economic conditions that facilitate the ability to make decisions and strategies for living in a household at low temperatures. People with self-confidence can act decisively and seek solutions to heat deprivation. This mindset also alleviates feelings of helplessness, frustration, and stress. Having control over resources or the trust of other family members facilitates autonomy in decision-making about warming strategies. For example, when fuel prices rise, independence helps people act more quickly, coping better with uncertainty. Earlier family learning also facilitates decision-making. In the cases of rural areas and medium-sized cities, it was common to find strategies under narratives of learning related to local identity and coping skills that span generations.

Finally, networks and access to services also shape the level of technical knowledge about home heating and insulation and define thresholds of acceptability. The social mobility of younger generations makes it possible to recognise other thermal comfort thresholds and learn about different technologies and materials. This knowledge opens people's expectations regarding the possibility of having higher temperatures in their homes. On the contrary, low awareness of alternatives increases the resignation of living in cold housing. Geographical conditions intensified this aspect in cities with difficult access to public services or programmes, such as rural or medium-sized cities with functional ties to large cities.

14.3.2 Equity dimension: economic thresholds shaping budget restrictions

14.3.2.1 Practices affecting sensitivity

Budgetary constraints appear as restrictions on heating expenditures or the inability to invest in home improvements or new heating technologies. The inability to cover the heating costs required to maintain a healthy minimum temperature in a poorly insulated building generates different responses that, in diverse forms, result in resignation. High heating costs lead families to choose between spending on health and food or heating, with the latter usually not being prioritised because it is not considered a primary expense. Resigning to the precariousness and deprivations of everyday life is crucial to the experience of vulnerability. They reduce sensitivities through over-adaptation to living conditions, resulting in the normalisation and de-problematisation of the state of heat deprivation. This situation is especially noticeable in older people with a low pension, where health expenses also become more significant, and in low-income households in rural areas.

The investments required for well-insulated housing are far beyond the financial capacity of families whose income is insufficient to cover heating costs. As a result, looking for further improvements to the house or the heating system is frequently absent from the interviewees' narratives. Even when asked about plans to refurbish the house, most people prioritised other improvements over upgrading the thermal conditions, which they regard as a privilege or luxury. Where pensions are low, there is also a fear of increased housing maintenance costs for older people. People are more likely to choose cheaper building materials perceived as robust and durable, even if their thermal insulation conditions, such as brick, are inadequate.

14.3.2.2 Practices affecting adaptive and responsive capacity

The foremost circumstance guaranteeing the capacity to respond is good household economic conditions, which ease the risks of heat deprivation. In many cases, households have savings, financial resources, and capital of various kinds to cope with such pressures. Other factors, however, enable overcoming the resignation experience when faced with budgetary constraints. Healthy interactions with the environment enable decision-making, negotiation, and formation of support networks. Similarly, people with mental and emotional health access develop a higher tolerance for frustration, allowing them to cope and negotiate better with these conditions. In the opposite case, people experiencing depression perceived a lower capacity to act, feeling forced to resign to their household and economic conditions.

Autonomy in housing relationships also shapes this ability, so we can see that people who own their own homes, even when faced with budget constraints, are described as having a more remarkable ability to deal with expenses within the limits of their situation. On the other hand, tenants were resigned to these costs and conditions, even if budgetary constraints were less severe.

14.3.3 Access dimension: physical thresholds and redundancy for heating alternatives

14.3.3.1 Practices affecting sensitivity

The sensitivity to abundance or scarcity in access to energy alternatives becomes visible in narratives about firewood. People commonly believed that firewood provided "better heat" or "longer heat". This perception is reinforced by accounts of the abundance of firewood in the environment in childhood memories, especially in rural areas, where people emphasise that firewood was much more abundant in earlier times. In some cases, switching from firewood to pellets or LPG, which can also be more expensive, increases the feeling of scarcity.

Something similar to the above occurs in the case of access to technologies since people perceive the most modern or less polluting technologies as being scarcely accessible, as those who have access to them can usually do so through government subsidies or due to better economic conditions. Market conditions have also shaped sensitivities. The pellet market has had episodes where it has not been sufficient to meet demand, so many households are reluctant to apply for subsidies to replace wood-fired heaters for fear of shortages. In the same trend, there is a widespread fear of a lack of technicians with sufficient knowledge to install thermal improvements in houses, so families decide not to choose this option for concern of worsening the situation.

14.3.3.2 Practices affecting adaptive and responsive capacities

Households can respond to deprivation by diversifying energy alternatives to maintain minimum temperatures with access to well-heated spaces and available fuels. In the Argentina case, the children spend much of their time at school, where they also have lunch and have access to proper heating conditions, which they perceive as an opportunity to avoid the low temperatures at home. For Chile, the opposite is the case, with children reporting feeling cold in their schools, which increases the perception of a lack of alternatives to living in low-temperature spaces.

Concerning access to fuels, in situations of deprivation of clean fuels, the supply of lower quality but more available fuel is perceived as a response to heating deprivation, even if its consequences on health quality are detrimental. Households do not necessarily substitute a traditional fuel for a modern one but strategically use multiple fuels and technologies that add to or are complementary to traditional ones. For example, households adopting new technologies with different fuels do not eliminate fuelwood consumption. In rural areas, fuelwood users complement heating by burning cow dung and pruning material.

14.4 Living with heating deprivation, the permanent negotiation

Using the three dimensions of energy poverty to observe practices that affect sensitivity and resilience to heating deprivation has allowed us to recognise how different thresholds shape vulnerability. As shown in Figure 14.3, we observed that the experience of vulnerability to heating deprivation is constantly modulating. We detected this permanent shaping and negotiation by identifying practices that affect sensitivity, adaptive capacity, and response capacity in different dimensions. These practices are multiple, dynamic, situated, and intertwined with varying living conditions. However, we believe it is possible to establish critical aspects that determine vulnerability thresholds in access, equity, and quality dimensions under minimum tolerated temperatures, maximum budget constraints, and minimum redundancy of heating resources, respectively.

Living conditions under these thresholds, in their respective dimensions, are characterised by:

i High exposure to low indoor temperatures (resulting in risks to household members' physical and mental health);
ii Inability to afford energy resources or services, heating, or insulation technologies;
iii Limited access to energy resources, heating technologies, or suitable thermal insulation.

The combination of these living circumstances and the on-going negotiations with them shape how vulnerable a household's residents feel. Focusing on the different dimensions allows us to recognise the problem as a dynamic process shaped by and shaping vulnerability (Eisfeld & Seebauer, 2022; Hargreaves & Longhurst, 2018). Analysing the thresholds denoting the vulnerability conditions in each dimension makes it possible to recognise the nature of these negotiations and their articulation with structural relations. Therefore, we break down what Kearns et al. (2019) have pointed out, "vulnerability as a cause and consequence of occupant behaviour is not only health-related but also social, including the nature of a person's social contacts and support, and their relationship status, particularly where these are volatile" (Kearns et al., 2019).

Both material strategies and cognitive responses characterise coping practices. Recognising the strategies allows for a diagnosis that assesses people's abilities to adapt to energy poverty without depoliticising the structural causes of these conditions (Middlemiss, 2020). Moreover, coping capacities depend on material strategies and attitudes, which are fundamental to sustaining self-esteem and dignity and reaffirming autonomy (Anderson et al., 2012; Clair & Baker, 2022). The "linked lives" effect, associated with the bonding and reciprocal influence between household members, also influences energy use. Thermal regulation practices are consistent with the relationship between household members and are simultaneously shaped by individual needs, solidarity, conflict, and consensus (Bolton et al., 2023). This dynamic configures a micro-level energy system adjusted to the internal logic of the household, to the definition of non-negotiable needs (such as those linked to health), and to ways of thinking and acting inherited from previous generations (Hargreaves & Middlemiss, 2020).

In any case, it is worth mentioning that not all adaptive and response capacities are expressions of resilience. While physical and psychological behaviours and accommodations seem to help cope with the problem, they can be maladaptive. This maladaptation means there are coping strategies that do not fulfil their purpose or have costs that affect other aspects of people's lives. Through maladaptation, habituation to environmental conditions and practices tends to harm well-being, ultimately preserve the condition, and even amplify the vulnerability of people's lives (Cardoso & González, 2019; Cardoso et al., 2022; Juhola et al., 2016). Vulnerability must account for physical and mental health and the economic and territorial contexts inhabited, as its consequences are individual and societal (Kearns et al., 2019). In this sense, orienting public policies towards reducing vulnerabilities by promoting technology and community solutions will help ease pressures and dependencies on environmental variations and global economies.

14.5 Conclusions

We have carried out an exercise to identify how vulnerability is formed from the sensitivity and response capacity to heating deprivation and the thresholds in the different dimensions that constitute energy poverty. Recognising these dynamics is relevant insofar as it allows for prioritising public policy measures and recognising the cases in which the population is even more vulnerable to energy poverty.

Applying a vulnerability lens allows us to understand the "lived experience" of those vulnerable to energy poverty and consequently diagnose its manifestation, which prioritises the statistics and allows for an insight into the problem beyond its techno-economic understanding (Hargreaves & Longhurst, 2018). We recognise the possibility of dialogue between the energy poverty framework and the energy vulnerability framework, which has facilitated this exercise.

The current socio-political crisis in the Southern Cone and the global socio-ecological crisis call for just actions that do not reinforce the contributing causal factors that instigated them in the first place. By observing the sensitivities and resilience that make up the vital experience of vulnerability in contexts of energy poverty, we get clues on how to face the current challenges from a territorially based and right-to-energy approach. This challenges the design of specific actions, from state initiatives to community-based efforts, considering financing mechanisms, co-participation, and innovation. Policies should include the provision of information, agile financing mechanisms, investment in energy infrastructure, targeted subsidies, and the co-participation of users. A just transition that addresses the affordability of energy services must consider vulnerabilities and the context of the environment and territories that shape them.

Notes

1 Which is derived from the plant *Ilex paraguariensis.*
2 Name given for Rubber hot water bottles in Chile.

References

Adger, W.N. (2006). Vulnerability. *Global Environmental Change*, 16(3), 268–281.

Anderson, W., White, V., & Finney, A. (2012). Coping with Low Incomes and Cold Homes. *Energy Policy*, 49, 40–52.

BCN (2021). Reportes comunales.

Bolton, E., Bookbinder, R., Middlemiss, L., Hall, S., Davis, M., & Owen, A. (2023). The Relational Dimensions of Renovation: Implications for Retrofit Policy. *Energy Research & Social Science*, 96, 102916.

Bustos, J.C. (2006). *Características climáticas del campo, Anexo Pilcaniyeu (Río Negro).* San Carlos de Bariloche, Argentina: Instituto Nacional de Tecnología Agropecuaria INTA, No. Comunicación Técnica N° 25.

Calvo, R., Álamos, N., Huneeus, N., & O'Ryan, R. (2022). Energy Poverty Effects on Policy-Based PM2.5 Emissions Mitigation in Southern and Central Chile. *Energy Policy*, 161, 112762.

Cardoso, M.B., & González, A.D. (2019). Residential Energy Transition and Thermal Efficiency in an Arid Environment of Northeast Patagonia, Argentina. *Energy for Sustainable Development*, 50, 82–90.

Cardoso, B., Miranda, A., Berrueta, V., Masera, O., & Gamboa, M. (2022). Longitudinal Analysis and Expected Evolution of Household Fuel and Stove Stacking Patterns in Rural Mexico. *Energy for Sustainable Development*, 70, 1–9.

CDT (2019). *Informe final de usos de la energía de los hogares de Chile 2018.* Chile: Corporación de Desarrollo Tecnológico & In-Data.

Centro de Ciencia del Clima y la Resiliencia (CR)2 (2018). "Guía de referencia para la plataforma de visualización de simulaciones climáticas". Centro de Ciencia del Clima y la Resiliencia (CR)2. (FONDAP 15110009).

Clair, A., & Baker, E. (2022). Cold Homes and Mental Health Harm: Evidence from the UK Household Longitudinal Study. *Social Science & Medicine*, 314, 115461.

Corona, E. (2016). Código de ética para la investigación, la investigación-acción y la colaboración etnocientífica en América latina. *Etnobiología*, 14(1), 33.

Cortés, A., & Amigo, C. (2022). Energy Culture and the Dynamics of Energy Poverty in South Chile: A Blind Spot for Decontamination Energy Efficiency Policies. *People, Place and Policy Online*, 16(1), 73–97.

Eisfeld, K., & Seebauer, S. (2022). The Energy Austerity Pitfall: Linking Hidden Energy Poverty With Self-Restriction in Household Use in Austria. *Energy Research & Social Science*, 84, 102427.

Estévez, A., Serrone, S., & Molinaroli, G. (2011). Vulnerabilidad psicosocial, resilencia y trauma: breve revisión conceptual. Presented at the III Congreso Internacional de Investigación de la Facultad de Psicología, La Plata.

González-Eguino, M. (2015). Energy Poverty: An Overview. *Renewable and Sustainable Energy Reviews*, 47, 377–385.

Gúber, R. (2004). *El salvaje metropolitano: reconstrucción del conocimiento social en el trabajo de campo.* 1a. ed. en Editorial Paidós. Buenos Aires: Paidós.

Hargreaves, T., & Longhurst, N. (2018). *The Lived Experience of Energy Vulnerability Among Social Housing Tenants: Emotional and Subjective Engagements.* Norwich, UK: Centre for Competition Policy, University of East Anglia, Working paper No. 2018–07.

Hargreaves, T., & Middlemiss, L. (2020). The Importance of Social Relations in Shaping Energy Demand. *Nature Energy*, 5(3), 195–201.

INE (2021). Conozca cuáles son las comunas 100% urbanas y 100% rurales del país [online]. INE – Instituto Nacional de Estadísticas. Available from: https://www.ine.gob.cl/sala-de-prensa/prensa/general/noticia/2021/09/13/conozca-cu%C3%A1les-son-las-comunas-100-urbanas-y-100-rurales-del-pa%C3%ADs [Accessed 3 February 2023].

IPCC (2022). Climate Change 2022: Impacts, Adaptation and Vulnerability. *Contribution of Working Group II to the Sixth Assessment Report of the Intergovernmental Panel on Climate Change*. Cambridge, UK and New York, NY: Cambridge University Press.

Juhola, S., Glaas, E., Linnér, B.-O., & Neset, T.-S. (2016). Redefining Maladaptation. *Environmental Science & Policy*, 55, 135–140.

Kearns, A., Whitley, E., & Curl, A. (2019). Occupant Behaviour as a Fourth Driver of Fuel Poverty (aka Warmth & Energy Deprivation). *Energy Policy*, 129, 1143–1155.

Mertens, M., Hermans, L., Van Oosterwijck, J., Meert, L., Crombez, G., Struyf, F., & Meeus, M. (2020). The Result of Acute Induced Psychosocial Stress on Pain Sensitivity and Modulation in Healthy People. *Pain Physician*, 23(6), E703–E712.

Middlemiss, L. (2020). Energy Poverty: Understanding and Addressing Systemic Inequalities. In: Galvin, R. (ed.) *Inequality and Energy*. Cambridge, Massachusetts: Academic Press. pp. 99–114.

Middlemiss, L., & Gillard, R. (2015). Fuel Poverty from the Bottom-up: Characterising Household Energy Vulnerability Through the Lived Experience of the Fuel Poor. *Energy Research & Social Science*, 6, 146–154.

Navarro-Espinosa, A., & Thomas-Galán, M. (2023). Firewood Electrification in Chile: Effects on Household Expenditure and Energy Poverty. *Energy Policy*, 173, 113337.

ONU (2007). Words into Action: A Guide for Implementing the Hyogo Framework. United Nations International Strategy for Disaster Reduction (UNISDR).

Pérez-Fargallo, A., Bienvenido-Huertas, D., Rubio-Bellido, C., & Trebilcock, M. (2020). Energy Poverty Risk Mapping Methodology Considering the user's Thermal Adaptability: The Case of Chile. *Energy for Sustainable Development*, 58, 63–77.

RedPE (2019). Acceso equitativo a energía de calidad en Chile. Hacia un indicador territorializado y tridimensional de pobreza energética. No. 5.

RedPE (2022). *Una mirada multidimensional a la pobreza energética en Chile*. Santiago, Chile: Red de Pobreza Energética (RedPE) & Generadoras de Chile, No. 1.

Reyes, R., Schueftan, A., Ruiz, C., & González, A.D. (2019). Controlling Air Pollution in a Context of High Energy Poverty Levels in Southern Chile: Clean Air but Colder Houses? *Energy Policy*, 124, 301–311.

Schueftan, A., & González, A.D. (2015). Proposals to Enhance Thermal Efficiency Programs and Air Pollution Control in South-Central Chile. *Energy Policy*, 79, 48–57.

Schueftan, A., Sommerhoff, J., & González, A.D. (2016). Firewood Demand and Energy Policy in South-Central Chile. *Energy for Sustainable Development*, 33, 26–35.

Strauss, A., & Corbin, J. (1998). *Basics of Qualitative Research: Techniques and Procedures for Developing Grounded Theory*, 2nd ed. Thousand Oaks, CA: Sage Publications, Inc.

Thomas, K., Hardy, R.D., Lazrus, H., Mendez, M., Orlove, B., Rivera-Collazo, I., Roberts, J.T., Rockman, M., Warner, B.P., & Winthrop, R. (2019). Explaining Differential Vulnerability to Climate Change: A Social Science Review. *WIREs Climate Change*, 10 (2), 1–18.

Urquiza, A., & Billi, M. (2020). Seguridad hídrica y energética en América Latina y el Caribe: definición y aproximación territorial para el análisis de brechas y riesgos de la población. CEPAL, | No. LC/TS.2020/138.

Urquiza, A., Amigo, C., Billi, M., Calvo, R., Labraña, J., Oyarzún, T., & Valencia, F. (2019). Quality as a Hidden Dimension of Energy Poverty in Middle-Development Countries. Literature Review and Case Study from Chile. *Energy and Buildings*, 204, 109463.

World Health Organization (2018). *WHO Housing and Health Guidelines*. Geneva: World Health Organization.

15 Female Voices of Energy Deprivation

The Lived Experience of Energy Vulnerable Women in North Macedonia and Austria

Ana Stojilovska and Mariëlle Feenstra

15.1 Introduction

The European Green Deal has opened a new chapter in the EU's energy transition. It aims to transform the process into a socially just one that takes into account disadvantaged locations and groups, including those affected by energy poverty. The Green Deal's most cited reference is its emphasis on leaving "no-one-behind." This concept is continuously being reinterpreted to consider various intersectional elements such as gender, ethnicity/migrant status, and others. Gender has received increased attention in energy poverty and energy justice academic debates. However, it has been disregarded in energy transition policies. The EU climate and energy transition policies, including the Green Deal, are gender-blind (Clancy & Feenstra, 2019; EIGE, 2020; European Commission, 2019). Research shows that women are disproportionately affected by energy poverty and energy injustices (Clancy & Feenstra, 2019; Feenstra & Özerol, 2021). Energy poverty, or the inability to satisfy the energy needs of the household, needs to be studied at an individual level (Robinson, 2019). Disregarding the gender dimension in the context of energy poverty can reinforce existing gender stereotypes (Tirado Herrero, 2020). Furthermore, energy-poor women are not a homogeneous group either (Listo, 2018). Social practice, a routinised type of behaviour involving bodily, mental, and emotional activities (Reckwitz, 2002), can expose gender roles in the household and beyond (Allen, 2022). There is a need to further explore the lived experience (Middlemiss & Gillard, 2015) of energy-poor women in households, to illustrate gendered household practices, energy use, and socio-cultural practices and how they reflect broader societal inequalities. The chapter contributes to developing the application of energy justice to energy poverty through a gender lens to inform about systematic gender inequalities. This is supported by two empirical case studies illustrating the lived experience of energy-vulnerable women and their coping strategies to manage their situation.

Energy injustice, defined as a lack of adequate access to sustainable, affordable, secure, and clean energy (Heffron & McCauley, 2014), has a clear gendered dimension. Women especially are struggling to have access to clean, sustainable, and affordable energy services, but this issue is less prominent in the Global North (Feenstra & Özerol, 2021). In the Global South, energy use is a highly gendered topic following a traditional division of labour that leaves women to attend to cooking and collecting fuelwood with all the associated health implications that arise from these activities (González-Eguino, 2015; Kaygusuz, 2011; Moniruzzaman & Day, 2020; Njenga et al., 2021). The link between energy poverty (or lack of energy access) and gender is global as expressed in the mutual reinforcement of SDG 7 (access to energy) and SDG 5 (gender equality). We observe that in Europe, a growing group of women suffer from energy deprivation, especially in the current and ongoing energy crisis with increasing energy prices for households. For example, not only are women more affected by energy poverty, but they also experience

DOI: 10.4324/9781003408536-18

greater severity of poverty, are prone to more persistent poverty, and face more barriers to lifting themselves out of poverty (Sánchez-Guevara Sánchez et al., 2020). Furthermore, power relations and institutions can discriminate against women, undermining their status and well-being (Pachauri & Rao, 2013). Due to societal, market, and personal lock-ins, households and women, in particular, engage in coping strategies that depend upon maximising energy savings, often underconsumption, and/or restricting oneself regarding other basic needs (Stojilovska et al., 2021). This kind of coping behaviour has adverse health impacts and entails an acceptance of these impacts (Stojilovska et al., 2023). To address and prevent these negative impacts and the widening of existing gender gaps in society, there is a call for policy efforts to be gender-sensitive and consider women's energy usage behaviour, needs, and rights (Chaudhry & Shafiullah, 2021; Räty & Carlsson-Kanyama, 2010; Vogiatzi et al., 2018).

To better understand gendered coping strategies to deal with energy service shortages, as well as the needs, experiences, and behaviours of women facing energy poverty, we approach the topic through an energy justice lens. We incorporate an intersectional gendered perspective, including elements like health, age, and ethnicity/migrant status, with a particular focus on coping mechanisms. We study energy poverty from a gender angle by considering the economic, biological, and socio-cultural perspectives of the issue (Clancy et al., 2017). Furthermore, we are guided by the understanding that studying energy poverty and its intersectionality is a question of recognition energy justice in which the needs of the energy users should be recognised (Feenstra, 2021a). We consider that other socio-demographic variables accompanying gender, such as health, material deprivation, migrant or minority status, and old age, can deepen the state of energy deprivation. We also incorporate the understanding that the low-carbon transition can magnify existing societal challenges and create new inequalities (Bouzarovski & Tirado Herrero, 2017; Clancy & Feenstra, 2019; Middlemiss, 2022) while considering that energy poverty is a result of distributive, recognition, and procedural injustices (Jenkins et al., 2016; Walker & Day, 2012). Finally, we operationalise our engendered approach to studying energy poverty as follows:

- Distributive gendered energy justice assesses how the socio-economic position of women shapes gendered socio-cultural practices.
- Procedural gendered energy justice explores how societal inequalities result in the limited agency of women in household decisions.
- Recognition gendered energy justice looks at how health, age, ethnicity/migrant status, and coping practices multiply the existing vulnerabilities of energy-poor women.
- Hidden gendered coping strategies reflect the societal and household injustices women in energy poverty experience.

This chapter aims to explore the experiences, challenges, and practices in the households of energy-vulnerable women in North Macedonia and Austria and to highlight how these issues are reflective of broader societal gender inequalities. We inspect these issues in two European countries with different path dependencies co-shaping energy poverty but similar gender equality indicators. We argue that energy-vulnerable women in both countries experience multiple inequalities at the household level, reflected in gendered coping strategies and socio-cultural practices. These gendered strategies and practices are informed by the disadvantaged socio-economic position and limited power of women and are magnified by other vulnerabilities such as health, age, and ethnicity. The experiences of female energy poverty at the household level only mirror existing gender inequalities in society.

15.2 Methods and materials

We explore gendered energy injustices at the household level through the lived experience of women in energy poverty in Austria and North Macedonia chosen following a maximum variation sampling (Miles et al., 2014). This means there is a large difference regarding the level of energy poverty as the key variable in the selected countries, as only 1.5% of households are in energy poverty in Austria in 2020 while 23.8% in North Macedonia, based on the EU-SILC indicator of the inability to keep households adequately warm (EUROSTAT, 2022). The difference in the extent of energy poverty can be explained by the post-socialist past of North Macedonia, which left it disadvantaged with low quality dwellings, increasing energy prices during the energy market liberalisation, and wide-spread material deprivation (Stojilovska, 2020). On the other hand, Austria has a strong social welfare system and liberalised energy markets, while material deprivation is low (Stojilovska et al., 2021). Austria and North Macedonia have, however, similar gender equality indicators based on the Global Gender Gap Index 2022 regarding economic participation and opportunity, educational attainment, and the health and survival of women (WEF, 2022). Austria scored better only regarding the political empowerment of women, leading to an overall 21st rank compared to the 69th rank for North Macedonia for 2022 (WEF, 2022). The socialist legacy of North Macedonia, similar to that of Bulgaria, which prioritised full labour participation and universal access to education, may have contributed to its good starting position in terms of gender equality (Feenstra, 2021b), despite its high levels of energy poverty and material deprivation. In sum, although the countries have different energy poverty outcomes and drivers, they illustrate the lived experience of energy-poor women and the conditions that perpetuate it. Finding commonalities in cases with significant differences (Miles et al., 2014), such as regarding the state of energy poverty, can foster the understanding of common experiences and drivers of female energy poverty in both European post-socialist and developed countries.

The case studies combine qualitative and quantitative data. A total of 219 online open-ended interviews with households nationwide in both countries were conducted using Limesurvey, which was sampled using purposive sampling. 300 phone surveys, including open-ended questions, with households in both countries in the capital cities were collected after they were randomly sampled. In both types of data, energy-poor and non-energy-poor households were included. For this chapter, the data were sub-sampled to include only the surveys and interviews given by female respondents, which included both single female and multiple-member households. This resulted in a total of 96 surveys in Vienna, 97 surveys in Skopje, and 37 interviews across North Macedonia. The sub-sampled data contains direct quotes from women in energy poverty, which provide valuable insights into their lived experiences. These personal stories offer a different kind of evidence because they have a different emphasis and can express emotions (Moezzi et al., 2017). The data was analysed following the steps of data condensation, data display, and conclusion drawing (Miles et al., 2014). We carried out a secondary analysis of primary empirical data to find answers to a new research question that differs from the question in the original studies and generate new knowledge (Hinds et al., 1997). That means we analysed the existing data on female respondents to explore gendered energy injustices and related coping strategies. The data was collected in 2017 in both countries, hence reflecting the status quo in lived energy poverty experiences before the COVID-19 pandemic and the energy crisis caused by the war in Ukraine in 2022. Because this data predates these major crisis events, it is reasonable to expect that the number of households in energy poverty has increased since then.

15.3 Results

Following our conceptual approach, this section is divided into four subsections to explore the gendered energy injustices of energy-vulnerable women in Austria and North Macedonia by applying the principles of distributive, procedural, and recognition justice. A separate section explores the hidden (gendered) coping strategies to mitigate energy deprivation. We use direct quotes to support the main findings about gendered energy injustices relevant to both countries.

15.3.1 Distributive gendered energy justice: determined by economic and socio-cultural perspectives

As a result of their socio-economic position in society, and involvement in gendered socio-cultural practices, women are more exposed to energy poverty. The key distributive gendered energy injustices driving female energy poverty in both countries include low incomes, the physical burden of using certain energy carriers, and the caretaker/caregiver role of women.

The link between energy poverty and gender can be explained by the insight that women more often have lower incomes than men, which impacts their ability to afford domestic energy costs. Evidence shows that single female pensioners on a minimal pension in Austria are a vulnerable group (Stojilovska, 2021). For example, one single female pensioner explains her inability to afford basic energy services: "*It is cold, I don't heat in April anymore in order to save, and now it's cold. It's too expensive, and my flat is cold.*" (no. 15/V, Austria). Another single female pensioner on a minimal pension uses fuelwood to keep her costs low, but still doesn't always have optimal warmth: "*I heat upon need when it's cold. I don't heat in the night; I don't like warmth in the night. I have bought a fuelwood stove, I like fuelwood.*" (no 37/V, Austria). The experiences of energy deprivation among single female pensioners receiving a minimal pension can be explained by the fact that they live alone and their minimal pension is very low. More female pensioners are affected by energy poverty due to the greater longevity of women and living alone at pensionable age (Clancy et al., 2017). The minimum pension in Austria is below the poverty line; thus, as a result, pensioners receiving a minimum pension are automatically income-poor (Stojilovska, 2021).

The link between energy poverty and gender also has biological aspects to consider. Elderly women, for instance, face greater challenges than men when it comes to the physical labour needed to cut and carry fuelwood, which limits their ability to use affordable fuel sources like fuelwood independently. An elderly widow in North Macedonia had to abandon using fuelwood due to her difficulties with the physical labour required, and as a result, she ended up using more expensive energy sources. She is using electricity for heating and explains: "*While my husband was alive, we heated with fuelwood, but he has passed away and I cannot chop the wood. I first pay the utility bills. I just paid and I'm left with 2000-3000 denar (33-49 EUR) for food (till the end of the month).*" (no.135/S, North Macedonia). Similar findings about the link between female biology and energy poverty in the literature show that women are more sensitive to ambient temperatures, especially when ageing because the ability of their bodies to self-regulate temperatures decreases (Clancy et al., 2017; Folkerts et al., 2022). Women prefer higher room temperatures than men and feel uncomfortably cold and uncomfortably hot more often than men (Karjalainen, 2007).

The socially constructed gender role of the caretaker can also explain how women are more exposed to energy poverty. In the capacity of this traditionally imposed domestic role of caretaker (Measham & Allen, 1994; Watkins, 2006), we found evidence of women sacrificing their comfort and health to keep their energy bills low. A retired woman living with her son uses electricity for heating: "*I turn on and off (the heating). During the day when I'm alone, I turn it on only for half an hour-hour, and when it's very cold. I need a higher temperature, I am old. It's difficult to pay for the electricity for heating. Therefore, we economise. It does not pay off to heat all the time when I'm alone. We heat more in the evening when we are both home with my son.*" (no. 95/S, North Macedonia). Research has shown that single-mother households living with children in Europe often lack sufficient financial resources, which makes them particularly vulnerable to energy poverty (Clancy et al., 2017). This is even observed when their child is an adult too but still lives with the mother.

15.3.2 Procedural gendered energy justice: reflected in the limited capacity to participate in household energy decision-making

Women are more affected by energy poverty due to a lack of sufficient resources. As a result, women have limited decision-making power or have experienced limited independence in pursuing their energy preferences. Procedural gendered energy injustices that contribute to female energy poverty include the limited agency of women to decide about energy matters in the household and their financial dependence on family members.

Some women have less decision-making power in the household regarding energy matters and depend on their male counterparts even when that means that they make a compromise and do not pursue their own energy needs. For example, an interviewee explains how she feels cold at home while her husband isn't: "*For me it's cold, but for my husband it's ok. This is a family question. I'm a bit cold. Personally, I need a higher temperature, but my husband wants to keep it lower. My husband takes care of financial matters.*" (no.138/V, Austria). It is well documented that women have been disadvantaged in the job market due to lower employment rates, part-time employment, provision of unpaid care, and even precarious jobs, resulting in lower incomes (Bouzarovski, 2022; EIGE, 2020; Robinson, 2019). These conditions can limit the position of women in the household regarding decisions with financial implications.

In some cases, women rely on their family or social circle for financial support or to meet their basic energy needs. For example, a single pensioner on a minimal pension depends on her daughter's help to use fuelwood: "*Fuelwood is cheap and I get it from friends. My daughter carries the fuelwood from the basement.*" (no. 38/V, Austria). Another interviewee, a single elderly woman without completed primary education, depends on the help of her grandchildren: "*I used to live in very poor housing. My nephew made two rooms in 2012 in which I live since. I don't have any income at all, I don't have a pension. My nephews pay for all of my costs. They supply and chop the fuelwood. I economise in order not to cause them costs.*" (no. 27/S, North Macedonia). In another example, a single female pensioner is dependent on her daughter to keep herself warm when she is ill: "*Last winter I was sick and I've spent part of the winter at my daughter's place and I finally got warm. Every cloud has a silver lining.*" (no. 62/S, North Macedonia). These examples illustrate how women within a family employ gendered coping strategies, which can be emotionally draining as women bear the burden of rationing energy use (Petrova & Simcock, 2019).

15.3.3 Recognition gendered energy justice: amplified by multiple vulnerabilities

Female energy poverty is amplified by the presence of poor health, old age, minority or migrant status, and coping practices. Observed recognition gendered energy injustices refer to the lack of recognition that women living in energy poverty often experience health issues, as their situation is often even more severe due to other vulnerabilities, such as illness, old age, being a migrant, or being a minority.

Gender, health, and energy poverty are closely linked (Aristondo & Onaindia, 2018; Lacroix & Chaton, 2015; Stojilovska et al., 2021). We found evidence that women in energy poverty experience health deterioration directly linked to the use of certain fuels or the lack of sufficient warmth. One elderly woman in a multi-member family had health implications as a result of using fuelwood to make fire and therefore had to abandon it: "*We would make the fire, she (the pensioner) would only add (logs) ... but she got a hernia (from lifting the logs). This was torture and my husband could not manage it anymore, he got back pain, so we said we would go for pellets.*" (no. 1/I/MK, North Macedonia). Another pensioner shared that she experiences adverse health effects as a result of using electric heating: "*The air is dry and it causes issues with my eyes, I have a dry eye.*" (no. 61/S, North Macedonia). A third respondent blames the exposure to cold indoor temperatures for having consequences for her health: "*Last winter I was freezing. But because of (the amount of) my pension, I have to economise. I have rheumatism.*" (no. 98/S, North Macedonia). Rheumatism can be exacerbated by living in energy poverty, for example, by exposure to cold indoor environments (Oliveras et al., 2021). These examples showing the harmful health impacts of energy poverty on women have also been documented elsewhere in Europe. Clancy et al. (2017) found that even in Europe, women are exposed to smoke from domestic fuelwood use due to their major responsibility for cooking.

People with illness and disability are at a higher risk of experiencing energy poverty (Healy & Clinch, 2004; Snell et al., 2015). It is not only that energy poverty contributes to adverse health outcomes, but pre-existing health conditions can increase the likelihood of being in energy poverty because some illnesses require high room temperatures, while others require dependence on electricity to power medical equipment (Simcock et al., 2021b). Having an illness that requires a warmer indoor temperature, can put the individual at either additional financial exposure or the risk of further health deterioration. A single pensioner explains: "*I am ill and I need a lot of warmth. I would prefer to have it warmer because of the illness, but that would be too expensive for me.*" (no. 88/V, Austria). Another pensioner living with her husband also prefers a higher indoor temperature due to health concerns but cannot afford it: "*We need it to be warmer, we are elderly. I have rheumatism. It's sad pensioners with higher education to live with restrictions.*" (no. 61/S, North Macedonia). Hence, their dependency on energy services is higher with more limited income to afford them.

There is a strong link between old age, health, and energy poverty (Shortt & Rugkåsa, 2007; Wright, 2004). Elderly women are significantly more vulnerable to energy poverty (EIGE, 2016). Similarly, being a minority or a migrant can additionally increase the likelihood of being in energy poverty (Bouzarovski et al., 2022; Stojilovska, 2021). Elderly women report that they need to juggle between affording energy bills, medicine, and food. A pensioner with her husband using electricity for heating explains their struggles: "*First we pay the electricity bills in order not to get disconnected, on the second place is medicine, and at the end, we manage somehow for food. I use the washing machine at night and on Sunday when there is a cheap electricity tariff.*" (no.5/S, North Macedonia). A single female pensioner of Roma ethnicity adds: "*What to prioritise (for paying) – bills or fuelwood or food or medicine?*" (no. 121/S, North Macedonia). Elderly women are more often in energy poverty due to their income gap compared to elderly men and also the demographic fact that women, in general, are living longer than men (Clancy et al., 2017). In addition to this, there is an increased need for warmth among the elderly due to an age-related decline in body temperature regulation and higher costs for necessities such as medicine. The situation requires developing coping strategies. Research suggests that the elderly in England experiencing energy poverty, for example, tend to use warmer clothes and economise on heating to afford their heating needs (Chard & Walker, 2016). Overall, the socio-economic and socio-cultural conditions that make women more energy vulnerable are doubled or tripled if combined with other intersecting vulnerabilities, such as old age, minority background, single-parent status, or disability (EIGE, 2020).

15.3.4 Coping strategies: the gendered way of dealing with material deprivation is invisible

The gendered coping strategies, which are the practices women undertake to deal with energy poverty, are deemed invisible, hidden, or unrecognised, although they mirror the societal and household injustices women in energy poverty experience. We have identified that the hidden gendered coping strategies include economising energy when performing gendered household tasks and gendered caretaking responsibilities. In addition, denying being in energy poverty is a hidden gendered coping strategy that can have mental health impacts.

Energy-poor women are heavily engaged in coping strategies aimed at self-restriction and minimising energy and other basic needs, especially regarding energy costs that they control, such as those while performing gendered household tasks. The aim is to be able to afford the energy bills, which includes having them below a certain threshold or prioritising their payment. Cooking, cleaning, and washing are more common responsibilities for women. Therefore, many women use electricity during the cheap electricity tariff to reduce costs that can increase their workload (Carlsson-Kanyama & Lindén, 2007). An example of the use of a cheap electricity tariff: "*Whenever possible, I use cheap electricity tariff for all chores needing electricity, such as washing or similar.*" (no. 50/S, North Macedonia). The interviewed women reported that they use domestic energy services, often in areas of care traditionally associated with women, to maximise their coping activities. In the literature, we can observe that some household chores, such as cooking, are viewed by women as invested in maternal emotion, and a central component of their childcare (Kaplan et al., 2020).

As caretakers, women take it upon themselves to protect the elderly or children from being exposed to cold homes or try to avoid a greater threat to themselves and the family, such as losing their home (if renting, e.g., in Vienna) or experiencing disconnection (which comes with additional costs and fears for dealing with a monopoly utility, such as in Skopje). This makes the respondents involved partake in an interplay of various coping strategies with other conflicting goals, such as satisfying the need for warmth and keeping energy costs low. A single female pensioner in Vienna of Hungarian background on a minimal pension explains that affording energy services is a must: "*One has to afford these things, they will become more expensive, so I economise on other things.*" (no. 71/V, Austria). An Albanian woman in a family with three children who use electric heating and can barely satisfy their basic needs clarifies the need to take care of her children: "*I cannot economise (the heating) because I have small children.*" (no. 3/MK, North Macedonia). In these coping practices by women, we can observe the element of carrying for and about the household, even at the expense of their own discomfort. The invisible work of care and household chores is socially constructed and expected of women, mothers, and grandmothers (Kaplan et al., 2020). The combination of the cognitive labour of family life – the thinking and organising of family members – and the emotional labour associated with this work, including the feelings of caring for and being responsible for family members, is creating an adverse mental load for women (Dean et al., 2022). This signals that governments should invest in caregiving infrastructure to reduce competing work and care demands (Dean et al., 2022).

Women often do not report that they are living in energy poverty but rather practice their hidden gendered coping strategies to manage the conditions of energy deprivation. In Figures 15.1 and 15.2, we can observe some of the most relevant coping strategies used by female respondents in Skopje and Vienna. These coping strategies are compared to whether

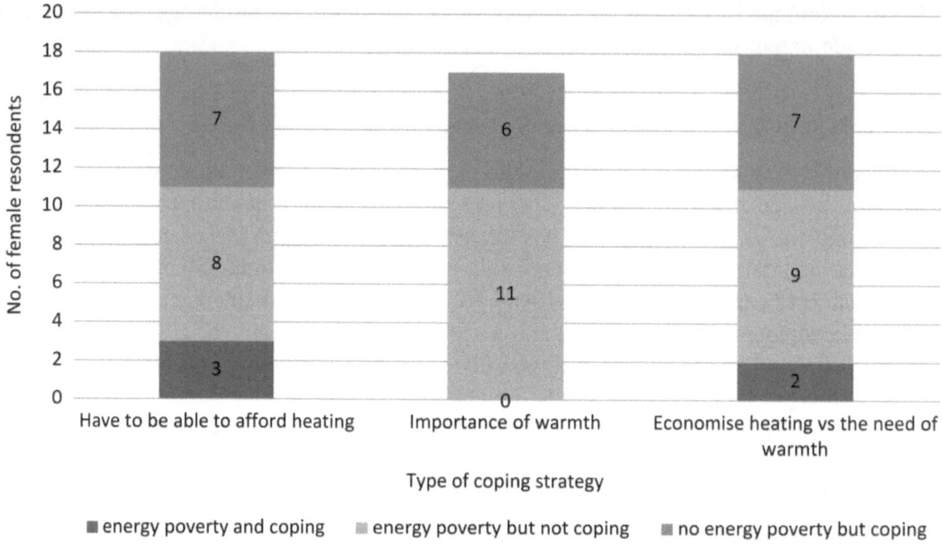

Figure 15.1 Interplay of energy poverty and coping strategies of female respondents in Vienna.
Source: Vienna survey dataset developed from (Stojilovska, 2021).

the female respondents self-report being in energy poverty, measured through their inability to keep their dwelling adequately warm. Both female respondents in Skopje and Vienna commonly prioritise warmth as a coping strategy. In the case of Vienna, female responders have indicated that it is as relevant to give priority to affording heat and express the tension arising between economising on heating and the need for warmth. In Skopje, female respondents reported tension between staying warm and affording the heating (pay vs. warmth tension);

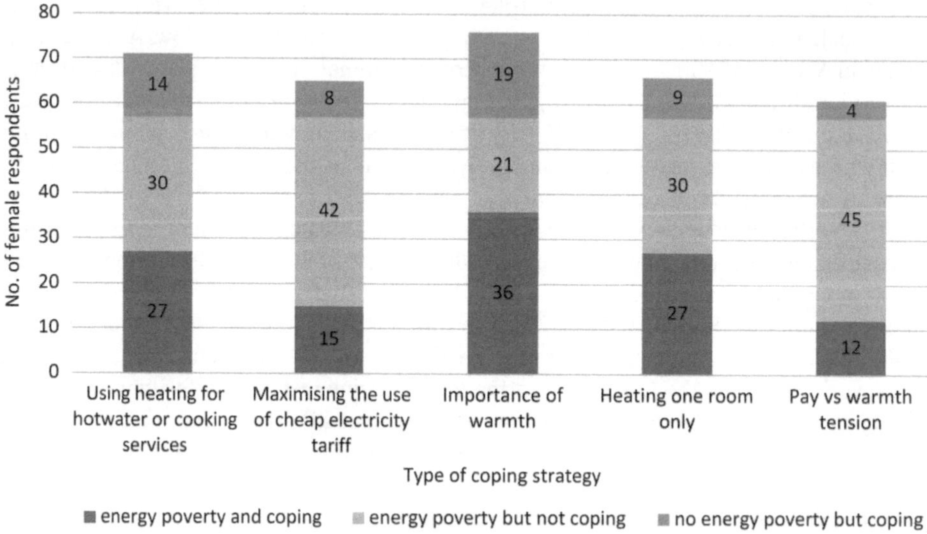

Figure 15.2 Interplay of energy poverty and coping strategies of female respondents in Skopje.
Source: Skopje survey dataset developed from (Stojilovska, 2021).

the use of heating for preparing hot water and cooking; maximising the use of the cheap electricity tariff; and heating one room only. There are three categories represented: women who reported energy poverty, and were engaged in coping, women who reported energy poverty but did not engage in coping, and the category of hidden energy poverty where women engaged in coping but did not report energy poverty. The phenomenon of hidden energy poverty is more pronounced in Vienna, possibly due to the overrepresentation of single female pensioners who prioritise energy bills over other essential needs. In Skopje, female respondents were more straightforward about their situation, but some were satisfied with the ability to minimise the energy costs related to their domestic duties. Figure 15.1 suggests that coping strategies in Vienna give female respondents a sense of control over their energy expenditures, allowing them to conceal energy poverty. Conversely, those who feel energy poor are less likely to be involved in coping mechanisms. In the case of Skopje in Figure 15.2, coping is an accompanying practice of co-experiencing energy poverty; however, not all energy-poor women are involved in coping strategies. Depending on the coping strategy, 6-25% of female respondents in energy poverty in Skopje do not acknowledge their energy poverty despite engaging in coping strategies that restrict energy usage and comfort.

These hidden coping strategies can be seen as gendered social practices aimed at managing material deprivation while performing gendered tasks such as managing household responsibilities and maintaining a sense of control over one's situation. Women experience a tension between satisfying their energy needs to carry out their domestic responsibilities and being able to afford them. This brings (female) energy poverty closer to material deprivation (Watson & Maitre, 2014). Experiencing energy poverty is linked to poorer mental health (Ballesteros-Arjona et al., 2022). This is because coping with energy poverty can adversely affect mental health (Mould & Baker, 2017). For instance, managing energy arrears can lead to anxiety (Fabbri, 2015). There is a stigma attached to those living in energy poverty, often expressed as embracement or shame (Hards, 2013; Longhurst & Hargreaves, 2019). Shame about energy poverty can lead individuals to keep their situation hidden and make coping strategies invisible. This may be explained by the underlying belief that deprivation is a result of the poor choices of individuals or households (Simcock et al., 2021a). Having to financially prioritise domestic energy services can leave women in energy poverty at risk of social exclusion as they may avoid social activities that may incur costs, such as inviting guests or other socialising activities (Maxim et al., 2016; Middlemiss et al., 2019).

15.4 Conclusions

We illustrated the experiences, challenges, and practices in the households of energy-vulnerable women in North Macedonia and Austria as a reflection of broader societal gender inequalities based on their lived experience. We argue that energy-vulnerable women in Austria and North Macedonia experience multiple inequalities at the household level reflected in gendered coping strategies and socio-cultural practices. These gendered strategies and practices are informed by the disadvantaged socio-economic position and limited power experienced by the majority of women and are magnified by other vulnerabilities such as health, age, and ethnicity. Female energy poverty at the household level only mirrors the existing gender inequalities in society.

We found similar gendered energy injustices that co-create energy poverty in a post-socialist country with high energy poverty, North Macedonia, and in a developed European country with a good standard of living, Austria. Although the extent of female energy poverty differs in the two case studies, we found similar experiences of distributive, procedural, and recognition gendered energy injustices, as well as hidden gendered coping strategies.

Women experience energy poverty due to low incomes, the physical burden of using outdated but affordable energy carriers, and their emotional and cognitive caretaker/caregiver role in the family. Women in energy poverty face limitations and dependence when it comes to making decisions about managing energy in their homes. The respondents are exposed to adverse health impacts as a result of living in energy poverty. If they are old, ill, minority, or migrant, energy-poor women's experiences of energy poverty are multiplied. Finally, gendered coping strategies often involve self-restriction and substandard satisfaction of basic needs. These strategies may not always be recognised as material deprivation by women and are thus rendered invisible. However, these strategies help women persist through their circumstances while performing gendered household tasks and caring for the family.

The energy transition provides an opportunity to align multiple social, health, and environmental policies with efforts to alleviate energy poverty (Stojilovska et al., 2022). To achieve a socially just energy transition, there is a need to recognise the female face of energy poverty. Exploring the often silenced stories of women facing energy poverty adds value to understanding the complexities of gendered material deprivation and broader domestic and societal inequalities. The case studies explored a pre-energy crisis context, and given the increasing energy prices and inflation across Europe, it is likely that the extent and severity of energy poverty and coping strategies have expanded. Finally, to render a socially just gendered approach to addressing energy poverty, it requires supporting the fair distribution of resources, tools, and opportunities for women, the meaningful participation of women and other vulnerable groups in the energy transition, and the recognition of gender as an indicator of social inequality.

References

Allen, I. K. (2022). Heated Attachments to Coal: Everyday Industrial Breadwinning Petro-Masculinity and Domestic Heating in the Silesian Home. In: Iwińska, K. & Bukowska, X. (eds.), *Gender and Energy Transition: Case Studies from the Upper Silesia Coal-Mining Region*. Cham: Springer International Publishing.

Aristondo, O., & Onaindia, E. (2018). Inequality of Energy Poverty between Groups in Spain. *Energy*, 153, 431–442.

Ballesteros-Arjona, V., Oliveras, L., Bolívar Muñoz, J., Olry de Labry Lima, A., Carrere, J., Martín Ruiz, E., Peralta, A., Cabrera León, A., Mateo Rodríguez, I., Daponte-Codina, A., & Marí-Dell'Olmo, M. (2022). What Are the Effects of Energy Poverty and Interventions to Ameliorate It on people's Health and Well-Being?: A Scoping Review with an Equity Lens. *Energy Research & Social Science*, 87, 102456.

Bouzarovski, S. (2022). Energy and Labour: Thinking Across the Continuum. *Progress in Human Geography*, 0, 03091325211051478.

Bouzarovski, S., Burbidge, M., Sarpotdar, A., & Martiskainen, M. (2022). The Diversity Penalty: Domestic Energy Injustice and Ethnic Minorities in the United Kingdom. *Energy Research & Social Science*, 91, 102716.

Bouzarovski, S., & Tirado Herrero, S. (2017). The Energy Divide: Integrating Energy Transitions, Regional Inequalities and Poverty Trends in the European Union. *European Urban and Regional Studies*, 24, 69–86.

Carlsson-Kanyama, A., & Lindén, A.-L. (2007). Energy Efficiency in Residences—Challenges for Women and Men in the North. *Energy Policy*, 35, 2163–2172.

Chard, R., & Walker, G. (2016). Living with Fuel Poverty in Older Age: Coping Strategies and Their Problematic Implications. *Energy Research & Social Science*, 18, 62–70.

Chaudhry, S. M., & Shafiullah, M. (2021). Does Culture Affect Energy Poverty? Evidence from a Cross-Country Analysis. *Energy Economics*, 102, 105536.

Clancy, J., Daskalova, V., Feenstra, M., Franceschelli, N., & Sanz, M. (2017). Gender perspective on access to energy in the EU. Brussels, European Parliament. Available from: https://www.europarl.europa.eu/RegData/etudes/STUD/2017/596816/IPOL_STU(2017)596816_EN.pdf

Clancy, J., & Feenstra, M. (2019). Women, Gender Equality and the Energy Transition in the EU. Brussels, European Parliament. Available from: https://www.europarl.europa.eu/RegData/etudes/STUD/2019/608867/IPOL_STU(2019)608867_EN.pdf

Dean, L., Churchill, B., & Ruppanner, L. (2022). The Mental Load: Building a Deeper Theoretical Understanding of How Cognitive and Emotional Labor Overload Women and Mothers. *Community, Work & Family*, 25, 13–29.

EIGE (2016). Gender and Energy. Available from: https://eige.europa.eu/sites/default/files/documents/ti_pubpdf_mh0116288enn_pdfweb_20170113163501.pdf

EIGE (2020). Beijing + 25: the fifth review of the implementation of the Beijing Platform for Action in the EU Member States. Available from: https://eige.europa.eu/publications/beijing-25-fifth-review-implementation-beijing-platform-action-eu-member-states

European Commission (2019). Communication from the Commission to the European Parliament, the European Council, the Council, the European Economic and Social Committee and the Committee of the Regions. The European Green Deal. Available from: https://eur-lex.europa.eu/resource.html?uri=cellar:b828d165-1c22-11ea-8c1f-01aa75ed71a1.0002.02/DOC_1&format=PDF

EUROSTAT (2022). Inability to keep home adequately warm – EU-SILC survey. Available from: https://ec.europa.eu/eurostat/databrowser/view/ilc_mdes01/default/table?lang=en

Fabbri, K. (2015). Building and Fuel Poverty, an index to Measure Fuel Poverty: An Italian Case Study. *Energy*, 89, 244–258.

Feenstra, M. (2021a). *Gender Just Energy Policy: Engendering the Energy Transition in Europe*. Enschede: University of Twente.

Feenstra, M. (2021b). Women as Change Agents of the Bulgarian Energy Transition. In: Mišík, M. & Oravcová, V. (eds.), *From Economic to Energy Transition Three Decades of Transitions in Central and Eastern Europe*. Switzerland: Springer Nature.

Feenstra, M., & Özerol, G. (2021). Energy Justice as a Search Light for Gender-Energy Nexus: Towards a Conceptual Framework. *Renewable and Sustainable Energy Reviews*, 138, 110668.

Folkerts, M. A., Bröde, P., Botzen, W. J. W., Martinius, M. L., Gerrett, N., Harmsen, C. N., & Daanen, H. A. (2022). Sex Differences in Temperature-Related All-Cause Mortality in the Netherlands. *International Archives of Occupational and Environmental Health*, 95, 249–258.

González-Eguino, M. (2015). Energy Poverty: An Overview. *Renewable and Sustainable Energy Reviews*, 47, 377–385.

Hards, S. K. (2013). Status, Stigma and Energy Practices in the Home. *Local Environment*, 18, 438–454.

Healy, J. D., & Clinch, J. P. (2004). Quantifying the Severity of Fuel Poverty, Its Relationship with Poor Housing and Reasons for Non-Investment in Energy-Saving Measures in Ireland. *Energy Policy*, 32, 207–220.

Heffron, R. J., & McCauley, D. (2014). Achieving Sustainable Supply Chains Through Energy Justice. *Applied Energy*, 123, 435–437.

Hinds, P. S., Vogel, R. J., & Clarke-Steffen, L. (1997). The Possibilities and Pitfalls of Doing a Secondary Analysis of a Qualitative Data Set. *Qualitative Health Research*, 7, 408–424.

Jenkins, K., McCauley, D., Heffron, R., Stephan, H., & Rehner, R. (2016). Energy Justice: A Conceptual Review. *Energy Research & Social Science*, 11, 174–182.

Kaplan, A., Sabbah-Karkabi, M., & Herzog, H. (2020). "When I Iron My Son's Shirt, I Feel My Maternal Role": Making Women's Invisible Work Visible. *Journal of Family Issues*, 41, 1525–1545.

Karjalainen, S. (2007). Gender Differences in Thermal Comfort and Use of Thermostats in Everyday Thermal Environments. *Building and Environment*, 42, 1594–1603.

Kaygusuz, K. (2011). Energy Services and Energy Poverty for Sustainable Rural Development. *Renewable and Sustainable Energy Reviews*, 15, 936–947.

Lacroix, E., & Chaton, C. (2015). Fuel Poverty as a Major Determinant of Perceived Health: the Case of France. *Public Health*, 129, 517–524.

Listo, R. (2018). Gender Myths in Energy Poverty Literature: A Critical Discourse Analysis. *Energy Research & Social Science*, 38, 9–18.

Longhurst, N., & Hargreaves, T. (2019). Emotions and Fuel Poverty: The Lived Experience of Social Housing Tenants in the United Kingdom. *Energy Research & Social Science*, 56, 101207.

Maxim, A., Mihai, C., Apostoaie, C.-M., Popescu, C., Istrate, C., & Bostan, I. (2016). Implications and Measurement of Energy Poverty Across the European Union. *Sustainability*, 8, 483.

Measham, F., & Allen, S. (1994). In Defence of Home and Hearth? Families, Friendships and Feminism in Mining Communities. *Journal of Gender Studies*, 3, 31–45.

Middlemiss, L. (2022). Who Is Vulnerable to Energy Poverty in the Global North, and What Is Their Experience? *WIREs Energy and Environment*, 11, e455.

Middlemiss, L., Ambrosio-Albalá, P., Emmel, N., Gillard, R., Gilbertson, J., Hargreaves, T., Mullen, C., Ryan, T., Snell, C., & Tod, A. (2019). Energy Poverty and Social Relations: A Capabilities Approach. *Energy Research & Social Science*, 55, 227–235.

Middlemiss, L., & Gillard, R. (2015). Fuel Poverty from the Bottom-up: Characterising Household Energy Vulnerability Through the Lived Experience of the Fuel Poor. *Energy Research & Social Science*, 6, 146–154.

Miles, M. B., Huberman, A. M., & Saldana, J. (2014). *Qualitative Data Analysis: a Methods Sourcebook*. Thousand Oaks, CA: Sage Publications.

Moezzi, M., Janda, K. B., & Rotmann, S. (2017). Using Stories, Narratives, and Storytelling in Energy and Climate Change Research. *Energy Research & Social Science*, 31, 1–10.

Moniruzzaman, M., & Day, R. (2020). Gendered Energy Poverty and Energy Justice in Rural Bangladesh. *Energy Policy*, 144, 111554.

Mould, R., & Baker, K. J. (2017). Documenting Fuel Poverty from the householders' Perspective. *Energy Research & Social Science*, 31, 21–31.

Njenga, M., Gitau, J. K., & Mendum, R. (2021). Women's Work Is Never Done: Lifting the Gendered Burden of Firewood Collection and Household Energy Use in Kenya. *Energy Research & Social Science*, 77, 102071.

Oliveras, L., Peralta, A., Palència, L., Gotsens, M., López, M. J., Artazcoz, L., Borrell, C., & Marí-Dell'Olmo, M. (2021). Energy Poverty and Health: Trends in the European Union Before and During the Economic Crisis, 2007–2016. *Health & Place*, 67, 102294.

Pachauri, S., & Rao, N. D. (2013). Gender Impacts and Determinants of Energy Poverty: Are We Asking the Right Questions? *Current Opinion in Environmental Sustainability*, 5, 205–215.

Petrova, S., & Simcock, N. (2019). Gender and Energy: Domestic Inequities Reconsidered. *Social & Cultural Geography*, 1–19.

Räty, R., & Carlsson-Kanyama, A. (2010). Energy Consumption by Gender in Some European Countries. *Energy Policy*, 38, 646–649.

Reckwitz, A. (2002). Toward a Theory of Social Practices: A Development in Culturalist Theorizing. *European Journal of Social Theory*, 5, 243–263.

Robinson, C. (2019). Energy Poverty and Gender in England: A Spatial Perspective. *Geoforum*, 104, 222–233.

Sánchez-Guevara Sánchez, C., Sanz Fernández, A., & Núñez Peiró, M. (2020). Feminisation of Energy Poverty in the City of Madrid. *Energy and Buildings*, 223, 110157.

Shortt, N., & Rugkåsa, J. (2007). "The Walls Were so Damp and Cold" Fuel Poverty and Ill Health in Northern Ireland: Results from a Housing Intervention. *Health & Place*, 13, 99–110.

Simcock, N., Frankowski, J., & Bouzarovski, S. (2021a). Rendered Invisible: Institutional Misrecognition and the Reproduction of Energy Poverty. *Geoforum*, 124, 1–9.

Simcock, N., Jenkins, K. E. H., Lacey-Barnacle, M., Martiskainen, M., Mattioli, G., & Hopkins, D. (2021b). Identifying Double Energy Vulnerability: A Systematic and Narrative Review of Groups at-Risk of Energy and Transport Poverty in the Global North. *Energy Research & Social Science*, 102351.

Snell, C., Bevan, M., & Thomson, H. (2015). Justice, Fuel Poverty and Disabled People in England. *Energy Research & Social Science*, 10, 123–132.

Stojilovska, A. (2020). Energy Poverty in a Subsistence-Like Economy: The Case of North Macedonia. In: Jiglau, G., Sinea, A., Dubois, U., & Biermann, P. (eds.), *Perspectives on Energy Poverty in Post-Communist Europe*. Routledge.

Stojilovska, A. (2021). *Synergies between Heating and Energy Poverty – The Injustice of Heat*. PhD, Budapest, Central European University.

Stojilovska, A., Dokupilová, D., Gouveia, J. P., Bajomi, A. Z., Tirado-Herrero, S., Feldmár, N., Kyprianou, I., & Feenstra, M. (2023). As Essential as Bread: Fuelwood Use as a Cultural Practice to Cope with Energy Poverty in Europe. *Energy Research & Social Science*, 97, 102987.

Stojilovska, A., Guyet, R., Mahoney, K., Gouveia, J. P., Castaño-Rosa, R., Živčič, L., Barbosa, R., & Tkalec, T. (2022). Energy Poverty and Emerging Debates: Beyond the Traditional Triangle of Energy Poverty Drivers. *Energy Policy*, 169, 113181.

Stojilovska, A., Yoon, H., & Robert, C. (2021). Out of the Margins, into the Light: Exploring Energy Poverty and Household Coping Strategies in Austria, North Macedonia, France, and Spain. *Energy Research & Social Science*, 82, 102279.

Tirado Herrero, S. (2020). Review of previous EU experiences Assessment from an energy poverty, gender, health and Mediterranean perspective. EmpowerMed. Available from: https://www.empowermed.eu/wp-content/uploads/2020/07/D1.3-Report-on-previous-experiences-Final-for-Web.pdf

Vogiatzi, C., Gemenetzi, G., Massou, L., Poulopoulos, S., Papaefthimiou, S., & Zervas, E. (2018). Energy Use and Saving in Residential Sector and Occupant Behavior: A Case Study in Athens. *Energy and Buildings*, 181, 1–9.

Walker, G., & Day, R. (2012). Fuel Poverty as Injustice: Integrating Distribution, Recognition and Procedure in the Struggle for Affordable Warmth. *Energy Policy*, 49, 69–75.

Watkins, H. (2006). Beauty Queen, Bulletin Board and Browser: Rescripting the Refrigerator. *Gender, Place & Culture*, 13, 143–152.

Watson, D., & Maitre, B. (2014). Is Fuel Poverty in Ireland a Distinct Type of Deprivation? *Economic & Social Review*. http://www.tara.tcd.ie/handle/2262/71985

WEF. (2022). Global Gender Gap Report 2022. Available from: https://www3.weforum.org/docs/WEF_GGGR_2022.pdf

Wright, F. (2004). Old and Cold: Older People and Policies Failing to Address Fuel Poverty. *Social Policy & Administration*, 38, 488–503.

Part III

Practices, Policies, and Sustainability

16 Transitioning Out of Vulnerability Using the Neighborhood Energy Center Model

Alison Kenner, Andrew Rosenthal, Briana Leone, Morgan Sarao, James Adams, and Abby Tzinberg

16.1 Introduction

Neighborhood Energy Centers (NECs) are a Philadelphia-based network of grant-funded offices that help households address energy vulnerability. In operation for decades, NECs have been described as "one-stop-shops" where households can apply for different forms of aid, receive budget counselling, and obtain referrals to other programmes. In this chapter, we explain why the NEC model is particularly well suited to address the structural and multifaceted nature of energy vulnerability, particularly through attempts to get at some of the root causes and intertwined issues in ways that are locally situated.

This study is based on a three-year ethnographic study[1] of energy vulnerability in Philadelphia, Pennsylvania (2019–2021). Data collection included 234 structured interviews between May 2020 and June 2021; interviews focused on subjects' ability to make utility bill payments and receive assistance. This was followed by 101 open-ended interviews in summer 2021. We also conducted participant observation at dozens of energy assistance events during this period, including public energy education workshops, national conferences and regional webinars for professionals working in the energy assistance sector, and monthly administrative meetings for Philadelphia's network of energy counsellors. This was followed by 20 key informant interviews with experts working in the energy assistance sector.

Our findings show that the NEC model is particularly robust because it is situated in micro-local ways (such as within community centres rather than state-level social service organisations), builds specialised expertise among energy counsellors, and leverages a broad network of adjacent programming that can help alleviate socioeconomic stressors that intersect with energy vulnerability. On the other hand, NECs also work within structural constraints that can drastically limit their impact. Some constraints include strict programme requirements that make the application process unforgiving and unable to recognise complex vulnerabilities; increased competition from utility companies looking to expand their own assistance infrastructure; and shifting accessibility needs, for both staff and clients during the pandemic.

In what follows, we first elaborate on the concept of energy vulnerability and what it looks like in Philadelphia. Next, we describe the NEC model, including its current network, application process, and activities. The following section elaborates on some of the above-described constraints that make the work of NECs more challenging and ultimately less effective, often because of the way that these constraints prevent access to assistance and counselling. We spend the final sections looking at what energy expertise looks like, in the context of energy vulnerability, and imagining how this expertise could translate into more robust programming and policy.

DOI: 10.4324/9781003408536-20

16.2 Background

16.2.1 Defining energy vulnerability

Sociologists Rosie Day and Gordon Walker define energy vulnerability as a household's inability to "achieve sufficient access to affordable and reliable energy services, and as a consequence is in danger of harm to health and/or well-being" (2013: 16). The term resembles but is distinct from energy poverty (Bednar & Reames, 2020) and energy insecurity (Hernández, 2016) in its emphasis on the locally contingent and temporal dynamics of access and harm, which are produced at the intersection of social, technological, and ecological spheres. Concretely, energy vulnerability emphasises that households may fall into and out of harmful or insecure conditions based on changes to household composition, acute and chronic health conditions, weather extremes related to climate change, public health crises, and disruptive technologies or policies. Despite its pervasive reach, many people who experience energy vulnerability may not recognise their predicament as such (Bouzarovski et al., 2013), including Philadelphia's own middle-to-low-income families who struggle to meet their utility bills on a monthly basis or periodically. This can make addressing the problem – at the scale of the household and also nationally – all the more difficult.

NECs are particularly well suited to address this phenomenon, as understood in this expanded definition, because of the way they are nested within neighbourhoods, communities, parent organisations, and a well-structured energy servicescape.

16.2.2 Energy vulnerability in Philadelphia

Philadelphia is a late industrial city (Kenner, 2020) that suffers from high rates of household energy burden, defined as the percentage of gross household income spent on energy costs. According to a study conducted by Greenlink Analytics, Philadelphia experiences energy burdens 86% higher than the national average; Philadelphian households spend on average around 6.7% of their income on energy. This phenomenon can be traced in part to the city's aged housing stock and infrastructural disinvestment in redlined neighbourhoods, which have produced widespread household energy inefficiency.[2] According to the US Census, 72% of Philadelphia's current housing stock was built more than 60 years ago, and 40% was built before 1940 (Tooher, 2014). Other factors contributing to household energy burden include high poverty rates in Philadelphia, with many households living on limited or fixed incomes. Research has also shown that households in predominantly Black neighbourhoods also experience higher energy burdens (Drehobl et al., 2020; Hernández, 2016; Jessel et al., 2019; Memmott et al., 2021).

Energy vulnerability in Philadelphia goes beyond energy burden in a number of key ways, however; it stems from the intersection of housing stock, inequitable income structures, and other compounded socioeconomic vulnerabilities, but can also be related to chronic or acute medical conditions (Hernández, 2013, 2016; Kenner et al., 2022), hotter summers due to climate change, regional and international energy politics, and growing technological demands that force households to have an increasing number of devices. Place matters, too, and neighbourhood location shapes households' energy environments (Petrova & Prodromidou, 2019).

16.2.3 Situating energy vulnerability in the welfare state

In the United States, assistance of all kinds, including Temporary Assistance for Needy Families and Supplemental Nutrition Assistance Programs, is tied to work in some form (Hacker, 2016).

Loss of employment can create dual economic vulnerability by exposing a household to the loss of welfare assistance that is critical in post-job loss periods (Falk et al., 2014). While energy assistance programmes such as the Low-Income Household Energy Assistance Program and the Weatherization Assistance Program are not tied to work requirements, they *are* tied to income guidelines (Berman et al., n.d.). To receive energy assistance, vulnerable households become dually dependent on earning a salary but not earning too much – where the state controls how the households spend the welfare assistance received (Daguerre, 2017a, 2017b).

NEC counsellors we interviewed often cited income guidelines as an inherent weakness in the assistance sector, giving voice to critiques that circulate in the scholarly literature (Haveman et al., 2015; Hernández & Bird, 2010; Prasad, 2011).

> All these programs have income requirements and that person, that working poor person, or the person on a fixed income, who their pension may be a little higher than somebody else, who may be a dollar over that income threshold, cannot receive the services.

Government assistance is tied to work here too, but differently than other federal or state programmes. The working poor, as noted by the counsellor above, are often excluded from assistance, even when they clearly need it.

With the widespread recognition that energy assistance programmes are an extremely limited approach to a deeply structural problem, the NEC model is designed to supplement education, resource distribution, weatherization and repair referral, and investment in renewable infrastructure.

16.3 NEC model

Philadelphia's NECs are a critical interstitial layer within the broader system working to address energy inequity, which includes more progressive energy transition efforts, legal and regulatory advocacy, and social services. In 2022, there were seventeen NECs: nine were identified as Community Development Corporations, seven were identified as social service organisations such as food pantries or senior centres, and one NEC provided energy assistance only. All but one NEC are nested within larger organisations that provide other forms of social service, including housing assistance, childcare, food pantries, referral programmes, and educational series; eleven NECs are also City of Philadelphia Neighborhood Advisory Committees – community organisations formally charged with connecting residents to City services. In 2021, some of the organisations that housed NECs also hosted vaccination clinics and heavily campaigned to provide accurate information about public health throughout the COVID-19 pandemic. In short, NECs work as a critical matrix of care for neighbourhoods, where people can access information on many types of governmental and private assistance, as well as other needed resources. The core activities NECs undertake can be divided into three categories: supporting applications for energy assistance, budget counselling, and conservation workshops.

16.3.1 Energy assistance

Energy assistance programmes help reduce energy burdens so that households are more likely to meet their energy needs. In Philadelphia, there are four primary forms of assistance available to households struggling with energy costs; these include federal and local grant programmes, weatherization programmes, and budget programmes made available through the utility companies.

Launched in 1981, LIHEAP (Low-Income Home Energy Assistance Program) is the principal federal energy assistance programme in the United States. In Pennsylvania, LIHEAP benefits help pay off heating costs. Benefits are distributed to households with low incomes (150% or below the Federal Poverty Guidelines, or 60% of the state median income), particularly those that have a high energy burden (percentage of income that goes to heating and cooling bills) and/or have members who are elderly, disabled, and/or have young children. A similarly structured private programme, the Utility Emergency Services Fund, is also available for Philadelphians and can be used to supplement LIHEAP and other forms of financial assistance.

Another way for households to cope with outstanding utility bills is to enrol in their local utility suppliers' assistance programmes. In Philadelphia, PECO (electricity supplier), PGW (gas supplier) (PGW, n.d.-a, n.d.-b), and the Philadelphia Water Department offer utility-sponsored programmes that tie customers to a cycle of bill payment that is meant to prevent their utilities from being shut off.

Utility bills can often increase due to inefficiencies in the home, a fact that was recognised at the federal level in 1976 with the creation of the Weatherization Assistance Program (WAP) (Energy.gov, n.d.; Fowlie et al., 2018). This programme seeks to help households address the previously discussed inefficiencies to help reduce utility bills and create a comfortable indoor living environment (Fowlie et al., 2018).

In addition to positive forms of assistance, the Pennsylvania Public Utility Commission also institutes an annual moratorium on utility shutoffs between December 1st and March 31st. During this period, income-qualified households (at or below 250% of the FPL) with outstanding bills cannot have their gas or electricity shut off between December 1st and March 31st, annually. This is intended to protect limited income households from losing heating during the winter months.

16.3.2 Budget counselling and referrals

When utility customers visit an NEC to learn about or apply for energy assistance, they also receive one-on-one budget counselling. As part of the process of determining whether a household qualifies for the above energy assistance programmes, an energy counsellor uses a budget template to help the client determine monthly expenses, which can be useful for households that are moving into a payment plan contract. This process helps clients see different expenses that may not be visible unless there is a query and things are marked on paper. Financial literacy and budgeting are not a practice formally taught in the United States, and many residents must figure out how to balance expenses without support.

16.3.3 Energy conservation workshops

The last core activity undertaken by NECs is energy education; each winter, staff members from ECA and Philadelphia's NECs teach dozens of energy conservation workshops around the city. These workshops are funded by Philadelphia Gas Works and Grandom Oil and help residents learn strategies to reduce their utility bills. NEC and ECA staff members also periodically facilitate off-season workshops that focus on cooling in the summer months or water conservation. Every workshop participant receives a free weatherization kit, which includes seasonally relevant conservation materials, such as rope and tube caulk, door sweeps, LED light bulbs, a surge protector, and plastic wrap for windows. Workshop facilitators give live demonstrations using kit items; each presentation also includes advice on lowering utility bills, information on federal and local assistance programmes, and the benefits of professional weatherization. The Q&A segments of the workshop allow the facilitator to get into specific details that are relevant to the community in attendance.

16.4 Cultivating energy expertise

The NEC network is managed by the Energy Coordinating Agency (ECA), a 35-year-old nonprofit that has been at the forefront of urban energy innovation. ECA is unique in that it specifically addresses energy inequity derived from structural conditions, such as redlining, lack of career opportunities, and aged housing stock, by providing various forms of energy assistance, weatherization, and green jobs training (Adams et al., 2022). This systems approach is also what sets Philadelphia's NEC network apart from other approaches; as Bednar and Reames argue, part of the problem in the United States is the lack of recognition, at the federal level, that energy poverty (or vulnerability) is a national issue requiring policy attention (2020).

NEC counsellors are trained energy experts. Monthly meetings, hosted by ECA, feature workshops and presentations from experts in different sectors of the energy servicescape. Counsellors bring this information back into their client work, where they are able to make referrals to other programmes and services. Some examples observed during fieldwork include solar energy programmes, community legal services, and basic systems repairs. Additionally, each summer, all NEC counsellors complete a week-long training to go over updates to UESF, LIHEAP, WAP, and other programmes.

Counsellors work together with clients to guarantee they receive the assistance they need. Many counsellors noted that a fundamental part of their work is making sure clients have all the documents they need before submitting their assistance applications, including the client's latest utility bill and pay stubs. Some counsellors also noted they go the extra mile for their clients, even adding hand-written letters to the applications to advocate for their clients' need for energy assistance.

Prior to the pandemic, the NEC model was highly dependent and modelled on the walk-in system; this system allowed anyone in the neighbourhood to easily access counselling and other energy literacy services. NEC counsellors help clients figure out whether they are eligible for assistance to help meet utility expenses or even retrofit their homes, if and when needed.

16.5 Structural limitations

While the NEC model offers myriad advantages compared to other approaches to energy vulnerability, there are also structural factors that prevent NEC from working. By highlighting these limitations, we hope we can inform organisations seeking to adopt this model on potential issues that may arise and offer advice on how to handle them.

Conservative income guidelines – which prevent many households from qualifying for federal and state programmes – are consistently announced as the greatest hurdle to addressing energy vulnerability. These guidelines are based only on gross income, not taking into account mortgage or rent costs. Medically vulnerable residents who are above the poverty line but most spend a large portion of their income on medical treatments may not qualify for assistance. In other words, most grant programmes do not take into account other financial burdens and expenses that households carry. One counsellor went as far as saying that their biggest struggle is working within power structures that are designed to help only the lowest of low-income households while moderately low-income and middle-class incomes, who still struggle to afford their energy bills, get overlooked. When asked what the hardest part of their job was, one NEC counsellor stated,

> I'd say just getting the powers that be to understand the needs that people have, the challenges that low income people have. And they only want to deal with the low, low income people and they don't have any consideration for the low income people or the middle income people, that blue collar worker that's trying every day.

Documentation is consistently named as the greatest barrier to energy assistance. Both NEC counsellors and utility customers shared stories about the excess and strict application materials. Paperwork needs to be up-to-date for applications to be successfully submitted. Missing paperwork or documents that are not from a specific time period – month or year, for example – meant that the application process could be delayed by 6–8 weeks. In other words, energy assistance applications have rigid time-based requirements for supporting documents; an utility bill and income documentation that did not match the time-period in question meant that the application was denied.

Staffing issues have been identified as a challenge for many NECs; while turnover or staff shortages are an issue common in social service industries, the pandemic exacerbated this dynamic. Most NECs have between one and four staff members, and depending on the size and location of the organisation, an energy counsellor could meet with anywhere from five to 25 clients a day. On high-traffic days, counsellors would often not be able to see everyone who came to the NEC, with some having to cut off entry after a certain number had signed up as walk-in clients. According to one counsellor, "In the past, about 8 years ago, they would have lines out the door for people waiting for services and would have to cap it at 50 people per day." Some clients desperately needed one-on-one help, however; some counsellors assist clients with reading their bills and even filling out applications, if a client has arthritis. Staffing issues have caused some NECs to cut back on programming as well, due to the time investment those programmes require. At the same time, with new forms of assistance offered during the pandemic, some counsellors were splitting their time between multiple programmes, such as rental assistance applications.

Staff shortages were not the only challenge brought on by or exacerbated by the COVID-19 pandemic. As large portions of social interaction, services, and cultural life shifted to remote and online platforms, already-vulnerable community members – such as seniors, people experiencing homelessness, and low-income families without broadband – became increasingly hard to reach. Many NEC clients are elderly, although this does vary from neighbourhood to neighbourhood. According to one energy counsellor, whose NEC serves a large senior population, even though most people have smartphones, "seniors still have trouble using them and acquainting themselves with this technology." Seniors struggled the most adapting to remote applications and appointments while also being more at risk for in-person appointments, causing many to simply not seek assistance. "Largely with the older folks that we work with, they were not in the least interested in coming in, even if it was just to meet in the parking lot and exchange paperwork. They were going to try to find some other kind of way, and were not necessarily plugged into technology." A large number of counsellors reported a stark decline in appointments, even when clients could opt for phone-based or in-person, socially distant appointments.

Numerous counsellors reported that their online outreach fell short in comparison to utility companies, which doubled down on efforts to get customers to apply for assistance directly through them as an energy provider. In some cases, this meant that customers were unable to apply funds across utility bills, which some NEC counsellors stated could have been handled more strategically. Younger clients, who are more tech-savvy we were told, will frequently apply for assistance through their utility company or through COMPASS (Pennsylvania's online portal for Health and Human Services programmes) instead of making an appointment with an NEC. While this may be more convenient, in such cases, the customers miss out on the adjacent services and programmes that NEC counsellors perform, including a more thorough check on every programme you apply for in an attempt to reduce energy burdens more than utility companies. They can also offer assistance outside not directly related to energy bills such as rental assistance. One NEC counsellor attributed their lack of appointments to "the fact that people can apply for LIHEAP online, and since younger generations are tech savvy they are applying for assistance themselves instead of coming in person." She also attributes it to utility companies conducting more outreach to clients, so individuals are applying on their own as opposed to going to NECs.

16.6 Solutions

The NEC's relative autonomy allows each organisation to find unique solutions to the structural problems and limited resources they face. For example, one understaffed NEC began providing energy counselling in groups, which allowed staff to help more people and leverage a group dynamic. According to one of the counsellors who implemented group counselling, "Sometimes we would have a group. We would do it in a group format. So, we would have a workshop as what we would call it, where people would come to fill out their applications all at once. And in that event, I would help with copying documentation, guiding people on how to fill out the applications. So, I would act a lot as if I was another intake specialist as well." The average group size at this NEC was 20 clients.

Since NECs are often nested within larger nonprofit organisations, many counsellors have found ways to leverage their parent organisation's resources. One counsellor who operates an NEC inside a housing preservation organisation stated that they collaborate across units within their organisation and leverage referrals to external programmes as well. Coalition work was identified as crucial. This counsellor explained that the housing preservation organisation that houses this NEC created a dumping advisory board, which is a group of individuals concerned about waste in the neighbourhood, the health of their environment, and the health of residents and their community. They are also a part of the North Philadelphia Community Collective, which assesses the impacts of the COVID-19 pandemic and reports on vaccination rates. The counsellor added, "These examples of collaboration and coalition building are indicative of strong and well organised social service networks in North Philadelphia, that includes social service organisations, other community groups, residents, and partners such as Temple University."

During the pandemic, several NECs reinvented themselves by deploying mobile services that would (1) allow them to conduct more outreach into Philadelphia communities and (2) allow NECs to reach those individuals who did not want to expose themselves to pandemic-related risks. These mobile services enabled counsellors to bring energy assistance information to non-mobile clients. One NEC used "door bags" which the counsellor explained was part of a collaboration: "I partnered with an organization called the Resource Pop-up Bus, and we bag up information, put it in a bag, the newsletter or flyers, whatever we're doing. We take the bags and put them on people's doors throughout the neighbourhood because everybody, I don't know if there's an assumption, but everybody's not connected, particularly our seniors."

With elderly clients feeling that in-person appointments were unsafe yet struggling to connect to remote appointments, one NEC began travelling to elderly living facilities to offer assistance directly. As one NEC counsellor puts it, "If people are not coming in, you got to find other ways to help them. I make appointments and go to the senior place, and I do their applications there."

Another NEC counsellor set up a small booth in the lobby of senior apartment centres, offering to sign up residents for LIHEAP. This counsellor had used a broad range of tactics to reach community members: social media, phone calls, flyers attached to free meals as part of their free meal programme, and calling past clients directly as outreach for energy assistance. She found that "setting up shop" in senior centres was the most effective means of signing people up for assistance "I definitely think going to senior housing buildings, kind of setting up shop in other places is probably going to be what really helps."

16.7 Conclusion

Energy vulnerability is not so much a singular issue as it is a complex and dynamic problem space, one that emerges at the nexus of many different kinds of (social, political, economic, infrastructural, epistemic, climatic, etc.) issues. Accordingly, the forms of assistance developed to address this kind of vulnerability must be similarly dynamic and wide-ranging enough to yield a flexible and responsive landscape of care (Bowlby, 2012). As we have shown, NECs (especially those housed within larger, more general assistance organisations) are well positioned to enhance a local energy-governance structure's dynamism and flexibility, serving, metaphorically, as the structure's "connective tissue" of care. That is, working between government and industry and both within and outside extant assistance structures, they have a remarkable capacity to bend their focus, to extend into new problem areas, and to reallocate vital resources.

Traditionally, however, these organisations have been simultaneously underutilised and overextended. That is, they are underutilised in that the NEC's unique position as a more flexible and locally attuned assistance infrastructure has yet to be adequately appreciated by decision makers. And they are overextended in that they tend to lack the staff and resources necessary to attend to the forms of energy vulnerability that they are best positioned to understand. In this chapter, we have attempted to redress these oversights, arguing that NECs provide a critical layer of flexibility within and between the otherwise well-stratified and ossified scales of energy assistance, developing a living network of deeply embedded and responsive care actors.

Notes

1 Ethnographic research is defined by descriptive analysis that is based on participant observation, a method where the researcher traditionally spends long periods of time conducting immersive fieldwork. See, for example, Marcus and Fischer (1986) and Fortun (2003).

2 Redlining emerged after the Great Depression, when the US federal government initiated low-interest lending (Jackson, 1987). In the lending process, federal institutions denied credit lending in neighbourhoods they considered hazardous and designated them as "red zones" (Jackson, 1987:362). These "red zones" were often populated by Black households (Hillier, 2003; Jackson, 1987). Studies defining and analysing redlining practices discuss its role in deepening neighbourhood segregation, racial divides, and inequalities (Hillier, 2003).

References

Adams, J., Kenner, A., Leone, B., Rosenthal, A., Sarao, A., & Boi-Doku, T. (2022). What Is Energy Literacy? Responding to Vulnerability in Philadelphia's Energy Ecologies. *Energy Research & Social Science*, 91, 102718.

Bednar, D. J, & Reames, T. G. (2020). Recognition of and Response to Energy Poverty in the United States. *Nature Energy*, 5(6), 432–439.

Berman, L., Brackbill, G., Edinger, L., Geesaman, K., Keaton, M., Marx, E., Pereira, R., & Sweet, J. (n.d.). The Low Income Home Energy Assistance Program (LIHEAP): Pennsylvania Advocates Manual 2021–2022 Edition. *The Pennsylvania Utility Law Project*. https://philalegal.org/sites/default/files/attachments/2022-06/PULP-2021-2022-LIHEAP-Advocates-Manual.pdf

Bouzarovski, S., Petrova, S., Kitching, M., & Baldwick, J. (2013). Precarious Domesticities: Energy Vulnerability Among Urban Young Adults. In K. Bickerstaff, G. Walker, & G. Bulkeley (eds.), *Energy Justice in a Changing Climate: Social Equity and Low-Carbon Energy*. Zed Books. London, UK. pp. 30–45.

Bowlby, S. (2012). Recognising the Time-Space Dimensions of Care: Caringscapes and Carescapes. *Environment and Planning A*, 44(9), 2101–2118.

Branigin, A. (2020). Black communities are on the 'frontline' of the COVID-19 pandemic. Here's why. *The Root*, March 31, viewed April 1 2020. https://www.theroot.com/black-communities-are-on-the-frontline-of-the-covid-19-1842404824

Bruch, S. K, Meyers, M., & Gornick, J. C. (2016). *Separate and Unequal: the Dimensions and Consequences of Safety Net Decentralization in the US 1994–2014*. University of Wisconsin-Madison, Institute for Research on Poverty.

Daguerre, A. (2017a). Conclusion: The American Welfare State in Comparative Perspective. In *Obama's Welfare Legacy*, 1st ed. Policy Press. pp. 113–118. https://doi.org/10.46692/9781447338345.008

Daguerre, A. (2017b). Eva Bertram (2015), The Workfare State: Public Assistance Politics from the New Deal to the New Democrats, Philadelphia: University of Pennsylvania Press. *Journal of Social Policy*, 46(4), 844–846. https://doi.org/10.1017/S0047279417000459

Drehobl, A., Ross, L., & Ayala, R. (2020). How High Are Household Energy Burdens. *An Assessment of National and Metropolitan Energy Burdens across the US*.

Energy.gov (n.d.). About the Weatherization Assistance Program. *Office of Energy and Community Energy Programs*. https://www.energy.gov/scep/wap/about-weatherization-assistance-program

Falk, G., McCarty, M., & Aussenberg, R. A. (2014). Work Requirements, Time Limits, and Work Incentives in TANF, SNAP, and Housing Assistance. *Congressional Research Service*.

Fortun, K. (2003). Ethnography In/Of/As Open Systems. *Reviews in Anthropology*, 32(2), 171–190.

Fowlie, M., Greenstone, M., & Wolfram, C. (2018). Do Energy Efficiency Investments Deliver? Evidence from the Weatherization Assistance Program. *The Quarterly Journal of Economics*, 133(3), 1597–1644. https://doi-org.ezproxy2.library.drexel.edu/10.1093/qje/qjy005

Hacker, J. S. (2016). America's Welfare Parastate. *Perspectives on Politics*, 14(3), 777–783.

Haveman, R., Blank, R., Moffitt, R., Smeeding, T., & Wallace, G. (2015). The War on Poverty: Measurement, Trends, and Policy. *Journal of Policy Analysis and Management*, 34(3), 593–638.

Hernández, D. (2013). Energy Insecurity: A Framework for Understanding Energy, the Built Environment, and Health Among Vulnerable Populations in the Context of Climate Change. *American Journal of Public Health*, 103(4), e32–e34.

Hernández, D. (2016). Understanding 'energy insecurity' and Why It Matters to Health. *Social Science & Medicine*, 167, 1–10.

Hernández, D., & Bird, S. (2010). Energy Burden and the Need for Integrated Low-Income Housing and Energy Policy. *Poverty & Public Policy*, 2(4), 5–25.

Hillier, A. E. (2003). Spatial Analysis of Historical Redlining: A Methodological Exploration. *Journal of Housing Research*, 14(1), 137–167. http://www.jstor.org/stable/44944777

Jackson, K. T. (1987). *Crabgrass Frontier: The Suburbanization of the United States*. New York: Oxford University Press.

Jessel, S., Sawyer, S., & Hernández, D. (2019. Energy, Poverty, and Health in Climate Change: A Comprehensive Review of an Emerging Literature. *Frontiers in Public Health*, 7(357), 1–19. https://doi.org/10.3389/fpubh.2019.00357

Kenner, A. (2020). Scrapping the Workshop of the World: Civic Infrastructuring and the Politics of Late Industrial Governance. *Engaging Science, Technology, and Society*, 6, 514–533.

Kenner, A., Sarao, M., Rosenthal, A., Leone, B., & Adams, J. (2022). Powering our health: How the pandemic impacted medically vulnerable households in Philadelphia. *The Energy Rights Project, Platform for Experimental Collaborative Ethnography*, last modified 14 October 2022. https://energyrights.info/content/powering-our-health-how-pandemic-impacted-medically-vulnerable-households-philadelphia

Marcus, G. E., & Fischer, M. J. (2014). *Anthropology as Cultural Critique: An Experimental Moment in the Human Sciences*. Chicago, IL: University of Chicago Press.

Memmott, T., Carley, S., Graff, M., & Konisky, D. M. (2021). Sociodemographic Disparities in Energy Insecurity Among Low-Income Households Before and During the COVID-19 Pandemic. *Nature Energy*, 6(2), 186–193.

PECO An Exelon Company (n.d.). Customer Assistance Program – Percentage of Income Payment Plan (CAP-PIPP). https://www.peco.com/MyAccount/CustomerSupport/Pages/CAPRate.aspx

Petrova, S., & Prodromidou, A. (2019). Everyday Politics of Austerity: Infrastructure and Vulnerability in Times of Crisis. *Politics and Space*, 37(8), 1380–1399. https://doi.org/10.1177/2399654419831293

PGW (n.d.-a). LIHEAP-Crisis-UESF Grants. https://www.pgworks.com/customer-care/liheap-crisis-uesf-grants?utm_campaign=liheap&utm_medium=seo&utm_source=google&utm_content=sem&gclid=Cj0KCQiAnuGNBhCPARIsACbnLzpbFSzlKwYdM8KpnxKFb5LizRgzte5Z-u-_mFhBMVQh5PzQoaipZZwaAo9REALw_wcB

Prasad, M. (2011). Tax 'expenditures' and Welfare States: A Critique. *Journal of Policy History*, 23(2), 251–266.

Rosie, D., & Walker, G. (2013). Household Energy Vulnerability as 'assemblage'. In K. Bickerstaff, G. Walker, & G. Bulkeley (eds.), *Energy Justice in a Changing Climate: Social Equity and Low-Carbon Energy*. Zed Books. pp. 30–45.

Tooher, N. (2014). Census: Most Philly Homes Built Before 1950. *The Philadelphia Inquirer*. 15 September. https://www.inquirer.com/philly/business/real_estate/zillow/20140915_ZILLOW_Census__Most_Philly_Homes_Built_Before_1950.html

17 Demand-side Flexibility to Address Household Energy Poverty in Sub-Saharan Africa

The Case of Burkina Faso and Madagascar

Marine Cornelis, Edmond VII Mballa Elanga, and Luc Richaud

17.1 Introduction

The European concept of energy poverty, which primarily concerns a household's inability to heat its living space, is largely unsuitable for Africa due to its hot and humid climate, where temperatures frequently fall between 22°C and 30°C. Instead, energy poverty on the continent is linked to the fact that people have little or no access to modern energy sources and services, such as electricity. On November 15, 2022, the world's population reached 8 billion people. With over 950 million people, Sub-Saharan Africa is the most electricity-poor region in the world[1]. In 2020, 568 million people had no access to electricity (out of 733 million globally), while one billion people relied on traditional biomass as their main source of energy for cooking (IEA et al., 2022). Less than half of the population has access to electricity, mainly in rural areas. Despite committed efforts, energy poverty persists and remains a severe problem.

Access to modern energy is at the core of the UN Sustainable Development Goals (SDGs), adopted in 2015. In particular, Goal 7 focuses on "Clean and affordable energy". It considers energy to be at the heart of all current and future social and economic challenges and opportunities. It suggests that access to modern energy is a fundamental human right, enabling other rights to be fulfilled such as health, well-being, education, social inclusion, and empowerment (Hesselman et al., 2021). It is critical to advance gender equity by helping women fulfil their professional and personal aspirations (e.g., storing food for eating and selling, recharging their phones, and watching television).

Based on the case of two Sub-Saharan African countries (Burkina Faso and Madagascar), this chapter examines the meaning and challenges of energy poverty and the responses provided by electrification programmes in these countries. It critically examines the relevant policies and programmes developed in these countries. The aim is to answer the question of *how to do better with the limited infrastructure available on the continent?* Better still, *can innovative tools like electricity-as-a-service and energy demand flexibility address household energy poverty?* Direct consumer participation is at the heart of the reflection, whether through energy-as-a-service or demand-side flexibility, that is, when a consumer (prosumer) adapts their consumption to production, price signals, or market incentives. Those tools aim to improve the existing access to electricity rather than add new layers.

This chapter focuses on provoking a conversation beyond the challenge of lacking access to electricity, raising the question of reliability and quality of supply in the African context. Given the extremely high prices and supply security issues in Europe, the experience of African countries, built on scarcity together with resilience logic and adaptation to economic and social constraints, makes it all the more relevant.

DOI: 10.4324/9781003408536-21

The discussion will make four stops. After methodological and conceptual clarifications, a brief presentation of the energy situation in Burkina Faso and Madagascar will be given. The challenges of energy poverty in these two countries and the responses provided by electrification programmes linked to digitalisation will then be analysed in the third portion. Finally, the fourth and last will attempt to show how the experience of African countries can be a source of inspiration for European countries facing increasing difficulties in accessing energy.

17.2 Methodological and conceptual clarifications

17.2.1 Methodological clarifications

This chapter is based on field and documentary data collected within the framework of the EU-funded EURICA project (project WP 15 – LEAP-RE EU Grant Agreement 963530). The LEAP-RE programme seeks to create a long-term partnership between African and European stakeholders in a quadruple helix approach involving government (programme owners and funding agencies), research and academia, the private sector, and civil society. It is organised into three pillars: external research funding and capacity-building activities; eight key projects; and management, coordination, monitoring, and evaluation.

EURICA is one of the key projects of the second pillar. To rapidly develop access to electricity services for African populations, EURICA mostly works on the backbone: existing infrastructure and networks. These must be reliable, decarbonised, decentralised, and intelligent (in the sense of "digital"). EURICA combines a top-down and bottom-up strategy to bring together individual electrification practices and support the development of transmission networks. This will address immediate and long-term challenges, such as maintenance and expansion.

The bottom-up approach, set up in Ambanja, Madagascar, interconnects existing small microgrids called "nanogrids". It aims to improve the quality of service and achieve a higher level of electricity access. It relies on an "electricity-as-a-service" model. The top-down approach in Ouagadougou, Burkina Faso, aims to improve the service in locations connected to the grid and make the area more resilient by promoting Distributed Energy Resources (DER). DER provides digital tools to facilitate their insertion into the grid and create a local flexibility market to reduce congestion problems with limited impact on the population. The project supports the take-off of data-driven methodology rather than time-consuming and costly physical network expansion. It enables local community engagement and creates local energy markets. This leads to flexibility management and improves mini-grid stability and supply reliability. All in all, it maximises local, sustainable, and cost-effective energy resource management and use.

For the project to gain insights into the views and experiences of specific populations, the authors conducted 67 in-person and remote interviews across Burkina Faso, Madagascar, and Cameroon[2] in the winter and spring of 2022. They used qualitative methods such as in-depth user and expert interviews, as well as participatory observation. The participants were selected according to their profiles and availability at the agreed times (e.g., availability had to match the travel schedules to Madagascar and Ouagadougou). The interviewees belonged to the direct and indirect networks of the project partners (contact persons). The authors' analysis aimed to uncover the underlying logic of interviewees and communities, focusing on induction from facts to ideas. The author used qualitative data analysis software and coding interview transcripts for coherent interpretation. While the authors' sample is not statistically representative, it aims to highlight and explain social realities linked to electricity consumption, integrating diverse stakeholders.

Table 17.1 List of interviewee categories

Categories of interviewees	*Cameroon*	*Burkina Faso*	*Madagascar*
Academia	2	2	
Agriculture/other businesses	3		
Business/Economic Interests (solar sector)		8	6
Consumers and end-users	9	3	6
Government Officials	4		1
International donors		4	1
Illegal distributors	3		
Local NGOs		1	2
Regulators	1		
Utilities/Distributors	4	5	
Other		1	1
Total	26	24	17

Source: The authors.

17.2.2 Conceptual clarifications

Given the context, the authors provide definitions to clarify what the debate is about.

17.2.2.1 "Flexibility" and decarbonising the energy sector

Today, the energy sector is responsible for roughly three-quarters of the greenhouse gases that cause climate change (heat, electricity, and transport) (Ritchie & Roser, n.d.). Therefore, it is essential to decarbonise this sector, that is for it to no longer rely on the combustion of fossil fuels, such as oil, gas, and coal. On the contrary, decarbonisation depends on using more energy from renewable sources, such as electricity produced by photovoltaic solar panels or wind turbines. However, electricity remains difficult to store, so it must be consumed when it is available in abundance. All around the world, grids and networks are the energy systems' backbones. They need to become more resilient to make space for renewables, absorb structural shocks, and implement energy efficiency much faster.

Thus, to allow renewable energies to reach their full potential and for networks to function correctly, three other Ds accompany decarbonisation: digitalisation, decentralisation, and democratisation of networks and production (Soutar, 2021).

Decentralisation is essential to developing renewable energies and corresponds to the situation where production is as close as possible to the final user. It avoids transporting electricity over very long distances, with all the associated costs and risks of loss. Transmission and distribution system operators are key players in enabling this. They are the entities responsible for transporting, distributing, and managing energy from the generation sources to the final user. They must constantly engage with regulators, policymakers, suppliers, and energy service providers (Brisbois, 2020). Digitalisation enables the monitoring of the flows and production of renewable energies and allows industries and energy-consuming appliances to adapt the time of consumption to the time of production. This is what is called demand-side management or flexibility.

Flexibility can be implemented automatically (in industry or with aggregators) or manually (i.e., by starting a washing machine or charging an electric vehicle when the solar panels are working at full capacity). Demand flexibility refers to the fact that consumers can also participate in the balance of the power system by intentionally changing their consumption (Herc et al., 2022). Therefore, the correlate is the direct participation of citizens – a phenomenon summarised as the "democratisation" of the energy sector, especially when the electricity they produce is shared with other users (Wahlund & Palm, 2022). It can increase the capacity for individual action, especially for those belonging to traditionally marginalised groups, such as women, indigenous communities, and vulnerable communities (Stephens, 2019).

17.2.2.2 Electricity-as-service

Energy-as-a-service (EaaS) is a business concept in which clients pay for an energy service without making any upfront capital commitment. EaaS models often take the form of a subscription for electrical equipment held by a service firm or energy use management to supply the required energy service.

The consumer benefits from avoiding direct power payments, costly updates for electrical equipment or software, or device administration while still enjoying the benefits of the gadget. EaaS, like other service models, may make better technology (such as energy devices and software) more available, benefiting customers, service companies, and the electrical grid (Cleary & Palmer, 2019).

17.2.2.3 Energy poverty

Energy poverty in Europe is defined as the lack of thermal comfort for households that cannot afford associated energy costs due to insufficient thermal quality (Bouzarovski et al., 2021). However, defining energy poverty in Africa is more complex due to varying socio-economic, urban, and climatic contexts. The lack of access to reliable, affordable, and sustainable energy services hinders development and makes Africa vulnerable to shocks. Cities are generally connected, whereas rural areas remain in the dark. Some areas of Sub-Saharan Africa do not need heating at all, while the need for cooling is much more urgent. Women, children, minorities, and indigenous groups are the most vulnerable (IEA et al., 2022). As a result, individual solutions such as diesel-powered generators or individual solar kits are proliferating.

The multi-tiered framework (MTF) has been developed to ensure access to quality and affordable energy services for development (Bhatia & Angelou, 2015; ESMAP, n.d.). Therefore, energy poverty in Africa refers to all situations where access to adequate, affordable, reliable, quality, safe, and sustainable energy services for development is not assured. This definition emphasises the importance of electricity quality, supply continuity (vs. untimely outages), and affordability (Anuga & Njenga, 2022).

Although initially European concepts, energy flexibility, decarbonisation, and energy poverty are crucial for Africa's human and economic development.

17.2.3 Energy situation and prospects in Burkina Faso and Madagascar

17.2.3.1 Energy situation and prospects in Sub-Saharan Africa

Regarding population growth, Sub-Saharan Africa now has over 800 million people. It is rising at more than 5 per cent annually (2.4 per cent in Madagascar, 2.7 per cent in Burkina Faso). By 2050, the African population might double (SEforALL Africa Hub, 2023; World Bank, n.d.). Fossil fuel costs are rising, subsidies are decreasing, and deforestation is increasing the cost of biomass. As a result, the cost of conventional alternatives to electricity (wood, candles, paraffin lamps, etc.) is rising even if the demands are mostly unchanged. This confluence of causes will likely increase already widespread poverty (IEA et al., 2022).

In parallel, the quick adoption of Western energy consumption norms (without most of the energy efficiency regulations), the adoption of the digital economy, and the accompanying appliances are driving the use of electricity for personal needs (from domestic appliances to smartphones, computers, servers, televisions, DVD players, radios, and Hifi). Innovative energy service models offered by private and public stakeholders are proliferating. For example, many African users still lack access to domestic energy while relying on their phones for everyday tasks. The growing popularity of electricity-as-a-service decentralised mobile phone charging stations illustrates such a phenomenon (Eun Hee et al., 2022; Victron Energy, n.d.).

As a result, embracing clean, modern energy sources is already a reality for many Africans. The continent could leapfrog from a situation of no energy access to abundant, emission-free sources, with the potential to be a springboard for society. However, as stated in SDG 7, social justice, particularly gender equality, must be prioritised in policies and activities. This will guarantee that communities' potential is fully realised and everyone benefits from higher living standards (Colantoni et al., 2021; SE4All, n.d.).

17.2.3.2 Energy poverty and access in Burkina Faso

Burkina Faso is one of the poorest and least electrified countries globally, as less than 20 per cent of the population receives access to electricity. Access to electricity is uneven and depends strongly on where people live; around 65 per cent of urban households are electrified compared to less than 5 per cent of rural households. Therefore, the national figure of 18.6 per cent hardly reflects reality and progress on the ground, despite some progress in the past decades. In 2010, the actual access rate to electricity was barely 13 per cent.

The country's energy demand is growing, driven by industrial consumption, transport needs, population growth, and urban domestic demand. Currently, the energy system overly relies on solid fuels, such as biomass. Non-solid fuels' access only reaches around 8 per cent nationally (SEforALL Africa Hub, 2023). Despite having tremendous solar potential and recent investments in these technologies, most electricity remains produced by fossil fuels (90 per cent in 2020), especially in diesel-fuelled power plants. Hydroelectricity accounts for only 10 per cent of production. Therefore, the country's population and economy are particularly vulnerable to variations in oil prices. Hence, the government heavily subsidises the electricity sector to limit the adverse effects of rising oil prices on electricity production (Alliance Sahel, 2020).

Pushed by international donors, Burkina Faso's government sets ambitious electrification targets regularly. The most recent objective is to achieve 95 per cent electricity access in urban areas in 2030 (SEforALL Africa Hub, 2023), with a threshold of 75 per cent by 2025 (Coulibaly, 2021). Access to clean cooking should also be extended to cover 100 per cent access in urban areas against 65 per cent in rural areas by 2030.

17.2.3.3 Energy poverty and access in Madagascar

Madagascar also experiences endemic poverty, and people are 42 per cent poorer today than in 1960, the year of Madagascar's independence (World Bank, 2018). Concerning energy specifically, the Malagasy population heavily relies on biomass. About 80 per cent of the island's energy consumption comes from plant and animal sources[3]. Madagascar's electrification level is among the lowest in Africa and the world. About 16.5 million people did not have electricity in 2019, with huge disparities. 80 per cent of the urban population had access, compared to 23.1 per cent of the rural population (IEA et al., 2022; World Bank, n.d.). The actual – or usable – levels might be lower. According to SE4All, at the end of 2018, 21 per cent of households in Madagascar had Tier 1 equivalent or higher electricity access per the MTF. Level 1 is the most basic level of access to electricity. It corresponds to task lighting and phone charging for at least four hours during the daytime and at least one hour in the evening.

Based on the social, cultural, economic, and energy situations in Burkina Faso and Madagascar and the challenges and significance of fuel poverty, we examine the electrification programmes in these countries and present proposed solutions from the EURICA project.

17.2.4 Innovative models to serve local needs: examples from Madagascar and Burkina Faso

17.2.4.1 Madagascar (Ambanja): serving local needs through a flexible "energy-as-a-service" model

Due to territorial complexity and the distance between large cities and remote hamlets from distribution lines, many rural Malagasy areas are completely excluded from national electrification plans. Private and non-profit organisations are stepping in, seeking solutions to enable households to benefit from minimal access to electricity. Solar kits and diesel generators are widespread but often insufficient, and their quality and durability are often questionable. Above all, these measures seem to do little to fill the gaps in electrification needs.

According to field workers interviewed by the authors in March 2022, one of the recurring problems is that many NGOs that seek to provide access to electricity aim to "dictate" their funding conditions to the donors' agenda (general public, government funding), especially what the donors think is needed. Local needs tend to be overlooked. For example, many donors will seek to electrify schools even though the pupils only attend during daylight. Meanwhile, they find it "inappropriate" for families to watch television at home in the evening, even though that is what they actually want.

Many users are gradually becoming dependent on those low-value solutions, which may be cheap to buy but need to be replaced often, leading to a waste of resources and money. For example, low-quality solar kits running on poorly efficient batteries may need to be changed often, and batteries may not be recycled. Another emblematic case is leasing solar kits (where people pay in instalments with loans at 20 per cent per year), which amounts to indebting people who already have very little. Besides, these solutions generally do not provide any social or long-term value such as training and employment that are essential to meet other development needs (Kinally et al., 2022). Furthermore, improper recycling processes expose people to health problems, such as lead poisoning (Gottesfeld et al., 2018). In short, even though they fulfil the immediate needs and demands of populations, these solutions are often socially and ecologically unsustainable.

Building on extensive empirical research and evidence, Nanoé, a partner in the EURICA project, follows the opposite logic. Local and international energy experts founded this Franco-Malagasy SME. It wants to sell a solution, not "sell a dream to someone financing a solution to a problem they do not understand". In the words of its cofounder, they "manage to sell something more specific and more useful in the long term" (interview with Nanoé's cofounder, 28 March 2022).

Nanoé has developed a "nanogrid" system, combining the pooling of solar panels and efficient batteries to interconnect houses and relying on an intelligent management system for electrical flows. The innovation, defined as a process of progressive expansion of the grid, or "lateral" in their terms, is based on linking these tiny networks to form a larger one, better able to support productive uses (such as water pumps or small machines) (Fondation Tuck et al., 2018).

But innovation is also about business practices. Rather than offering to supply a certain volume of kilowatt-hours at a certain price, Nanoé sells different packages of electricity services, fulfilling the power needs of the appliances that users are equipped with (such as light bulbs, mobile phone charging, TVs, or even fridges). This business model, known as "energy-as-a-service", was inspired by what exists in other sectors, such as telecommunication or video streaming.

This service-based business model provides customers with affordable, quality products daily. Tariffs and packages are made very simple and flexible for users; rather than talking about power, they are explained in terms of equipment (see Figure 17.1). For example, the "economy" package connects two bulbs. Installation costs MGA 10–30,000 (EUR 2.1 to 6.3)[4], and the daily rate is MGA 650 per day (EUR 0.14) or MGA 4,000 (EUR 0.84) per week. Households have to buy the two bulbs and the cable to connect them for MGA 60–80,000 (EUR 12.6 to 16.8), but Nanoé provides a discount of MGA 50,000 (EUR 10.5). This discount is applicable regardless of the initial package. If, for example, a household chooses the "multimedia +" or "premium" packages, which include the power to connect a

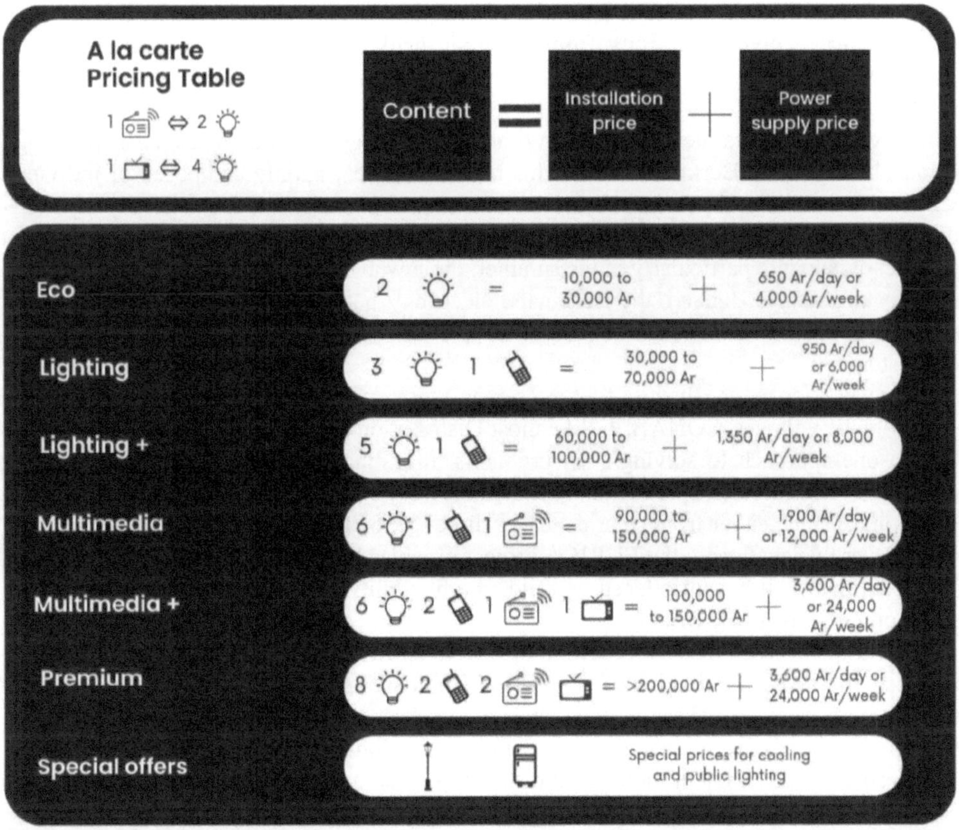

Figure 17.1 Nanoé's various offers and fixed fees.

Source: EURICA team (2022).

12V television set (MGA 300,000 (EUR 63.02) for a 19″ model and 650,000 (EUR 136.54) for a 32″ model), families will also benefit from a MGA 50,000 discount. In the latter case, the installation costs more than 200,000 MGA (EUR 42), and the weekly fee is 24,000 MGA (EUR 5.04). Nanoé guarantees a maximum daily performance of 5 hours and takes care of the maintenance of networks and equipment.

As a social enterprise, Nanoé reinvests all its profits in community development. It relies on a network of nanoentrepreneurs, that is, local young people trained as technicians and sales representatives selling, installing, and managing the nanogrids.

Nanoé's solution meets technical requirements for sustainable electricity supply thoroughly monitoring the material and technology sourcing, delivery, and recycling chains. It addresses economic and social needs through job creation, training, and affordable pricing. All in all, by pooling costs and promoting community collaboration, Nanoé enables more households to access greater comfort and maximises the impact of electrification efforts.

17.2.4.2 Demand-side flexibility to overcome power outages in Burkina Faso

As the multi-tier framework emphasises, the quality of electricity matters as much as its availability – and the absence of blackouts is one of the benchmarks (Padam et al., 2018). In Burkina Faso, the weak grid presents a fundamental challenge to better and broader access to electricity and makes it difficult to adopt energy from renewable sources. Even though the study area receives electricity from the main grid, the quality is not sufficient – but digital flexibility solutions could provide a sustainable solution to improve the quality of electricity.

This example follows the case of the Azimmo 2000 neighbourhood in the heart of Ouagadougou, the capital of Burkina Faso. In this residential area, mainly composed of individual houses, most dwellings are connected to the national grid (managed by the vertically integrated utility SONABEL), but suffer from recurrent power outages. Those outages are caused by the high level of demand, particularly in the summer, the low availability of supply, and the overall instability of the grid. Indeed, the Burkinabè electricity network and infrastructure have not been sized to handle this substantial energy increase and the fast pace of new grid connections (new customers).

Those overloads reduce the grid lifetime's expectancy and the customer's supply quality due to out-of-range voltages. SONABEL, like most Distribution System Operators (DSOs), mostly relies on one approach to solving such problems: investing in the electricity grid, which is highly capital-intensive. In parallel, many customers have been investing in their own electricity generation capacity, either from solar panels or diesel generators.

To answer this challenge, the EURICA project organised "Azimmo Ouaga 2000" as a local energy community. It has introduced demand-side flexibility into the local electricity environment to optimise available resources.

This solution is based on several technologies that allow physical devices to "communicate" through smart meters, the Internet of Things (IoT) digitisation and electronic observation of the local electricity network. These tools automate the power distribution that the grid can effectively provide during the day and between consumers according to their needs and means. The network observability system monitors the status of these constraints to detect them, forecast them, and request the activation of the necessary flexibility to mitigate them. By fulfilling the conditions, the project creates an intelligent and flexible local network, which allows for a better distribution of uses and avoids untimely outages by adjusting the network to the constraints in real-time (see Figure 17.2).

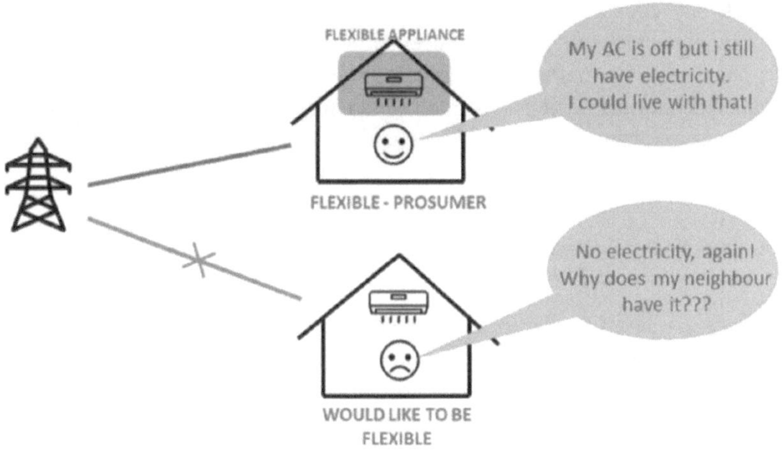

Figure 17.2 Flexible user. Small changes in individual and local communities' behaviour can help flatten the local demand to reduce peak loads and congestion of low-voltage assets and avoid outages linked to this issue.

Source: EURICA (2023).

The cases of Madagascar and Burkina Faso show that it is possible to optimise electrical energy consumption by mobilising intelligent strategies thanks to innovative business models such as electricity-as-a-service and flexibility (or demand-side response) mechanisms.

17.2.5 Doing more with less: build on flexibility and energy-as-a-service experiences in Africa for European countries

17.2.5.1 Historical context

It is difficult, if not impossible, to explain specific energy consumption logic (energy waste, energy resilience, or energy flexibility) without trying to understand the history of consumption in the world, which is primarily influenced by Western consumption logic. Consumption is an "analyser of society" (Desjeux, 2006). It illustrates the way of thinking about the world, the power relations between social groups, and the relationship between the people and their environment. It is also a marker of social belonging and differentiation (class, gender, generation, and culture) and, therefore, of social distinctions. Therefore, it is a "total social fact" to use Marcel Mauss' formula insofar as it allows us to grasp the world in its entirety (Olivier, 2008).

Concisely, consumption has gone from a craft mode where people consumed out of necessity, that is, to satisfy their basic needs, to a compulsive way of consumption-oriented, in addition to necessity, towards pleasure and luxury. With the development of salaried employment, urbanisation, and its artefacts, we no longer necessarily produce and consume to satisfy basic needs but also increasingly satisfy contemporary desires for luxury. Conspicuous consumption and wastefulness gradually come to give meaning to everyday consumption and become a way of life – the way of life of the modern ("developed") person.

The behaviours of the actors illustrate the consequence of socialisation and habits acquired by the populations progressively in the course of history, encouraged in this by the structuring of Western societies where mechanisation (and thus energy consumption) has become indispensable. Energy consumption is sustained by a desire for ever greater comfort. Over the decades in Europe, electricity consumption has grown exponentially, even if it has stabilised or even fallen due to gains in energy efficiency in recent years[5]. In such a context, changing the energy consumption patterns of populations that equate their well-being with a form of consumption focused on material comfort becomes difficult. It is necessary to consider this reality because it is a question of moving away from a set of mechanisms, some of which are cyclical and most of which are structural, and integrating others more oriented towards sufficiency.

17.2.5.2 How to do better with less in Europe: the experience of flexibility and electricity-as-a-service in African countries

The brief analysis above allows us to distinguish between Western and African (Sub-Saharan Africa) logics of electrical energy consumption. While it is established that consumption patterns in Africa are changing and increasingly adapting to those of European countries, it should be noted that African societies have long been structured by precariousness and uncertainty in all areas. Resilience is more of the population's existential necessity than a voluntary choice. "Doing better with less" thus appears to be a survival strategy adapted to the structural reality of the environment and the various economic situations.

Based on the cases of Burkina Faso and Madagascar, the experiences of service-based electricity and flexibility in certain African countries could contribute to promoting a more sustainable consumption logic of electrical energy in Europe. The main idea is to move away from the traditional consumption logic (always more) to an "intelligent" consumption logic based on resource sharing. However, the latter is strongly correlated with the development of digital technologies. Therefore, this approach is part of the quadruple helix of the "four Ds of the energy transition": decarbonisation, digitalisation, decentralisation, and democratisation of energy.

One of the strengths of flexibility and its corollary, smart grids, which could be of interest to Western countries, is the ability to activate or deactivate devices depending on the situation. In the context of the EURICA project, flexibility is defined as the ability to activate or deactivate a device at the right time to be helpful for the consumer and the energy network to offset peaks in supply or demand. The principle for the flexibility providers is to shift prosumers' electricity consumption depending on its availability when demand is forecast one day ahead by the Grid Observability system. It enables us to do more with less by improving the network while allowing users to pay less for a better-quality service. In Madagascar's case, energy availability is guaranteed for a certain number of hours. While this limiting prospect is unlikely in a European context where people have become accustomed to an unlimited supply of quality electricity, it does show that the rationale is, primarily, to address specific needs that serve the people.

17.3 Conclusion

The various energy supply crises that have shaken the world in recent years have finally established the need for a change of course regarding energy consumption habits. This chapter aimed to show the importance of service-based consumption patterns and system flexibility. Based on case studies, we have shown that these measures allow for a better distribution of resources by focusing on specific needs and for the grid to adapt to a lack of production, surplus consumption, etc. Smart grids, that is, the digitalisation and modernisation of electricity networks, are a means of achieving this objective. This encourages electricity consumers (individuals or professionals) to become increasingly involved and adapt their consumption patterns to the actual electricity production. The cases studied in Burkina Faso and Madagascar reveal the benefits of intelligent systems.

Notes

1 Africa's total population is 1.4 billion, or 18 per cent of the world's population (United Nations Department of Economic and Social Affairs, 2022).
2 At this stage, the project will not roll-out a technological solution in Cameroon, but the study of the local situation serves as a benchmark for Burkina Faso and Madagascar. Institut Supérieur Dale Kietzman in Douala, Cameroon, is a partner of the project.
3 in particular, firewood accounts to 68 per cent of the share, charcoal 10 per cent and other biomass 2 per cent), 17 per cent on petrol (transport), 2 per cent on electricity (hydropower and diesel power plants) and 1 per cent on coal (energypedia, n.d.).
4 MGA 1.00 = EUR 0.00021 (7 April 2023).
5 In 2020, the EU's primary energy consumption was 1 236 megatons of oil equivalent (Mtoe), a sharp drop from 1 353 Mtoe in 2019. Mainly due to the COVID-19 crisis, this drop allowed the EU to surpass its 2020 target (1 312 Mtoe). Primary energy consumption has thus fallen in all Member States between 2019 and 2020. The average decrease in the EU amounts to 8.7 per cent. Large falls were recorded in Spain (−12.5 per cent) and France (−11.4 per cent). In comparison, Lithuania showed a slight decrease of 1.6 per cent (European Environment Agency, 2022).

References

Alliance Sahel (2020). DÉPLOIEMENT DES ÉNERGIES RENOUVELABLES AU BURKINA FASO AVEC LA CENTRALE SOLAIRE DE ZAGTOULI [WWW Document]. Alliance Sahel.

Anuga, S., & Njenga, N. (2022). *Why Does an African Interpretation of Energy Poverty Matter?* European University Institute (Florence) U. https://data.europa.eu/doi/10.2870/229795

Bhatia, M., & Angelou, N. (2015). Beyond Connections. Energy Access Redefined. World Bank (Washington, DC). http://hdl.handle.net/10986/24368

Bouzarovski, S., Thomson, H., & Cornelis, M. (2021). Confronting Energy Poverty in Europe: A Research and Policy Agenda. *Energies (Basel)* 14. https://doi.org/10.3390/en14040858

Brisbois, M.C. (2020). Decentralised Energy, Decentralised Accountability? Lessons on How to Govern Decentralised Electricity Transitions from Multi-Level Natural Resource Governance. *Glob Transit* 2, 16–25. https://doi.org/10.1016/j.glt.2020.01.001

Cleary, K., & Palmer, K. (2019). Energy-as-a-Service: A Business Model for Expanding Deployment of Low-Carbon Technologies. Resources for the future (Washington, DC). Issue Brief 19-09. https://media.rff.org/documents/IB_19-09_EaaS.pdf

Colantoni, L., Franza, L., Sofia Sarno, G., Bompard, E.F., Corgnati, S.P., Grosso, D., & Leone, P. (2021). Africa's Energy Future. *Energy Leapfrogging Potential in Four African Countries*. Rome/Turin: Istituto Affari Internazionali (IAI).

Coulibaly, N. (2021). Burkina Faso : « Nous voulons limiter la dépendance énergétique à l'égard de la Côte d'Ivoire et du Ghana » [WWW Document]. Jeune Afrique. URL https://www.jeuneafrique.com/1129486/economie/burkina-faso-limiter-la-dependance-energetique/ (accessed 2.1.23).

Desjeux, D. (2006). *La consommation.* Presses Universitaires de France. https://doi.org/10.3917/puf.desje.2006.01

energypedia (n.d.). Madagascar Energy Situation [WWW Document]. energypedia. URL https://energypedia.info/wiki/Madagascar_Energy_Situation (accessed 2.2.23).

ESMAP (n.d.). Multi-Tier Framework for Energy Access (MTF) [WWW Document]. ESMAP. URL https://www.esmap.org/mtf_multi-tier_framework_for_energy_access (accessed 2.1.23).

Eun Hee, L., Woo Young, K., Kang, H., & Han, K. (2022). Still lacking reliable electricity from the grid, many Africans turn to other sources. Afrobarometer (Accra) Dispatch No. 514 |.

European Environment Agency (2022). Primary and final energy consumption in Europe [WWW Document]. European Environment Agency (Copenhagen). URL https://www.eea.europa.eu/ims/primary-and-final-energy-consumption (accessed 2.2.23).

Fondation Tuck, algorus, Nanoé (2018). Electrification latérale: vers un nouveau modèle d'électrification pour l'Afrique. Fondation Tuck (Rueil-Malmaison).

Gottesfeld, P., Were, F.H., Adogame, L., Gharbi, S., San, D., Nota, M.M., & Kuepouo, G. (2018). Soil Contamination from Lead Battery Manufacturing and Recycling in Seven African Countries. *Environmental Research* 161, 609–614. https://doi.org/10.1016/J.ENVRES.2017.11.055

Herc, L., Pfeifer, A., Duić, N., & Wang, F. (2022). Economic Viability of Flexibility Options for Smart Energy Systems with High Penetration of Renewable Energy. *Energy* 252, 123739. https://doi.org/10.1016/j.energy.2022.123739

Hesselman, M., Tirado-Herrero, S., Smith, M., Cornelis, M., Antepara, Í, Bajomi, A., Barrella, R., Cao, U., Chatterjee, S., Cuerdo, T., Dobbins, A., Frigo, G., Fuller, S., Feenstra, M., Gouveia, P.J., Guyet, R., Kizilcec, V., Lennon, B., González, I., Robinson, C., Robić, S., Smilevska, N., Varo, A., Yoon, H., & Živčič, L. (2021). Moving forward on the right to energy in the EU Engagement Toolkit. European Energy Poverty Agenda Co-Creation and Knowledge Innovation (ENGAGER 2017-2021). URL https://www.engager-energy.net/wp-content/uploads/2022/02/ENGAGER_Right-to-Energy-Toolkit_FINAL.pdf (accessed 8.1.23).

IEA, IRENA, UNSD, World Bank, WHO (2022). Tracking SDG 7: The Energy Progress Report. Washington, DC.

Kinally, C., Antonanzas-Torres, F., Podd, F., & Gallego-Schmid, A. (2022). Off-Grid Solar Waste in Sub-Saharan Africa: Market Dynamics, Barriers to Sustainability, and Circular Economy Solutions. *Energy for Sustainable Development* 70, 415–429. https://doi.org/10.1016/J.ESD.2022.08.014

Olivier, N. (2008). Marcel Mauss, Essai sur le don. Forme et raison de l'échange dans les sociétés archaïques. *Lectures*. https://doi.org/10.4000/lectures.520

Padam, G., Rysankova, D., Portale, E., Koo, B.B., Keller, S., & Fleurantin, G. (2018). Beyond Connections Energy Access Diagnostic Report Based on the Multi-Tier Framework Beyond Connections ETHIOPIA. World Bank. Washington, DC

Ritchie, H., & Roser, M. (n.d.). Emissions by sector [WWW Document]. Our World in Data. URL https://ourworldindata.org/emissions-by-sector (accessed 2.1.23).

SE4All (n.d.). Sustainable Development Goal 7 (SDG7) | Sustainable Energy for All [WWW Document]. URL https://www.seforall.org/sustainable-development-goal-7-sdg7 (accessed 3.4.22).

SEforALL Africa Hub (2023). Burkina Faso [WWW Document]. SEforALL Africa Hub. URL https://www.se4all-africa.org/seforall-in-africa/country-data/burkina-faso/ (accessed 2.1.23).

Soutar, I. (2021). Dancing with Complexity: Making Sense of Decarbonisation, Decentralisation, Digitalisation and Democratisation. *Energy Research & Social Science* 80, 102230. https://doi.org/10.1016/j.erss.2021.102230

Stephens, J.C. (2019). Energy Democracy: Redistributing Power to the People Through Renewable Transformation. *Environment: Science and Policy for Sustainable Development* 61, 4–13. https://doi.org/10.1080/00139157.2019.1564212

United Nations Department of Economic and Social Affairs (2022). Population Division (2022). World Population Prospects 2022: Summary of Results., United Nations. New York.

Victron Energy (n.d.). Giraffe charges phones [WWW Document]. Victron Energy. URL https://www.victronenergy.com/blog/2021/11/19/giraffe-charges-phones/ (accessed 2.3.23).

Wahlund, M., & Palm, J. (2022). The Role of Energy Democracy and Energy Citizenship for Participatory Energy Transitions: A Comprehensive Review. *Energy Research & Social Science* 87, 102482. https://doi.org/10.1016/J.ERSS.2021.102482

World Bank (n.d.). Madagascar | Data [WWW Document]. World Bank. URL https://data.worldbank.org/country/madagascar (accessed 2.1.23).

World Bank (2018). The World Bank Madagascar-Least-Cost Electricity Access Development Project-LEAD (P163870) Combined Project Information Documents/Integrated Safeguards Datasheet (PID/ISDS) Appraisal Stage|.

18 Assessing Energy Poverty and Potential Retrofit Scenarios in Residential Stock in Tirana, Albania

Sokol Dervishi, Gliti Mazniku, Alessia Tafani, and Ina Dervishi

18.1 Introduction

Rapid and unplanned urbanization of cities has projected harmful consequences The city structure has changed informally regarding urban form, building construction and materials (Pee & Pan, 2022). Due to these developments, energy poverty (EP) is considered one of the most significant challenges, requiring immediate attention on a global scale energy poverty by increasing the impact on the well-being of people and the environment (Middlemiss et al., 2019; Sareen et al., 2020; Simcock et al., 2021).

It encompasses households' inability to use the necessary energy services for thermal comfort, cooking, or illumination (Ortiz et al., 2021). Numerous definitions of EP exist, all attributed to its complexity and multi-dimensionality. Bouzarovski (2013) defined it as difficulties accessing or affording adequate domestic energy services (space heating, cooling, etc.) due to financial, regulatory, or technical constraints.

"Lived experience energy poverty" refers to the subjective experience of individuals or households who are unable to afford adequate heating, cooling, or lighting in their homes, leading to discomfort, health issues, and reduced quality of life. It encompasses the social, economic, and environmental factors that contribute to a person's inability to access or afford energy services, such as low-income, high-energy prices, poor housing quality, and limited access to energy-efficient technologies (Bell & Lane, 2018). This term emphasizes the importance of understanding energy poverty from the perspective of those who are directly impacted by it and highlights the need for solutions that address the unique challenges faced by vulnerable communities (Bouzarovski & Petrova, 2015).

Research on energy poverty has increasingly recognized the importance of the lived experience perspective and has used various qualitative methods such as interviews, focus groups, and ethnography to explore the subjective dimensions of energy poverty (Bell & Lane, 2018; Bouzarovski & Petrova, 2015). This approach has highlighted the diversity of experiences and coping strategies among energy-poor households and the social and cultural factors that shape these experiences (Bouzarovski et al., 2012). According to Boardman (2012), a household suffers from energy poverty if it needs to spend more than 10% of its income on total fuel. By this definition, the concept of energy poverty expands in dimension, encompassing more than the economic capability of "low-income households" (Llera-Sastresa et al., 2017; Zhang et al., 2021). Many researchers have studied Energy Poverty in respective European states, providing a valuable guideline to follow. Papada and Kaliampakos (2016) assessed the significant vulnerability of Greek households to energy poverty in the middle of a severe economic crisis. The findings showed that 58% of Greek households are energy poor, and among families under the poverty threshold, the energy poverty rate exceeds 90%. Karásek and Pojar (2018) conclude

DOI: 10.4324/9781003408536-22

that household expenditures on energy consumption had recently increased, and the number of households in energy poverty had reached 16% in the Czech Republic.

A practical approach to reducing energy poverty has proven to be the retrofit methods applied to households suffering from this problem. In a study conducted by Ardente et al. (2011), it was indicated that the most significant benefits of energy consumption reduction were the improvements in envelope thermal insulations, lighting, and glazing. Omar et al. (2015) concluded that retrofitting techniques in the envelope could reduce energy consumption by up to 28%. Reducing energy consumption in the building sector is essential for achieving sustainability goals and mitigating the impacts of climate change. However, achieving energy efficiency goals should not come at the expense of human comfort and building identity. It is critical to develop climate and intervention strategies that consider the unique characteristics of each building and the needs of its occupants (Dervishi & Mahdavi, 2013a, 2013b; Mahdavi & Dervishi, 2011a, 2011b). One approach to reducing energy consumption in buildings is through passive design strategies, which prioritize using natural light, ventilation, and insulation to reduce energy demand for heating and cooling. Passive design can be integrated into retrofits of existing buildings, resulting in significant energy savings while maintaining or enhancing indoor comfort (Dervishi & Mahdavi, 2013a, 2013b; Mahdavi & Dervishi, 2013).

On the other hand, the accurate calculation of irradiance values on various surfaces is crucial in simulation applications, and it depends heavily on the sky models used in different climatic conditions. Numerous studies have addressed this issue, emphasizing the importance of using appropriate sky models to improve the accuracy of simulations (Dervishi & Mahdavi, 2012, 2013a, 2013b; Mahdavi & Dervishi, 2010, 2011a, 2011b; Mahdavi et al., 2009; Orehounig et al., 2014). For instance, Reinhart and Perez (2004) proposed a simplified Perez sky model for solar radiation simulations suitable for locations with cloudless or partially cloudy skies. Similarly, Haberl et al. (2014) developed a new sky model based on the statistical analysis of high-resolution sky images that can be used for solar energy system simulations in any location worldwide. Kummert and Lhomme (1996) suggested an isotropic sky model that considers the hemispherical distribution of solar radiation. This was more accurate than other traditional sky models for buildings in northern latitudes.

A review of the literature shows how various researchers have explored retrofitting historic buildings, especially in Europe. Several theories have been proposed in this field, some focusing on possible interventions to reduce energy consumption (Ardente, 2011) and others on the importance of traditional buildings' historical values (Bao et al., 2018; De Feo & Brano, 2021). Preserving traditional buildings not only contributes to the cultural heritage of a region but also has economic, social, and environmental benefits. These buildings often have unique architectural features reflecting local culture and history. By preserving them, we can maintain the historical identity of a region and promote tourism, which in turn can boost local economies (Padeiro et al., 2018).

Several studies have investigated energy poverty, energy and thermal performance, and potential scenarios in buildings in Albania, but limited research has utilized comprehensive building simulation models to address these issues (Dervishi et al., 2012; Göçer & Dervishi, 2015; Picari & Dervishi, 2019). Dervishi et al. (2022) studied the thermal performance of traditional residential buildings constructed in Albania between the 16th and 19th centuries. Retrofitting the building fabric improved energy performance by up to 46.3% and thermal comfort by up to 7.2°C, with a payback period of 7.9 years. Similarly, Belba and Dervishi (2018) analysed the thermal performance of five traditional housing buildings in the Kukes region. They found that insulation improvements in building fabrics could reduce annual energy consumption by up to 35%. Resuli and Dervishi (2015) used a combination of questionnaires and computational simulations to explore the energy performance of Italian Houses in Tirana and propose possible retrofit interventions. Specifically, a specific combination of improvement measures could significantly reduce the annual heating energy by up to 30%.

On the other side, the energy performance simulation has been studied in other building typologies. Dervishi and Karamani (2020) explored the energy performance and thermal comfort conditions of an existing industrial building in Durres, Albania. The study analyses the building envelope retrofitting via building performance simulation. The results suggest that improvements and insulations in building fabrics and ventilation regimes could reduce the annual energy consumption by up to 19% and the monthly summer temperature by up to 1.5°C. Breçani et al. (2019) studied the simulation-assisted energy performance and thermal comfort conditions of selected agricultural structures and alternative thermal retrofit measures in three climatic regions in Albania. The alternative scenarios incorporate thermal insulation of the walls and a change in building geometry. The results suggest that changes in building morphology could reduce annual energy consumption by up to 25%. Breçani and Dervishi (2019) studied the thermal performance of bunkers in Albania via detailed computational simulations and possible retrofitting scenarios. The results show that insulation does not affect the heat flux through the outer walls. Furthermore, the results establish that the same building consumes 44–145% more energy when located above ground instead of underground. Bano and Dervishi (2021) evaluate the greenery's effectiveness in reducing energy loads and improving indoor thermal comfort. Several design variables, including WWR (window-to-wall ratio), LAI (leaf area index), and plant height, are selected for a comprehensive whole-building energy performance analysis. The impact of vertical greenery effectiveness on a fully glazed building scenario considerably reduces 4.7°C indoor temperature and 9°C radiant temperature in the summer, ranging from 14% up to 34% energy savings. Kalaci and Dervishi (2014) focus on the adaptive reuse of a historical housing building and examine the challenges in incorporating a sustainability framework into the adaptive reuse of the building, viewed more holistically, integrating social, economic, environmental, and urban political policies.

The literature review analysed highlights the importance of energy poverty and energy retrofitting, especially in Albania. This research can provide an original contribution addressed by the following knowledge gaps in the literature and identify potential areas for further improvement in simulation.

- Limited studies have applied a detailed building simulation model based on comprehensive onsite input (energy bills, questionnaires, temperature and humidity data loggers, building, HVAC, occupancy and ventilation schedules, and cost estimation) in the housing sector in Albania.
- Limited studies have assisted in comparing and evaluating the thermal and energy performance of housing typologies located in Mediterranean climates, highlighting the differentiation of performance and needs for each typology.

Therefore, the study's main objective is to evaluate energy poverty in Tirana, Albania, and develop potential energy retrofit scenarios to improve energy performance. The chapter aims to develop a comprehensive methodology based on an analytical and quantitative approach that would significantly contribute to the energy retrofitting field in the Mediterranean region within the case study of Albania. Specifically, the present study explores comprehensive research on the prevalence of energy poverty in housing stock in Tirana, highlighting the vulnerability of households and providing potential passive retrofitting measures for its eradication. The long-term goal of this research is to analyse existing energy efficiency in buildings and develop a simulation-based retrofitting scheme for these buildings.

18.2 The Albanian context

18.2.1 Prevalence of energy poverty

Albania is subject to the constant challenges of urban development and those presented by continuous economic, socio-cultural, and environmental transformation. Focusing on the development during these past 30 years, a radical change in society's socio-cultural and lifestyle models is noticed. This new model is developed on the existing environment and infrastructure, which need to be improved (Pojani, 2010). Most of the available residential stock needs to respond to contemporary building criteria, thus providing a higher quality of life. More than half of the existing building stock in Albania was built between 1945 and 1990, which corresponds to 83.7% of the residential stock (Triantis, 2018). During this period, the government initiated a massive construction programme to meet the growing demand for housing. Most of the housing stock built during this period consisted of buildings designed and built quickly and inexpensively to meet the pressing housing needs with low thermal efficiency. As a result, many need to be better insulated and equipped with efficient heating systems, leading to high energy consumption and waste. According to a report published by the World Bank (Review, 2021) that categorizes countries based on their GNI (gross national income) and GDP (gross domestic product), Albania is ranked as an "upper middle income" country in the GNI list. However, Albania is ranked 40th out of 49 countries on the GDP list.

The primary household energy source is electricity (45%), followed by biomass (38%). Furthermore, Albania has a high unemployment rate of 19%, which is continuously growing. Many households face difficulties in paying the bills and need more heating to reach thermal comfort (Instat, 2023), resulting in the high severity and prevalence of insufficient access to energy services in Albania. Although far from attention, according to studies being performed, Albanian households will be even further endangered by energy poverty, caused by a lack of available government funding to implement beneficial monetary policies for the householders (Mustafaraj, 2014). From a long-term perspective, it is essential to ensure that buildings constructed after 1991 are retrofitted, as they will be responsible for around 43% of the final energy consumption in the sector (Taipi & Ballkoci, 2017).

On the other side, recently, Albania's Ministry of Infrastructure and Energy compiled two national laws regulating energy efficiency and buildings' energy performance. Law no. 124/2015, "Energy Efficiency," (2015) outlining the measures that must be taken to improve energy efficiency across the country, focusing on industry, transportation, and public buildings. Law no. 116/2016, "Energy Performance of Buildings," (2016) sets minimum energy performance standards for new and existing buildings and requires building owners to obtain an energy performance certificate. These laws aim to ensure that Albania complies with Directive 2010/31/EU (2010) by promoting renewable energy sources, reducing energy consumption, and improving the energy efficiency of buildings.

18.3 Methodology

18.3.1 Overview

The research methodology implemented for studying housing typologies in Tirana involves five distinct stages. The first stage is focused on identifying a representative group of housing typologies for the city, with a specific emphasis on those that are more prevalent and relevant for the study. The chosen typologies include Prefabricated Buildings (PRT), Silicate Brick Buildings

(SBT), and Informal House Units (INT). In Section 3.3, detailed information on the construction time, typology, geometry, and window-to-wall ratio (WWR) is collected during this stage, while in Section 3.4, a questionnaire is administered to the residents of the selected buildings. This questionnaire is designed to gather information about several aspects of their living conditions, such as the temperature comfort in their houses, the type of heating and cooling systems used, their annual electricity bills, and how their economic situation influences their heating and cooling habits.

In the third stage, simulation models are developed to collect data on local climate, occupancy, and heating and cooling schedules for one or more representative buildings from each typology. The models are then calibrated by comparing their results with actual bills obtained in the previous stage to establish a baseline for assessing the impact, need, and potential for reducing energy demand in the buildings. The fourth stage involves exploring different retrofitting scenarios to improve the sustainability of residential buildings. The selected buildings and retrofitting scenarios are evaluated based on indoor thermal comfort, energy demand reduction, and overall financial profitability for the homeowners. Finally, the fifth stage involves analysing the study results and developing recommendations based on the findings. These recommendations may include guidelines for design and construction standards as well as energy efficiency measures to improve the energy performance of the studied housing typologies in Tirana.

18.3.2 Climate

The climate of Tirana, defined by the Köppen climate classification like Cfa, is Mediterranean, with mild, rainy winters and hot, sunny summers. The average precipitation is about 1,266 millimetres per year. The city receives the most rainfall in the winter, from November to March, and less in the summer, from June to September. Springs and summers are hot, reaching over 20°C from May to September. Temperatures vary throughout the year, from an average of 6.3°C in January to 23.8°C in July. January is the coldest month with an average high temperature of 9.7°C and an average low temperature of 2.9°C. August is the warmest month with an average high temperature of 29.2°C and an average low temperature of 18.4°C During autumn and winter, from November to March, the temperature drops but is not lower than 6.7°C. The city receives approximately 2,500 hours of sun, making it one of the sunniest cities on the European Continent. Figure 18.1 illustrates the annual temperatures and average solar irradiance on the horizontal plane of Tirana.

18.3.3 Site selection

The main focus of the chapter is on three distinct types of housing in Tirana, which have been chosen based on their prevalence at the city-wide level and their susceptibility to low construction standards. This selection approach aims to increase the significance and relevance of the study's outcomes, as they may impact a larger number of households. The housing typologies that have been identified for the city of Tirana are Silicate Brick Buildings [SBT], Precast Concrete Buildings [PRT], and Informal House Units [INT].

18.3.3.1 Prefabricated Building (referred to as PRT)

The multi-family building was constructed as part of modular projects and prefabrication of panels between 1960 and 1980 to reduce costs and time for the production and assembly of

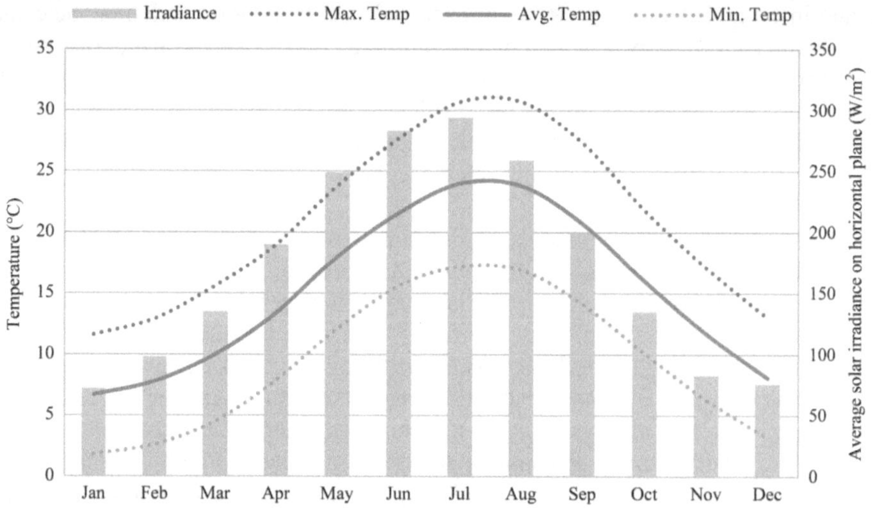

Figure 18.1 Annual temperatures and average solar irradiance on a horizontal plane for the city of Tirana.
Source: Extracted from Meteonorm.

housing units across the country (see Figure 18.2). The building's design consists of a vertical circulation node where the block of stairs is positioned in the centre. The rectangular floor plan has two sides left empty to allow it to be attached to other sections. The building has six apartments on each level, a mixture of one-bedroom and two-bedroom apartments. All apartments have two-way orientations and indirect ventilation. The panels used in the building were reinforced and poured into metal moulds before being treated with relevant technology. The external panels are 22 cm thick reinforced concrete, which also acts as a structural component and provides thermal insulation through the foam concrete layers. The layer of foam concrete is 14 cm thick while the partition panels dividing the functional areas are 14 cm thick reinforced vertical concrete. The building's roof and internal floors are horizontally reinforced concrete slabs.

18.3.3.2 Silicate Brick Building (referred to as SB)

This building, erected in 1970, is a multi-family structure containing 15 apartments located across five levels above ground. Access to each level is provided by staircases that lead to three apartments on each floor, consisting of two 2+1 apartments and one 3+1 apartment (as illustrated in Figure 18.3). Some of the apartments on the ground floor are currently serving commercial and service functions. The building's construction system features retaining walls made of red bricks, while the external walls, left unembellished on the outside, are plastered on the inside and 24 cm thick. Partition walls, which are 14 cm thick, are constructed using red bricks and serve as retaining walls. The flat, reinforced concrete slabs form the roof and internal floors.

18.3.3.3 Informal Building (referred to as IN)

The informal housing typology is prevalent in the informal areas of Tirana. This residential building has two levels and a square plan, featuring four rooms and two toilets (refer to Figure 18.4). The building's organization comprises a vertical circulation hub, with the staircase block inside the structure. The building's supporting columns are constructed of reinforced concrete, while the building envelope is made of aerated bricks plastered

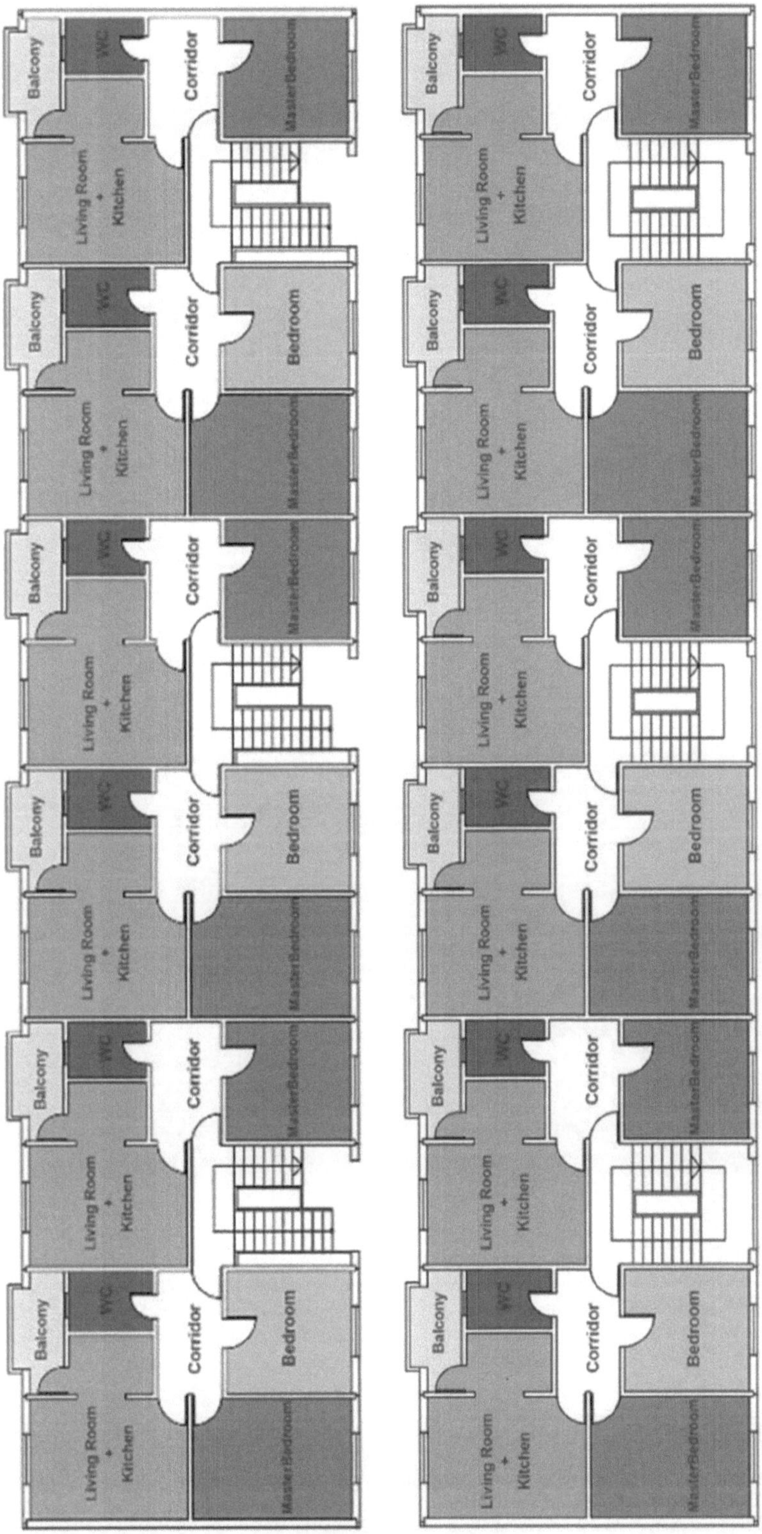

Figure 18.2 Building layout plans with associated functional spaces for [PRT] and [PRP] (ground and typical floor).

Source: Image courtesy of the authors.

Figure 18.3 Building layout plans with associated functional spaces for [SBT] and [RBP] (ground and typical floor).

Source: Image courtesy of the authors.

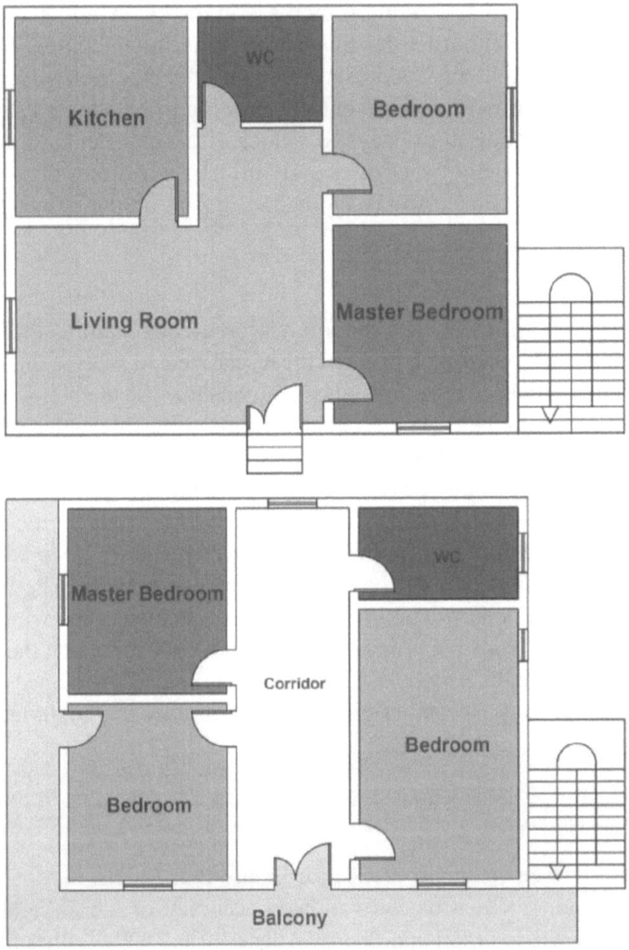

Figure 18.4 Building layout plans with associated functional spaces for [INT] (ground and first floor).

Source: Image courtesy of the authors.

on both sides, measuring 25 cm. The internal partitions also feature aerated bricks plastered on both sides, measuring 15 cm. The roof and interior floors of the building are constructed of horizontally reinforced concrete slabs. For more detailed technical information regarding the chosen typologies for Tirana, Albania, please refer to Table 18.1. As the table shows, electricity is the primary heating mode for the selected housing typologies while the households are highly sensitive to price changes in the electricity markets. Any increase in

Table 18.1 Technical information of the selected typologies for Tirana, Albania

Building code	*Floor Nr*	*Building area (m^2)*	*WWR (%)*	*Apartment Nr*	*Density (people/m)*	*Heating system*	*Cooling system*
PRT	5	2,190	20.18	30	0.05	Electricity	–
SBT	5	1,313	18.7	15	0.05	Electricity	–
INT	2	149	5.51	–	0.03	Electricity	–

electricity prices could lead to a substantial increase in the household's energy bills, which could be a significant financial burden for households with low incomes. This sensitivity to market prices also implies that any increase in demand for electricity during peak periods may lead to a strain on the power grid, potentially leading to blackouts or other disruptions in the power supply. As such, finding ways to reduce electricity demand in the housing typologies through energy efficiency measures and building retrofits is crucial for improving the households' financial resilience and reducing the impact on the power grid.

18.3.4 Questionnaire

The questionnaire was designed to gather important information regarding the factors that affect energy consumption in each building. It enabled the researchers to analyse the social, economic, and cultural aspects that influence the inhabitants' relationship with the building. The survey targeted randomly apartment or house owners and inhabitants of the three selected building typologies in Tirana. The questionnaire comprised 26 questions and was divided into four sections: personal information section, energy poverty section, building information section, and energy consumption section. The personal information section collected data on the respondents' age, gender, education level, and occupation. The energy poverty section assessed the respondents' ability to pay for energy bills and their perception of energy poverty. The building information section gathered data on construction time, typology, geometry, and window-to-wall ratio (WWR). Finally, the energy consumption section aimed to collect data on the respondents' energy consumption patterns and behaviours. The questionnaire was developed and implemented over a two-month period from December 22, 2019, to February 20, 2020. During this period, 66 valid questionnaires were obtained from the targeted survey respondents.

18.3.5 Computational simulation

The DesignBuilder interface version 6 (2020) for EnergyPlus (2020) is chosen to develop the simulation. The selection is based on the validated accuracy of the DesignBuilder software procedure developed by the International Energy Agency. The local weather file for the various climatic contexts in Southeast Europe is generated through Meteonorm 7.3 (2016) software. To enhance the reliability and impact value of this parametric exploration, each model generated in the simulation tool is first calibrated. Data from site surveying, documentation, questionnaires, and electricity and biomass bills are used as input. Thus, the input parameters of occupancy schedules, local climate, heating and cooling schedules, manufacturing operation times, and building systems availability are set based on the actual case scenarios of the inhabitants of each building.

The data presented in Table 18.2 provide an overview of the actual construction details. A calibration process was carried out to match the energy consumption values obtained from the actual energy bills with the computational simulation output to ensure accuracy. The simulations were conducted over a one-year period, and the monthly energy demand was measured from the energy bills and converted to kWh m^{-2} to enable comparison. The results showed that the energy demand for each building was 9.5 Lek per kWh. This data was used to establish base-case scenarios and retrofit scenarios. The calibration process was deemed successful when the energy consumption values from the energy bills were close to the output values from the computational simulations. The energy demand for each building was measured and calibrated over one year and converted to kWh m^{-2} for comparison purposes. For biomass, the conversion equivalency was 1 kWh, equal to 4,500 L or 0.4 kg/h of wood.

Table 18.2 Description of the typologies (BR, PR, IN, respectively) and associated U-value assumptions regarding the relevant building components

Code	*U-Value* ($W\,m^{-2}\,K^{-1}$)	*Description*
BR	$U_{outer\ wall} = 2.804$	Outer Wall (0.27 m): Calcium Silicate Brick (0.25 m), cement sand render (0.02 m), no insulation.
	$U_{inner\ wall} = 2.632$	Inner Walls (0.16 m): Cement sand render (0.02 m), Calcium Silicate Brick (0.12 m), Cement sand render (0.02 m).
	$U_{roof} = 4.440$	Roof (0.17 m): Reinforced concrete (0.15 m), Cement sand render (0.02 m), no insulation.
	$U_{window} = 5.894$	Single Glazing, Clear, 3 mm
PR	$U_{outer\ wall} = 1.161$	Outer Wall (0.24 m): Reinforced concrete (0.04 m), Cast concrete lightweight dry (0.14 m), Reinforced concrete (0.04 m), plastered (0.02 m), no insulation.
	$U_{inner\ wall} = 3.839$	Inner Walls (0.14 m): Plaster (0.02 m), Reinforced concrete (0.04 m), Plaster (0.02 m).
	$U_{roof} = 4.440$	Roof (0.17 m): Reinforced concrete (0.15 m), Cement sand render (0.02 m), no insulation.
	$U_{window} = 5.894$	Single Glazing, Clear, 3 mm
IN	$U_{outer\ wall} = 0.958$	Outer Wall (0.27 m): Cement sand render (0.02 m), Brick-aerated (0.25 m), Cement sand render (0.02 m), no insulation.
	$U_{inner\ wall} = 1.250$	Inner Walls (0.16 m): Cement sand render (0.02 m), Brick-aerated (0.12 m), Cement sand render (0.02 m).
	$U_{roof} = 4.440$	Roof (0.17 m): Reinforced concrete (0.15 m), Cement sand render (0.02 m), no insulation.
	$U_{window} = 5.894$	Single Glazing, Clear, 3 mm

18.3.6 Scenarios

The first retrofit scenario, [SC01], involves making improvements to the thermal insulation of the walls. The second scenario, [SC02], involves making improvements to the roof insulation, while the third scenario, [SC03], focuses on improving the glazing. Scenario [SC04] is a combination of the best options from scenarios [SC01] and [SC03], while scenario [SC05] combines [SC02] with [SC03]. The sixth scenario, [SC06], involves applying the best thermal insulation for the walls, roof, and windows. The seventh scenario, [SC07], aims to increase natural ventilation from 5 to 10 ac/h, while scenario [SC08] involves increasing the WWR by 15%. In Table 18.3, Tirana's scenarios [SC01-SC08] are illustrated, along with the U-value assumptions for the relevant building components for each typology. These scenarios are important in evaluating the impact of different passive interventions on energy consumption reduction and thermal performance. By simulating each scenario using the DesignBuilder interface, the energy consumption and thermal performance results can be compared to the base case scenario, [BC], to assess the effectiveness of each intervention. This information can be useful in identifying the most effective interventions for reducing energy consumption and improving thermal performance in Tirana's building stock.

18.4 Results and discussion

18.4.1 Questionnaire results

A survey was administered to 66 homeowners residing in the selected building types in Tirana. Most of the homeowners fell into the age group of 45–65, with the remainder belonging to the age group of 36–45. When asked about their monthly income, it was discovered that 50% of

Table 18.3 Retrofitting scenarios of the typologies (BR, PR, IN, respectively) and associated U-value assumptions regarding the relevant building components

Code	Scenario	U-Value ($W\ m^{-2}\ K^{-1}$)	Description
SC 1	Wall Insulation	U_{Walls} = 0.469 (SBT) U_{Walls} = 0.591 (PRT) U_{Walls} = 0.436 (INT)	EPS Expanded Polystyrene (5 cm)
SC 2	Roof Insulation	U_{Roof} = 0.301	XPS Extruded Polystyrene (10 cm) + 3 cm Screed + 1 cm Roofing Felt
SC 3	Window Glazing	$U_{glazing}$ = 2.665	Double, Clear Glazing 6 mm/13 mm Air
SC 4	Wall Insulation + Double Glazing		
SC 5	Roof Insulation + Double Glazing		
SC 6	Wall + Roof Insulation + Double Glazing		
SC 7	Natural Ventilation (ac/h) = 10		
SC 8	WWR increased by 15%		

the homeowners in the [PRT] type, 30% of those in [SBT], and 28% of those in [INT] had an income of less than 380 euros, which is considered low average income. According to data from the World Bank for 2021, the median income in Albania will be $5,188 per year, or $432 per month. The second most frequent response for income fell in the range of 120–300 euros, with 10% in [SBT], 50% in [PRT], and 20% in [INT]. The findings mentioned above are summarized in Figure 18.4.

Regarding the concept of energy poverty, it was found that only a fraction of householders in Tirana had knowledge of it. A crucial aspect of the survey was the building-related questions. In terms of house insulation, a significant proportion of residents in [SBT] (45%), [INT] (40%), and [PRT] (25%), reported having no insulation in their homes. For [SBT], 45% claimed to have only wall insulation, whereas [INT] (45%), [PRT] (25%), and [SBT] (11%) have both wall and roof insulation. In the energy consumption section of the survey (refer to Figure 18.4), it was found that all the apartments and houses in Tirana use electricity for heating. Residents of [PRT] and [SBT] reported the highest levels of temperature discomfort during the winter or summer, but even [INT] residents claimed to experience discomfort frequently. When it comes to difficulties with electricity bill payments, a significant proportion of respondents reported having problems. However, most of them managed the situation by cutting other expenses, avoiding late payments, and resolving issues with electricity bill payments.

The questionnaire results presented in Figure 18.5 provide valuable insights into the different aspects of the three building typologies in Tirana. The data collected from the householders reveals interesting trends and patterns in relation to monthly income, building insulation, window glazing, window frame, energy used for heating, thermal comfort, electricity spent for space heating/cooling, and electricity bill payment.

18.4.2 Energy consumption

Figure 18.6 presents a comparison of the annual energy use for heating, cooling, and total energy consumption among the actual real case (RC), base case (BC), and eight different scenarios (SC 01–SC 08) for three building typologies: Prefabricated Building [PRT], Silicate Brick

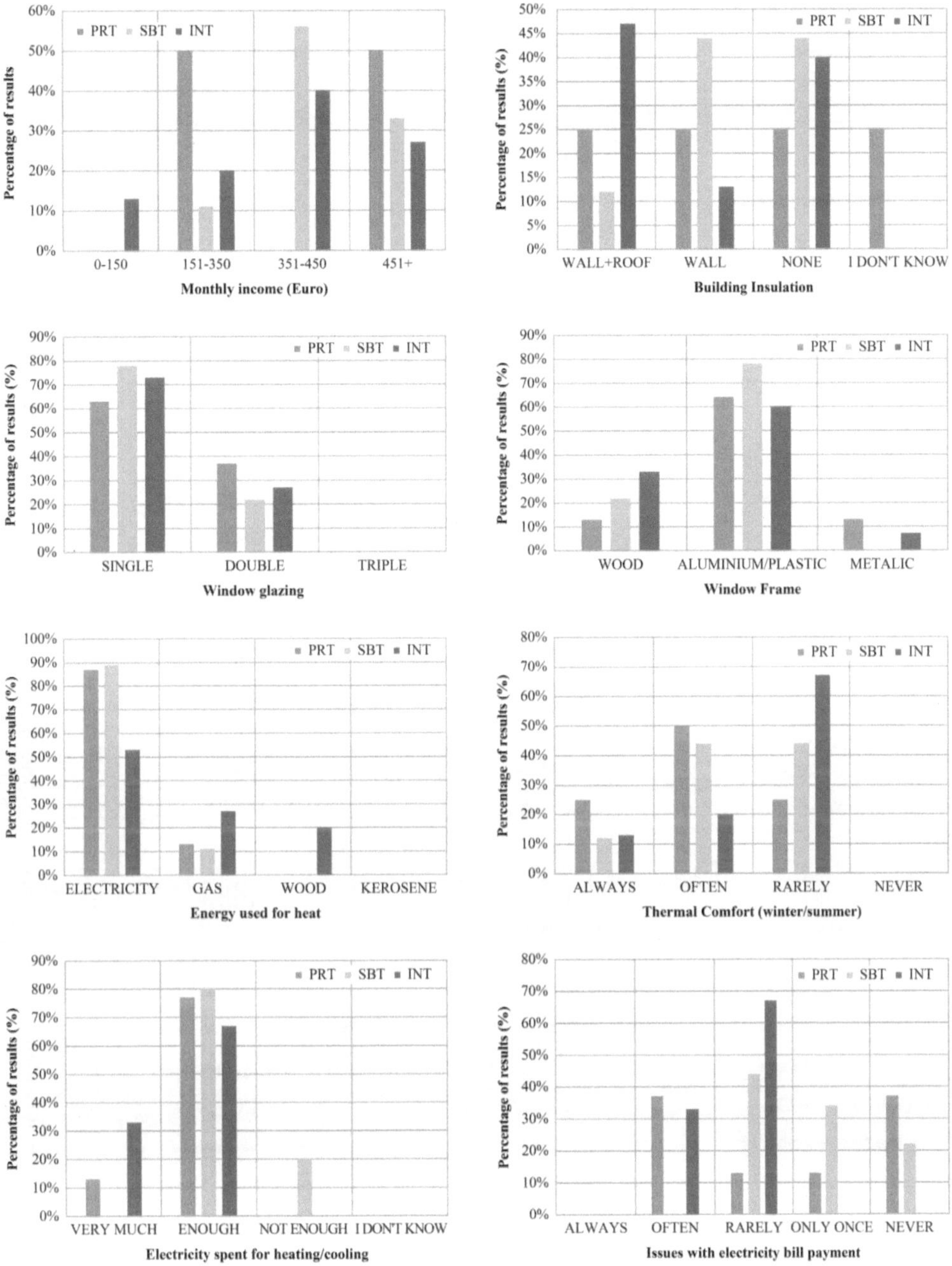

Figure 18.5 The results of questionnaires (%) of Prefabricated Building [PRT], Silicate Brick Building [SBT], and Informal Building [INT] typologies related to monthly income, building insulation, window glazing, window frame, energy used for heating, thermal comfort, electricity spent for space heating/cooling and electricity bill payment.

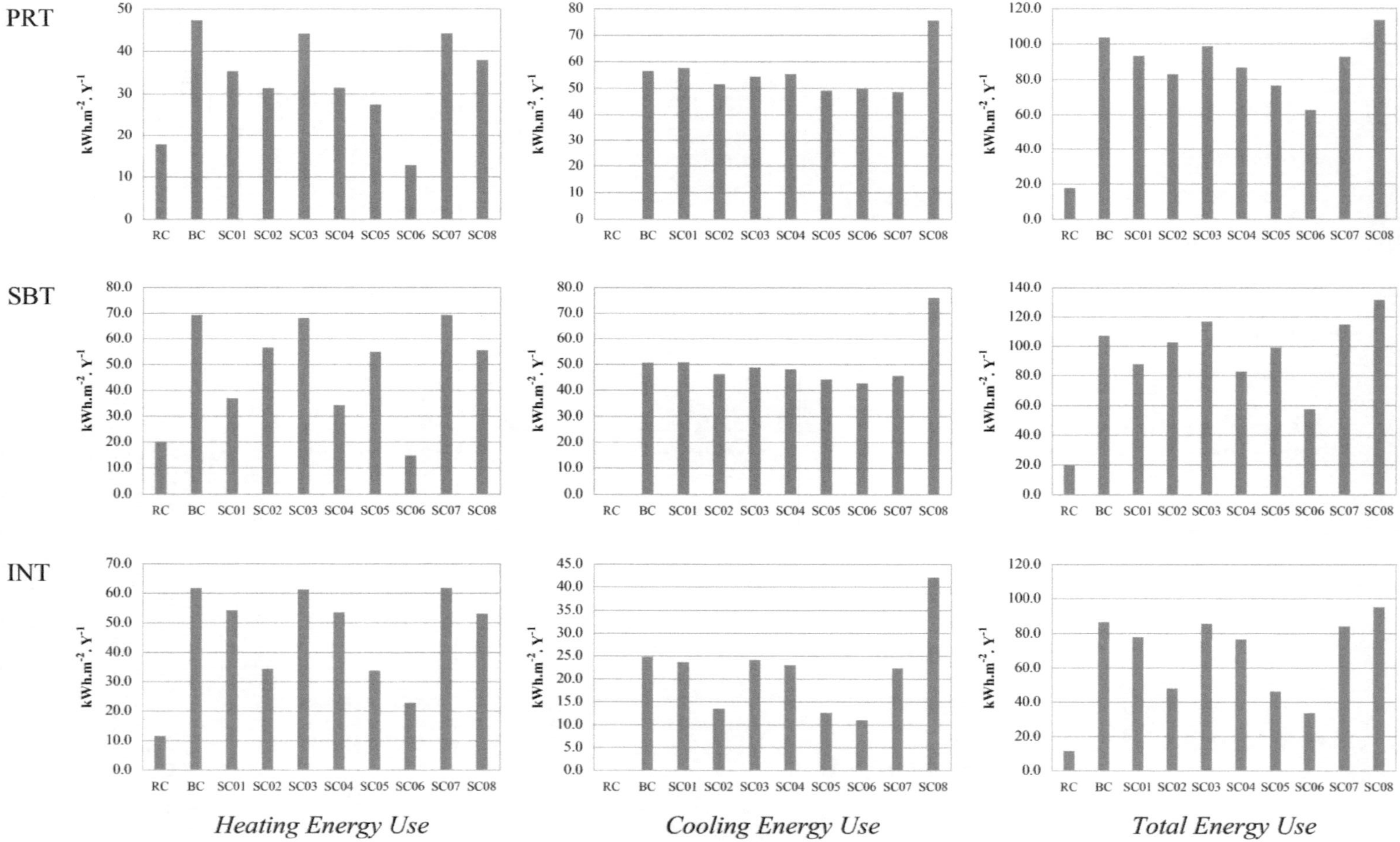

Figure 18.6 Computational evaluation of heating, cooling, and total energy use for Prefabricated Building [PRT], Silicate Brick Building [SBT], and Informal Building [INT] typology throughout one year in Tirana. Comparison of annual total energy demand for the real case (RC), base case (BC), and the eight scenarios (SC 01–SC 08).

Table 18.4 Energy reduction (%) of the refurbishment scenarios for total energy use

PR		*BC*	*SC 1*	*SC 2*	*SC 3*	*SC 4*	*SC 5*	*SC 6*	*SC 7*	*SC 8*
PRT	kWh m^{-2}	103.9	93.2	82.9	98.6	86.8	76.5	62.6	92.5	113.5
	% Reduction	–	**10.3%**	**20.2%**	**5.1%**	**16.5%**	**26.4%**	**39.7%**	**11.0%**	**−9.2%**
SBT	kWh m^{-2}	107.4	87.8	102.5	117.1	82.7	99.0	57.7	115.0	131.5
	% Reduction	–	**18.2%**	**4.6%**	**−9.0%**	**23.0%**	**7.8%**	**46.3%**	**−7.1%**	**−22.4%**
INT	kWh m^{-2}	96.6	78.0	47.9	85.5	76.8	46.4	33.8	84.1	95.4
	% Reduction		**19.3%**	**50.4%**	**11.5%**	**20.5%**	**52.0%**	**65.0%**	**12.9%**	**1.2%**

Building [SBT], and Informal Building [INT]. In Table 18.4, the percentage of energy reduction resulting from each scenario's implementation compared to the base case scenario is displayed. The graphs in Figure 18.5 reveal that the actual energy consumption scenario falls short of the required energy consumption based on each building's conditions, as indicated by the base case scenario [BC]. This indicates that the occupants of the considered buildings live in energy poverty when they resort to underconsumption strategies. The simulation results indicate that modifications to the building fabric can improve each building typology's energy performance. Specific combinations of improvement scenarios display high thermal performance, with the most effective being the first three scenarios (SC 6), followed by wall and glazing replacement (SC 5), and finally, wall and ceiling insulation (SC 2).

For Prefabricated Building [PRT], the results on energy consumption fluctuate between 103.9 kWh m^{-2} Y^{-1} (base case) and 62.6 kWh m^{-2} Y^{-1} (SC 6). Specifically, SC 6 reduces the annual total energy demand by up to 39.7%. The energy consumption of Silicate Brick Building [SBT] fluctuates between 107.4 kWh m^{-2} Y^{-1} (base case) to 57.7 kWh m^{-2} Y^{-1} (SC 6). Scenario six (SC6) reduces the annual total energy demand to 46.3%. Energy consumption in Informal Building typology [INT] fluctuates between 96.6 kWh m^{-2} Y^{-1} (base case) and 33.8 kWh m^{-2} Y^{-1} (SC 6). Scenario six (SC6) reduces the annual total energy demand to 65.0%.

The energy consumption for Prefabricated Buildings [PRT] varies from 103.9 kWh m^{-2} Y^{-1} (base case) to 62.6 kWh m^{-2} Y^{-1} (SC 6), reducing up to 39.7% in the annual total energy demand. Silicate Brick Building [SBT] also shows fluctuations in energy consumption, ranging from 107.4 kWh m^{-2} Y^{-1} (base case) to 57.7 kWh m^{-2} Y^{-1} (SC 6) reduces the annual total energy demand by 46.3%. Informal Building [INT] shows the most significant reduction in energy consumption, with variations ranging from 96.6 kWh m^{-2} Y^{-1} (base case) to 33.8 kWh m^{-2} Y^{-1} (SC 6), reducing the annual total energy demand by 65.0%. The sixth scenario (SC6) is the most effective in reducing energy consumption for all building typologies.

Energy-poor buildings are one of the most important topics, where the objective of decreasing the final energy demand and achieving energy efficiency must be attained. Nevertheless, choosing a type of support that will be feasible regarding the household's income is necessary. At the same time, the support must be practical for the shortest payback period possible and a decrease in energy demand. A programme focusing on energy poverty decrease should be based on the requirements of the affected households. It should provide sufficient support, aiming primarily at decreasing household energy expenditures while achieving higher comfort.

18.5 Conclusions

The study's main objective is to evaluate energy poverty in Tirana, Albania, and develop potential energy retrofit scenarios to improve energy performance. The residential stock in Tirana is made up of poorly constructed buildings with no insulation or thermal comfort, leading to low-quality buildings and energy poverty. The study's selection criteria for the building typologies aim to represent Tirana's different types of buildings. The authors aim to contribute to developing policies and strategies to address energy poverty in Tirana and improve living conditions for residents. The study's findings can provide policymakers and other stakeholders with insights into practical ways to tackle energy poverty in similar settings.

Eight easily implemented scenarios are explored, focusing on the building envelope retrofit. The retrofit packages result in more success than others in dependence on the improved elements and their impact on a particular floor or building. Given Tirana's climatic conditions, houses require heating and cooling during the year. The best scenario for Tirana (lowering energy consumption and achieving thermal comfort) is SC06. It combines the wall (5 cm) with roof insulation and double glazing. The SC06 scenario has a more significant impact in the [SBT] typology (46.3% energy demand reduction), followed by the [PRT] typology (39.7% energy demand reduction) and [INT] typology (65.0% energy demand reduction). The typology that requires higher energy consumption to achieve thermal comfort is the [SBT] typology, followed by the [PRT] typology and [INT] typology.

The study evaluates eight scenarios focusing on building envelope retrofit, which is relatively easy to implement. The success of each retrofit package depends on the elements being improved and their impact on specific floors or buildings. Tirana's climate requires heating and cooling throughout the year, making energy consumption a critical issue. Among the eight scenarios, SC06 is the most effective in achieving thermal comfort and reducing energy consumption. It involves wall insulation (5 cm), roof insulation, and double glazing. The impact of SC06 is most significant in the SBT typology, reducing energy demand by 46.3%, followed by the PRT typology (39.7% energy demand reduction) and INT typology (65.0% energy demand reduction). Among the three typologies, the SBT requires the highest energy consumption to achieve thermal comfort, followed by PRT and INT. Overall, the study highlights the importance of building envelope retrofit to reduce energy consumption and address energy poverty in Tirana.

The impact of various factors such as climatic conditions, building fabric, orientation, window-to-wall ratio (WWR), and construction technique on energy use has been identified through simulation results. Each of these factors plays a crucial role in determining the overall energy performance of a building. For instance, the climatic conditions in a particular region can significantly influence the heating and cooling required to maintain thermal comfort indoors. Similarly, the building fabric, such as insulation in walls and ceilings, can help reduce the energy demand for heating and cooling.

Moreover, the orientation of a building and the WWR can impact the amount of solar radiation entering the building and, consequently, affect the heating and cooling needs. Finally, the construction technique used in building design can also impact energy demand, with more energy-efficient designs typically resulting in lower energy use. The simulation results highlight the need to consider these factors when designing and retrofitting buildings to improve their energy performance and achieve thermal comfort for occupants.

To overcome the challenges associated with residential retrofit practices in Albania, a comprehensive approach involving multiple stakeholders is required. The government must lead in developing policies and initiatives that encourage building owners and residents to invest in energy-efficient retrofitting measures. In addition, building owners and residents must be educated about the benefits of energy-efficient retrofitting and the various options available. This can be done through public awareness campaigns, training programmes, and workshops that provide information on retrofitting techniques, products, and services. Addressing energy poverty and promoting energy-efficient retrofitting in Albania requires a multi-faceted approach involving the government, building owners, residents, and other stakeholders. With the right policies, incentives, and education, it is possible to transform Albania's building stock and improve the quality of life for its citizens.

References

Ardente, F., Beccali, M., Cellura, M., & Mistretta, M. (2011). Energy and Environmental Benefits in Public Buildings as a Result of Retrofit Actions. *Renewable and Sustainable Energy Reviews*, 15, 460–470

Bano, P., & Dervishi, S. (2021). The Impact of Vertical Vegetation on the Thermal Performance of High-Rise Office Building Facades in the Mediterranean Climate. *Energy and Buildings*, 236, 110761.

Bao, Y., Li, B., & Li, Y. (2018). Research on the Preservation and Reuse of Traditional Architecture Based on Sustainable Development. *Journal of Cleaner Production*, 172, 4414–4423.

Belba, R., & Dervishi, S. (2018). *Thermal Performance Analysis of Traditional Housing in Albania.* Syracuse, NY, USA: Syracuse University Libraries. pp. 1485–1490.

Bell, K., & Lane, R. (2018). Energy Poverty: An Overview. *WIREs Energy and Environment*, 7(1), e288.

Boardman, B. (2012). Fuel Poverty. In *International Encyclopedia of Housing and Home*. San Dieso. pp. 221–225.

Bouzarovski, S. (2013). Social justice and climate change: Addressing energy poverty at the European scale. Spring Alliance's Conference "Go Green, Be Social". pp. 1–8.

Bouzarovski, S., & Petrova, S. (2015). A Global Perspective on Domestic Energy Deprivation: Overcoming the Energy Poverty–Fuel Poverty Binary. *Energy Research & Social Science*, 10, 31–40.

Bouzarovski, S., Petrova, S., & Sarlamanov, R. (2012). Energy Poverty Policies in the EU: A Critical Perspective. *Energy Policy*, 49, 76–82.

Breçani, R., Belba, R., & Dervishi, S. (2019). Simulation assisted energy optimization of abandoned agricultural facilities: An adaptive reuse approach. In Building Simulation Conference Proceedings, vol. 4. International Building Performance Simulation Association. pp. 2554–2561.

Breçani, R., & Dervishi, S. (2019). Thermal and Energy Performance Evaluation of Underground Bunkers: An Adaptive Reuse Approach. *Sustainable Cities and Society*, 46, 101444.

De Feo, G., & Brano, V. L. (2021). The Historic Urban Landscape Approach and the Value of Cultural Heritage for Sustainable Development: An Italian Case Study. *Sustainable Cities and Society*, 68, 102787.

US Department of Energy. (2020). EnergyPlus | EnergyPlus. US Department of Energy's. Retrieved from https://energyplus.net/

Dervishi, S., & Karamani, S. (2020). Exploring the energy performance of an existing industrial building. In E3S Web of Conferences, vol. 181. EDP Sciences.

Dervishi, S., Lacaj, E., & Vathi, R. (2012). Urban Heat Islands (UHI) Mitigation in Densely Urban City of Tirana, Albania: Materials, Energy, Comfort. *International Journal of Business & Technology*, 1(1), 48–57.

Dervishi, S., & Mahdavi, A. (2012). Computing Diffuse Fraction of Global Horizontal Solar Radiation: A Model Comparison. *Solar Energy*, 86(6), 1796–1802.

Dervishi, S., & Mahdavi, A. (2013a). A simple all-weather sky radiance model. In Proceedings of BS 2013: 13th Conference of the International Building Performance Simulation Association. pp. 916–921.

Dervishi, S., & Mahdavi, A. (2013b). A Simple Model for the Derivation of Illuminance Values from Global Solar Radiation Data. *Building Simulation*, 6(4), 379–383.

Dervishi, S., Pashako, F., Dushaj, X., & Dervishi, I. O. (2022). Energy Performance Optimization of Traditional Housing in Mediterranean Climate. *Journal of Building Engineering*, 45, 103423.

DesignBuilder Software Ltd. (2020). DesignBuilder. Retrieved from https://designbuilder.co.uk/software/product-overview

European Parliament. (2010). Directive 2010/31/EU. *Official Journal of the European Union*, L153/13(18.6.2010), 13–35.

Göçer, Ö., & Dervishi, S. (2015). The use of building performance simulation tools in undergraduate program course training. In 14th International Conference of IBPSA – Building Simulation 2015, BS 2015, Conference Proceedings. International Building Performance Simulation Association. pp. 2781–2786.

Haberl, J.S., Norton, E.A., & Deline, C. (2014). High-Resolution Solar Irradiance and Daylight Simulations Using Integrated Sky and Daylighting Analysis. *Solar Energy*, 108, 29–41.

Instat (2023). http://www.instat.gov.al/

Kalaci, E., & Dervishi, S. (2014). Implementation Challenges to the Adaptive Reuse of a Heritage Building in Tirana, Albania. In 2nd ICAUD International Conference in Architecture and Urban. Tirana, Albania: Epoka University.

Karásek, J., & Pojar, J. (2018). Programme to Reduce Energy Poverty in the Czech Republic. *Energy Policy*, 115, 131–137.

Kummert, M., & Lhomme, W. (1996). A Simple Model to Compute the Clear Sky Radiation on Horizontal and Tilted Surfaces. *Solar Energy*, 56(3), 239–251.

Law no. 116/2016, "Energy Performance of Buildings".

Law no. 124/2015, "Energy Efficiency".

Llera-Sastresa, E., Scarpellini, S., Rivera-Torres, P., Aranda, J., Zabalza-Bribián, I., & Aranda-Usón, A. (2017). Energy Vulnerability Composite index in Social Housing, from a Household Energy Poverty Perspective. *Sustainability (Switzerland)*, 9(5), 691–720.

Mahdavi, A., & Dervishi, S. (2010). Approaches to Computing Irradiance on Building Surfaces. *Journal of Building Performance Simulation*, 3(2), 129–134.

Mahdavi, A., & Dervishi, S. (2011a). A Comparison of Luminous Efficacy Models Based on Data from Vienna, Austria. *Building Simulation*, 4(3), 183–188.

Mahdavi, A., & Dervishi, S. (2011b). Exploring the energy performance of simulation-powered lighting and shading systems controls in buildings. In Proceedings of Building Simulation 2011: 12th Conference of International Building Performance Simulation Association. pp. 772–778.

Mahdavi, A., Schuss, M., Suter, G., Metzger, S., Camara, S., & Dervishi, S. (2009). Recent advances in simulation-powered building systems control. In IBPSA 2009 – International Building Performance Simulation Association 2009. pp. 760–766.

Meteonorm. (2016). Meteonorm: Irradiation data for every place on Earth. Retrieved from http://www.meteonorm.com/downloads

Middlemiss, L., Ambrosio-Albalá, P., Emmel, N., Gillard, R., Gilbertson, J., Hargreaves, T., & Tod, A. (2019). Energy Poverty and Social Relations: A Capabilities Approach. *Energy Research and Social Science*, 55, 227–235.

Mustafaraj, E. (2014). Practices, barriers and challenges of risk management implementation in Albanian construction industry. In 3rd Annual International Conference on Civil Engineering, At Athens, Greece, vol. 6(39), pp. 17–28.

Omar, O., Sabsaby, Y., & Amro, R. (2015). Zero Energy University Buildings Energy Performance Evaluation of Faculty of Architectural Engineering in Tripoli's Branch. *Architecture and Planning Journal*, 23(1), 1–9.

Orehounig, K., Dervishi, S., & Mahdavi, A. (2014). Computational Derivation of Irradiance on Building Surfaces: An Empirically-Based Model Comparison. *Renewable Energy*, 71, 185–192.

Ortiz, J., Jiménez Martínez, M., Alegría-Sala, A., Tirado-Herrero, S., González Pijuan, I., Guiteras Blaya, M., & Canals Casals, L. (2021). Tackling Energy Poverty Through Collective Advisory Assemblies and Electricity and Comfort Monitoring Campaigns. *Sustainability (Switzerland)*, 13(17), 9671.

Padeiro, S., Bragança, L., & Mateus, R. (2018). Retrofitted Traditional Buildings: Energy Efficiency Versus Cultural Heritage Preservation. *Energy and Buildings*, 169, 32–47.

Papada, L., & Kaliampakos, D. (2016). Measuring Energy Poverty in Greece. *Energy Policy*, 94, 157–165.

Pee, L. G., & Pan, S. L. (2022). Climate-Intelligent Cities and Resilient Urbanization: Challenges and Opportunities for Information Research. *International Journal of Information Management*, 63. 102446.

Picari, V., & Dervishi, S. (2019). Analysis of urban heat Island phenomenon and mitigation strategies for Tirana, Albania. In Building Simulation Conference Proceedings (Vol. 6, pp. 3786–3793). International Building Performance Simulation Association.

Pojani, D. (2010). Tirana. *Cities*, 27(6), 483–495.

Pu, Y., Wang, Y., & Wang, P. (2022). Driving Effects of Urbanization on City-Level Carbon Dioxide Emissions: from Multiple Perspectives of Urbanization. *International Journal of Urban Sciences*, 26(1), 108–128.

Reinhart, C.F., & Perez, R. (2004). New Simplified Models for Assessing the Solar Radiation Distribution in the Urban Environment. *Solar Energy*, 77(2), 159–169.

Resuli, P., & Dervishi, S. (2015). Thermal Performance of Cultural Heritage Italian Housing in Albania. In *Energy Procedia*, vol. 78. Elsevier Ltd. pp. 753–758.

Review, W. P. (2021). GDP Ranked by Country 2021.

Sareen, S., Thomson, H., Tirado Herrero, S., Gouveia, J. P., Lippert, I., & Lis, A. (2020). European Energy Poverty Metrics: Scales, Prospects and Limits. *Global Transitions*, 2, 26–36.

Simcock, N., Frankowski, J., & Bouzarovski, S. (2021). Rendered Invisible: Institutional Misrecognition and the Reproduction of Energy Poverty. *Geoforum*, 124, 1–9.

Taipi, E., & Ballkoci, V. (2017). Capital Expenditure and Firm Performance Evidence from Albanian Construction Sector. *European Scientific Journal, ESJ*, 13(28), 231.

Triantis, L. (2018). The Post-Socialist Restitution of Property as Dispossession: Social Dynamics and Land Development in Southern Albania. *Land Use Policy*, 71, 584–592.

Zhang, Z., Shu, H., Yi, H., & Wang, X. (2021). Household Multidimensional Energy Poverty and Its Impacts on Physical and Mental Health. *Energy Policy*, 156, 112381.

19 From Energy Poverty to Vulnerability

A Discourse Analysis of the European Union's National Energy and Climate Plans

Ahmed Noaman El-Sherbini and Ingmar Lippert

19.1. Introduction

Great efforts to study EP have been undertaken since 1975–1985 (Primc et al., 2019, 727; Stojilovska et al., 2022, 1). Yet, many claim that research on EP – especially in Europe– is "insufficient", "less developed", or has "a distinct gap in knowledge" (e.g., Thomson and Snell 2013, 564; Okushima 2017, 1159; Ntaintasis et al. 2019, 160). Additionally, the political economies and the different day-to-day experiences of EP are "poorly understood" within Europe (Bouzarovski et al. 2012, 78). Further research is needed for a better understanding of the complexity of EP, which remains considered "largely under-researched" (Stojilovska et al. 2022, 1).

We set out from observing some research on EP treating the phenomenon mono-dimensionally, as if it could be identified by a single cause or measured by a specific singular indicator (Sovacool et al. 2012; Scarpellini et al. 2015). However, other research suggests EP is identifiable in varied ways, exists in different forms, or is caused by different conditions within different contexts (Castaño-Rosa et al. 2019, 36–37; Ntaintasis et al. 2019, 160; 162; 163; 168). Supposedly, that multidimensional characteristic of EP makes it very hard to measure, and any attempt to mono-dimensionally define EP is most likely to fail in many contexts. Moreover, neither common nor official standardised ways of assessing EP in Europe exist (Castaño-Rosa et al. 2019, 37). However, attempts to metricise EP for governance towards green and just transitions critically hinge on how EP is understood (Lippert and Sareen 2023). Following Castaño-Rosa et al. (2019, 37) and Bouzarovski et al.'s research and policy agenda (2021, 7), we engage two of their points. First, they reflect increasing attention towards understanding, analysing, and defining the concept of "*vulnerable consumer*" as a way to grasp EP; second, they point to the notion of vulnerability but do not make the relationship to EP clear.

In this chapter, employing a document-based discourse analysis of EU member states' policy documents, we problematise conceptualisations of EP. Next, we shed light on the historical emergences and contemporary negotiations of EP in Europe. Subsequently, we present the methods and concepts underlying our analysis. This allows us to ascertain policy imaginaries of the relationship between EP and vulnerability.

19.2 Scholarly discourse of Energy Poverty

Energy Poverty is a term that refers to the incapability of a household/consumer to either access or afford certain basic energy services (Okushima 2017, 1159; Castaño-Rosa et al. 2019, 36–37; Bouzarovski et al. 2021, 02). The related term *Fuel Poverty (FP)* creates a dilemma in relation to EP (Bouzarovski et al. 2012, 76; Thomson and Snell 2013, 564; Okushima 2017, 1159; Castaño-Rosa et al. 2019, 36). Some authors use both terms to indicate the same phenomenon,

DOI: 10.4324/9781003408536-23

using them interchangeably (Bouzarovski et al. 2012, 76; Boemi et al. 2017, 167–168). Others suggest that EP describes the lack of access to energy services and is more developing-countries oriented, while FP indicates social/material issues and focuses on developed countries (Bouzarovski et al. 2012, 76; Ntaintasis et al. 2019, 160; Castaño-Rosa et al. 2019, 36–37). In short, the scholarly discourse seems open, even open to whether EP/FP is one phenomenon or several phenomena. We zoom in by exploring the historical emergences and current use of these terms.

19.2.1 Temporal-spatial emergences of the terms Energy Poverty, and Fuel Poverty within Europe

Historically, the terms EP and FP first appeared in literature between 1975 and 1985 (Moore 2012, 1; Papada and Kaliampakos 2016, 157). According to Castaño-Rosa et al. (2019, 37), it was Isherwood and Hancock who first introduced the term EP in 1979, as a result of the inflation in fuel and energy prices that followed the 1973–1974 petroleum crisis. Moore (2012, 1) claims that Bradshaw and Hutton were the first to introduce FP back in 1983.

Geographically, the term FP originates in the United Kingdom (UK) and Ireland, where it is still widely applied (Boemi et al. 2017, 168). Originally, FP was mainly referring the state of being "*depriv[ed ...] of domestic energy*" (Boemi et al. 2017, 168). In contrast, the term EP is more commonly applied across continental Europe, especially in central and eastern Europe (Boemi et al. 2017, 167–168; Ntaintasis et al. 2019; 160).

Boardman (1991) is widely considered the first author to formulate a definition of FP, considering a household as fuel-poor when its expenditure on basic energy services exceeds 10% of its total income (Moore 2012, 01; Thomson and Snell 2013, 564; Primc et al. 2019, 730; Ntaintasis et al. 2019, 162). This definition by Boardman is widely mobilised as the 10% rule and is considered the "ground-zero", on which manifold definitions of EP/FP build (Papada and Kaliampakos 2016, 157; Sareen et al. 2020, 30).

Zooming into Boardman's *Fuel Poverty: From Cold Homes to Affordable Warmth* (1991), we find the problem of FP has deep roots. Defining it in terms of a percentage of a household's income is merely the tip of the iceberg (Boardman 1991, 230). Boardman does not introduce the 10% definition until about the last fifth of the book, and she introduces it as a "first [*quantitative definition*] of FP". However, she acknowledges it should be neither seen as an absolute nor a grounded definition (Boardman 1991, 205).[1] Boardman lays out a complex network of intertwined factors, such as historical/economic/social events, health impacts/status, demographic status/statistics, governmental/social actors, weather/housing stock conditions, national subsidies, inflation rates, and domestic fuels/fuel prices; all these factors are temporally and spatially contextualised (Britain, early 1960s to late 1980s) (Boardman 1991, 1; 5; 12–14; 16–21; 24–25; 230).

This brief historical background on how EP/FP was imagined and understood illustrates that among both terms EP and FP, and the phenomenon/phenomena they refer to, a range of relations exist. These relations change through time and across scholarly and expert conversations. In sum, across these relations, it remains open whether EP/FP is a singular phenomenon, a multidimensional one, or two distinct phenomena. Therefore, historically, we have identified EP/FP as a complex concept.

19.2.2 The EP/FP-Complex within present Europe

Shifting to the early 21st century, we show that such complexity also characterises the contemporary situation of EP/FP in Europe. For that, consider the following definition samples, illustrating the ambiguity and the multiple ways of rendering EP/FP governable.

- *General EP*: One way to define EP is as a condition, in which a household cannot access energy services, rendering it incapable of achieving the socially and materially necessitated level of comfort (Boemi et al. 2017, 167; Castaño-Rosa et al. 2019, 36; Sareen et al. 2020, 26; Bouzarovski et al. 2021, 2).

- *FP, and different national contexts*: Thomson and Snell (2013, 564) describe FP as a by-product of a complex interaction of, for example, low income, insufficient domestic energy efficiency, the absence of savings, and living in rented accommodations. Furthermore, Thomson and Snell (2013, 564) differentiate among three FP definitions across three national contexts (i.e., Ireland, France, and the UK). In Ireland, FP is either the inability to afford adequate warmth at home or being unable to achieve it due to the home being energy-inefficient (Thomson and Snell 2013, 564). In France, FP occurs when a person/household faces difficulties having the required energy supply at home, which covers the elementary energy needs; either these difficulties are financial or related to housing conditions (Thomson and Snell 2013, 564). In the UK, FP is a case of lacking sufficient financial resources, which are enough to keep the home adequately warm (Thomson and Snell 2013, 564; Castaño-Rosa et al. 2019, 37).
- *EP, and different thresholds*: González-Eguino (2015, 379) defines EP as a state, in which a household suffers an "absence of sufficient choice" to access "adequate, affordable, reliable, high-quality, safe and environmentally benign energy services". González-Eguino (2015, 380) adds that EP can also be defined in terms of technological, physical, or economic thresholds. Technological EP is the lack of access to modern energy services (e.g., electricity but not biomass) (González-Eguino 2015, 380). Physical EP refers to the case of a household being unable to meet the minimum energy consumption threshold for covering its basic necessities (González-Eguino 2015, 380). Economic EP occurs when a household's expenditure on energy services surpasses a certain threshold of its total income (González-Eguino 2015, 380).
- *EP, FP*: According to Castaño-Rosa et al. (2019, 36–37), FP does not only refer to a household that is deprived of heating/cooling but also of hot water, electricity, or any other energy services that are essential for the household. While EP is related to the geospatial status of the household, referring to areas with higher poverty levels (e.g., old parts of towns, suburbs, socially excluded zones, rural areas), in which many buildings/housing stocks lack energy-efficient infrastructure/equipment (Castaño-Rosa et al. 2019, 36–37).
- *Hidden EP*: In Italy, EP comes in two forms, depending on the situation; the first is the inability of a household to afford high energy bills, and the second is when a household is forced to constrain its energy consumption to avoid having high energy bills in the first place (Betto et al. 2020, 3). When a household is "forced to under-consume" its energy needs to avoid high energy bills, this particular case is called "hidden energy poverty" (Betto et al. 2020, 1–3).

Castaño-Rosa et al. (2019, 37) state that despite the countless definitions and the variety of indicators of EP/FP that "were developed in Europe", neither a standardised definition nor an indicator was unified across the EU. The more literature on the phenomenon is reviewed, the more complex, ambiguous, and elusive defining EP/FP becomes (Thomson and Snell 2013, 568; Primc et al. 2019, 727; Sareen et al. 2020, 28).

According to Bouzarovski et al. (2012, 78), delimiting the phenomenon of EP/FP to a single cause (e.g., low income, inefficient buildings, high prices of energy) ignores a multitude of other technical and socio-economic causes of a complex phenomenon. Additionally, Primc et al. (2019, 728) claim that the factors affecting EP/FP are "diverse and sometimes contradictory". Therefore, for "explaining" the phenomenon of EP/FP, "consider[ing]" the "interaction" of multiple causes is necessary. Furthermore, Pachauri and Spreng (2011, 7497) suggest that

EP/FP is difficult to measure because of diverse factors that "influence" it, such as "economic growth", "social and infrastructure development", "environmental factors", and "behavioural factors", which can also interact, causing variations in both the form and the magnitude of EP/FP. Mitigating EP/FP as a multidimensional phenomenon might benefit from attempts to define it through the consideration of a range of indicators that are relevant for a specific national, subnational, regional, or even household context (Nussbaumer et al. 2012, 234).

In sum, this section reveals that the research literature widely considers EP/FP ambiguous, complex, and multidimensional. We find two terms (EP and FP), and each has its own plethora of definitions. Both terms came to exist within specific temporal-spatial contexts. We wonder then how the policy discourse reflects or reconfigures this definitional plethora. But before applying discourse analysis to the NECPs, in the next section, we introduce some basic concepts upon which our framework and methodology are established.

19.3 Data, concepts, and methods

To explore how the policy discourse relates to EP, we require empirical material, analytical concepts, and/or a methodology. To render these transparent, we next introduce the corpus, the empirical material we analyse. Then, we address the concepts of discourse and the perspective of discourse analysis. This allows us to introduce SKAD as our framework and finally specify our methodology.

19.3.1 The corpus: NECPs

The NECPs are integrated and established plans that cover the ten-year period (2021–2030) of how the EU-27 member states intend to meet their energy and climate objectives. Each of the 27 NECPs addresses themes like energy efficiency, renewable energy, greenhouse gas emissions reductions, or research and innovation. For our analysis, we use the English-language versions of the 27 plans.

We chose the NECPs for two reasons. First, the NECPs approach EP/FP as a policy target, a problem that is yet to be monitored or mitigated, which differs from other EP/FP documents. Second, we treat the NECPs as the "state of the art" understanding of the distributed EP/FP governance apparatus, specifying not only EP/FP in general but also the specific natures of the phenomenon and its indicators across 27 different national contexts.

19.3.2 Discourse and discourse analysis

A Discourse is an "ensemble of ideas, concepts, and categorisations, which are produced and transformed in a specific set of practices", which construct an aspect of truth about a particular social reality (Leipold et al. 2019, 447). In that sense, discourse is a *medium* through which knowledge about a particular social reality can be constructed and produced (Marvasti 2004, 107–108; Rosenthal 2018, 206; Leipold et al. 2019, 447).

Keller et al. (2018, 20) define discourse as a practice, that regulates formulating a statement in response to a problem. Empirically, this practice can be accessed with a corpus containing communication evidence (e.g., speech, written/printed texts, visual representations, etc.) (Keller et al. 2018, 19–20; 26; 31).

Discourse Analysis critically interprets the language in a corpus, that is, a selected body, of particular texts, to identify patterns of how a specific issue (e.g., EP/FP) is rendered thinkable and understandable (Keller et al. 2018, 26; 169). As Rosenthal (2018, 205) puts it, "The term 'discourse analysis' should not be understood as referring to a specific method, but rather as a '*research perspective*'", which we present in detail in the next two sub-sections (Framework and Methodology).

19.3.3 Our framework

For analysing the NECPs, we employ the discourse analysis approach SKAD. It allows designing and adapting strategies for interpreting the data and analysing how a specific topic is known, including, for instance, its processing through classifications (Keller 2011, 48; Keller et al. 2018, 5–6; 27–28; 74). This capacity of SKAD is key for our analysis of the NECPs because their technical policy language employs a range of implicit lists and classifications to render EP/FP known and, thus, governable.

To analyse a discourse, SKAD recommends the use of the Foucauldian notion of "dispositifs", as discourses "exist through and in [dispositifs]" (Keller 2011, 49; 58; Keller et al. 2018, 4). Keller et al. (2018, 4) define the dispositif as the "infrastructure established by social actors or collectives to resolve a particular situation". SKAD specifies two different types of "dispositifs": the first refers to the set of all entities that construct/enable/generate the discourse, while the second is the "discursive outcome", that is, the set of effects that intervene within the discourse (Keller 2011, 56; Keller et al. 2018, 27; 162).

According to Keller (2011, 46; 55–56; 64) and Bacchi (2012, 4–5; 7), discourse does not appear out of nothing; it is rather "[*brought to life*]" through processes of a variety of interactions, which are historically situated in social fields. To analyse such processes implies turning the discourse into a problem that offers historically and politically specific solutions. Analysts have addressed such problems with a variety of case studies, including, for instance, on discourses of ecological modernisation (Hajer 1995). In such studies, we find that discourses construct the phenomenon they are concerned with. SKAD builds, *inter alia*, on Foucauldian analyses, for which we mobilise the concept of *Problematisation.*

To "*problematise*" a phenomenon is to study its history, relations, and politics, which led to its appearance, and to access both temporal and spatial spaces, within which it came to exist (Bacchi 2012, 6–7). Bacchi (2012, 1) adds that "problematisation" or "thinking problematically" is a theoretical approach that has two components. First, when analysing a specific problem, the aim is to explore how this issue is discussed from different viewpoints within particular historical period(s), and under particular conditions. Second, the analyst must explore how the phenomenon is thought of and why it is turned into a problem.

19.3.4 Methodology

We mobilise the concept of discourse as a theoretical device for ordering the NECPs, in which language on EP/FP is used, and through and by which a context of EP is structured. For this, we build on Prior (2003) to argue for using documents as key forms of text (data). According to Prior (2003, 51), "texts [do not] merely represent some aspect of the world, [but they] are also involved in making the world. In part, they constitute the world". Hence, any documents concerned with EP/FP, however they contextualise the phenomenon, do not just reflect an aspect of it; they also participate in constituting and forming the phenomenon and its discourse(s).

Using SKAD's interpretation of dispositif, for our study, the "particular situation" to be "resolved" is the EP/FP phenomenon. The "established infrastructure" is the collection of all the documents concerned with the phenomenon. In this chapter, we specifically focus on the NECPs, which represent the second dispositif in Keller's typology. We treat the NECPs as

EU-level "national statements", which are concerned with EP/FP.[2] We specifically focus on the variations within the terms EP and FP, exploring the ways this problem is textually approached within the NECPs.

19.4 Discourse analysis of EP/FP within the NECPs

The NECPs present the respective member states' plans concerning climate change, energy infrastructure, and transition in highly technical policy language. Their structure typically involves many sections and subsections.

Almost all the NECPs emphasise EP/FP under the subsection "Internal Energy Market". Then, given 27 different internal energy markets, we can expect 27 different market constructions, different political circumstances, different consumption conditions, different suppliers, different policies and policymakers, etc. Hence, from this formalistic vantage point, we can potentially expect –at least– 27 different forms of EP/FP.

In terms of the terminology used, we find that the majority of NECPs refer to the phenomenon only as EP. The exceptions are four NECPs (Belgium's, France's, Ireland's, and Spain's) that refer to the phenomenon with both EP and FP. Let us turn from mentioning EP/FP to how these are approached. Here are four examples:

> Definition of energy poverty
>
> (heading in Austria NECP 2019, 97)

> 'A household is considered energy poor if its income is below the at-risk-of-poverty threshold and, at the same time, it has to cover above-average energy costs.'
>
> (Epigraph under the heading in Austria NECP 2019, 97)

> A comprehensive state policy oriented towards solving the problem of energy poverty is planned to be developed.
>
> (Poland NECP 2019, 44)

> Italy has no official definition of energy poverty. … the phenomenon of energy poverty (EP), [is] understood to mean the inability to purchase a minimum energy basket of goods and services …
>
> (Italy NECP 2019, 116)

> Energy poverty is not an isolated phenomenon; it is an integral part of overall poverty.
>
> (Belgium NECP A 2019, 309)

Analysing the patterns of how and whether EP/FP has been defined in the 27 NECPs reveals a wide range of variations. As shown in Figure 19.1, the NECPs of countries in *Group A* identify EP/FP as a specific phenomenon, with an official definition and specific measures to mitigate that phenomenon. NECPs in *Group B* identify EP/FP as a problem/phenomenon that is yet to be defined, monitored, and mitigated, with plans to do so. And the NECPs of *Group C* claim that EP/FP is a phenomenon that has no official definition yet is recognised/discussed under

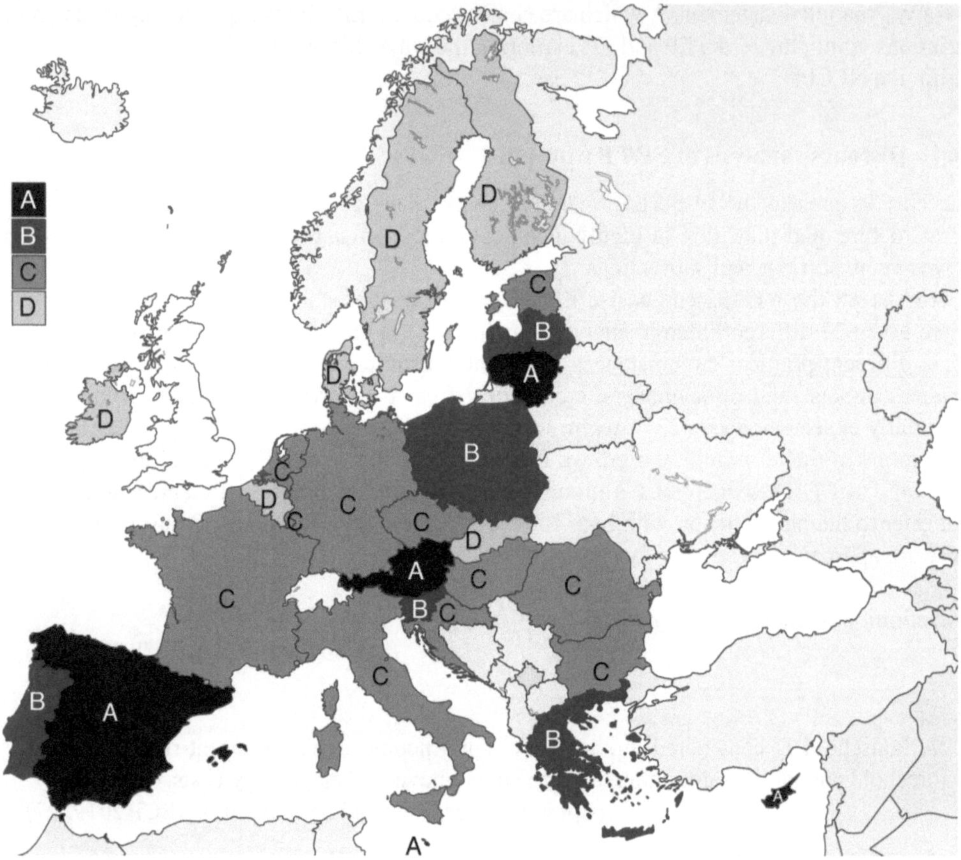

Figure 19.1 Variations of identifying EP/FP within the EU.

Source: Adapted by Ingmar Lippert from https://commons.wikimedia.org/w/index.php?title=File:European_Union_Enlargement.svg&oldid=715430364, CC by-sa/4.0 license. [Country Group A: Austria, Cyprus, Lithuania, Malta, Spain; B: Latvia, Poland, Portugal, Slovenia, Greece; C: Croatia, Germany, Hungary, Luxembourg, The Netherlands, Italy, Bulgaria, Czech Republic, Estonia, France, Romania; D: Belgium, Denmark, Finland, Ireland, Slovakia, Sweden].

particular conditions. Finally, NECPs of *Group D* identify EP/FP not as an independent phenomenon but as an effect of general poverty.

Zooming into individual NECPs shows that EP/FP is evoked with a range of textual actions, such as EP/FP being "defined", "indicated", "measured", "caused", "mitigated", and "tackled". To illustrate, here are two cases of evoking EP via listing "causes" of the phenomenon:

> Energy poverty is caused by four main problems: energy inefficiency, high energy prices and low household incomes, as well as lack of consumer awareness.
>
> (Lithuania NECP 2020, 45)

> Energy poverty is caused by a combination of three basic factors at household level: energy prices, energy efficiency and household income.
>
> (Czech Republic NECP 2019, 69)

In these two cases, EP was related to several causes. Analysing such causes across the 27 NECPs indicated five patterns through which EP/FP is approached: (1) *financial issues/difficulties* (e.g., low income, over-expenditure, affordability, arrears on energy bills, energy tariffs, high energy prices); (2) *vulnerabilities/vulnerable groups/consumers* (e.g., lack of energy services, specific socio-economic vulnerabilities/issues, struggling with energy supply/disconnection, energy saving misbehaviour/lack of awareness); (3) *low/bad-energy efficient buildings/housing* (e.g., need for renovations, poor insulation, insufficient heating); (4) *over/under/misconsumption of energy*; and (5) *general poverty*.

To illustrate the variations within such patterns, consider a "generally poor" household. This would indicate that this household is facing "financial difficulties", entailing a form of socio-economic "vulnerability", which can be mitigated through different types of "financial/social support" policies.

"Vulnerability", we suggest, is also reflected in the "energy consumption" pattern. Throughout the 27 NECPs, when EP/FP was contextually related to "energy consumption", they evoked "underconsumption" (which is caused by "financial vulnerability"). Households in "hidden energy poverty" are forced to "under-consume" their essential energy needs. NECPs also mentioned that forms of "misbehaviour" towards "energy use" or "energy consumption" are caused by a household's "lacking awareness" in that regard. And if a household lacks certain required knowledge regarding their energy needs, services, supply, or consumption, then this constitutes another dimension of "vulnerability".

"Energy efficiency of buildings/houses" constitutes a strong discursive pattern that we find nearly throughout all the 27 NECPs. This pattern appears in diverse forms and conditions, including "old buildings", which were built in certain period(s) of time, and are in need of "renovations". Or through specific types of households, which are inhabited by specific "socio-economic groups/individuals", who need "financial support" for a proper "heat/energy supply", or for "renovating" their houses. This pattern also emerges when "renewables" are evoked to "improve energy efficiency", or to minimise "energy costs". Or through "vulnerable/low-income groups" who live in homes characterised as "lacking modern energy standards buildings/houses". We find that any instances of the pattern "energy efficiency of buildings/households" reflect a specific "vulnerable socio-economic group", or reflect the "vulnerability" of specific buildings themselves, as these buildings need to be "renovated", or upgraded on a technical level due to variations of technical "vulnerabilities".

19.5 From EP/FP to vulnerability

Across our analysis, we find patterns in the NECPs that support defining EP/FP in terms of "vulnerabilities", or as a phenomenon that is an effect of some form(s) of "vulnerability". The "vulnerability" is ascribed to either social entities (e.g., consumers or households) or non-human infrastructures, especially buildings, in which some "vulnerable" social entity might live, or a combination of human/non-human entities. Thus, we find a techno-political textual pattern that understands EP/FP as a consequence of some form of vulnerability (rather than as a self-contained phenomenon or as sub-ordinary to "general poverty").

Vulnerability provides a wider and richer variation than general poverty. Vulnerability can include a technical dimension to express the technical vulnerability of a building, for example, which does not necessarily indicate a state of financial poverty/deprivation. Vulnerability can also express the "behavioural" and "awareness" dimensions of the phenomenon. As for a social group to "lack the required/needed awareness" about the "most efficient" or "best energy/financial saving" practices towards their energy services, this could entail a form of "social/cultural vulnerability". Our analysis aligns with Bouzarovski et al.'s (2021, 7) theorisation of energy vulnerability: "It could include income levels, the share of energy expenditure of disposable income, the energy efficiency of homes, critical dependence on electrical equipment for health reasons, age or other criteria".

In Bouzarovski et al.'s (2021) analysis of the NECPs, the term *Vulnerability* plays a significant role. In their concluding remarks, they identify "an acute need to [*rethink theorisations*] of the causes of energy poverty so as to expose the wider injustices and policy failures that lead to deprivation and vulnerability" (Bouzarovski et al. 2021, 14). Our analysis shows that the contemporary policy language in the NECPs supports a theorisation of EP as an effect of vulnerabilities, rather than as merely something that is vaguely related to vulnerabilities.

We argue that our reading of the NECPs provides indirect support for the concept of "contextualised identification" by Sareen et al. (2020). They define contextualised identification as a strategy that allows for identifying and measuring individual EP at the household or doorstep level.

This orientation directs theorising EP/FP towards addressing "*Who is energy/fuel poor?*" or "*How, and why are they energy/fuel poor?*" instead of answering "*How is EP/FP measured/defined?*" (cf. Pachauri and Spreng 2011, 7499; Primc et al. 2019, 734). This implies that the more individual (on national/sub-national/regional/neighbourhood/household levels) cases of EP/FP are allocated, the more EP/FP becomes a multidimensional phenomenon (Moore 2012, 1; Papada and Kaliampakos 2016, 163; Ntaintasis et al. 2019, 160; Sareen et al. 2020, 28; Bouzarovski et al. 2021, 5; 6). For a better diagnosis of EP/FP we need to set out from identifying who is the "*Vulnerable Consumer*", and the reasons for their vulnerability, and then consider the measures to combat/mitigate that specific kind of vulnerability (cf. Castaño-Rosa et al. 2019, 37; Betto et al. 2020, 1; Sareen et al. 2020, 32; Bouzarovski et al. 2021, 6). We thus argue that the NECPs' framing of EP/FP exemplifies well EP/FP as a phenomenon that is being "[*socially shaped*]", with a "[*varied*]" and "[*multidimensional*]" identity (Sareen et al. 2020, 2).

19.6 Conclusions

This chapter sets out from a literature review, analysed as the scholarly discourse of energy poverty. This indicated a multitude of conceptualisations of EP/FP within Europe, through history, and to the present, in a variety of contexts. Still, some research on the phenomenon seeks to establish a unifying definition or a singular indicator for EP/FP that can be applied across diverse contexts. Analysing critiques of such approaches suggested that the phenomenon of EP/FP is well described in terms of ambiguity, complexity, and multidimensionality. The literature further indicated an emerging discussion of EP/FP in relation to vulnerabilities. Against this backdrop, the core of the chapter consists of exploring how the technical policy language of the European Union member states' National Energy and Climate Plans defines EP/FP and how that language relates to vulnerabilities.

Our analysis of the NECPs indicates a variety of EP/FP concepts within the 27 national contexts. Across these concepts, we identify vulnerabilities as key. Specifically, we identified five patterns of causes of EP/FP, all of which we traced to forms of vulnerability. Recognising forms of vulnerability explains the ambiguous, complex, and multidimensional characters of EP/FP. Noteworthy is that the NECPs are not strong at conceptualising EP/FP explicitly in terms of vulnerabilities; instead, the technical policy language leaves vulnerability mostly implicit.

Furthermore, our analysis now allows to address the shared problematisation in the literature that claims EP/FP is poorly understood, under-researched, and suffers from a knowledge gap. Against attempts to establish a unified definition or singular method for assessing EP/FP, our analysis of the NECPs indicates that EP/FP is imagined as occurring in diverse forms. These diverse forms cannot be reduced to a singular condition. Our research suggests that the NECP's technical policy language is in principle able to approach EP/FP as a multidimensional and complex phenomenon. However, we observe that only a few NECPs use that language to achieve such recognition.

While our analysis of the NECPs, thus, confirms identifying EP/FP as an ambiguous, complex, and multidimensional problem, promises of the recognition of these characteristics and the multiplicity of EP/FP as effectively establishing the capacity for policy action require caution. That is, we recommend caution because EP/FP can be effectively discursively addressed as an umbrella term that allows heterogeneous and even contradictory phenomena of EP/FP to be ignored by a coalition of actors that performs policy action on EP/FP. However, the analysis of the NECPs also underlines that research now can draw on these policy documents themselves to call for research and monitoring resources to study energy poverty as a multidimensional phenomenon that reflects and produces a variety of vulnerability forms, which can vary greatly within a single national, including sub-national, context. Identifying EP/FP in such a manner might provide critical insight that can inform and problematise future policies and plans to mitigate the phenomenon.

Notes

1 For further details on the technicalities and the calculations, check Boardman (1991, 199–219).
2 Note that our literature section on the scholarly discourse on EP has been constructed based on the SKAD discourse analysis of the first dispositif in Keller's typology. That analysis focuses on scholarly documents that are EP/FP related. In that analysis, we investigated the emergences of EP/FP, the claims regarding the appearance of both terms EP and FP, the different contexts and conditions underlying these emergences, and the different socio-economic and historic dimensions of the phenomenon, specifically addressing the ways EP/FP is rendered defined and measurable. For details on this methodology, see El-Sherbini (2022).

References

Bacchi, Carol. "Why Study Problematizations? Making Politics Visible." *Open Journal of Political Science* 02, no. 01 (2012): 1–8. https://doi.org/10.4236/ojps.2012.21001

Betto, Frida, Patrizia Garengo, and Arturo Lorenzoni. "A New Measure of Italian Hidden Energy Poverty." *Energy Policy* 138 (March 1, 2020): 111237. https://doi.org/10.1016/j.enpol.2019.111237

Boardman, Brenda. *Fuel Poverty: From Cold Homes to Affordable Warmth.* 95-211049+01. Great Britain: Belhaven Press, 1991.

Boemi, Sofia-Natalia, Spyros Avdimiotis, and Agis M. Papadopoulos. "Domestic Energy Deprivation in Greece: A Field Study." *Energy and Buildings* 144 (June 1, 2017): 167–174. https://doi.org/10.1016/j.enbuild.2017.03.009

Bouzarovski, Stefan, Saska Petrova, and Robert Sarlamanov. "Energy Poverty Policies in the EU: A Critical Perspective." *Energy Policy*, Special Section: Fuel Poverty Comes of Age: Commemorating 21 Years of Research and Policy, 49 (October 1, 2012): 76–82. https://doi.org/10.1016/j.enpol.2012.01.033

Bouzarovski, Stefan, Harriet Thomson, and Marine Cornelis. "Confronting Energy Poverty in Europe: A Research and Policy Agenda." *Energies* 14, no. 4 (January 2021): 858. https://doi.org/10.3390/en14040858

Castaño-Rosa, Raúl, Jaime Solís-Guzmán, Carlos Rubio-Bellido, and Madelyn Marrero. "Towards a Multiple-Indicator Approach to Energy Poverty in the European Union: A Review." *Energy and Buildings* 193 (June 15, 2019): 36–48. https://doi.org/10.1016/j.enbuild.2019.03.039

Dagoumas, Athanasios, and Fotis Kitsios. "Assessing the Impact of the Economic Crisis on Energy Poverty in Greece." *Sustainable Cities and Society* 13 (October 1, 2014): 267–278. https://doi.org/10.1016/j.scs.2014.02.004

El-Sherbini, Ahmed Noaman. *"Discourse Analysis on Energy Poverty within the Context of the European Union's National Energy and Climate Plans"*. Master thesis at Brandenburg University of Technology (2022). Retrieved February 6, 2023 via https://www.researchgate.net/publication/368301922

González-Eguino, Mikel. "Energy Poverty: An Overview." *Renewable and Sustainable Energy Reviews* 47 (July 1, 2015): 377–385. https://doi.org/10.1016/j.rser.2015.03.013

Hajer, Maarten. *The Politics of Environmental Discourse: Ecological Modernization and the Policy Process*. Oxford: Oxford University Press, 1995.

Keller, Reiner. "The Sociology of Knowledge Approach to Discourse (SKAD)." *Human Studies* 34, no. 1 (2011): 43.

Keller, Reiner, Anna-Katharina Hornidge, and Wolf J. Schünemann, eds. *The Sociology of Knowledge Approach to Discourse: Investigating the Politics of Knowledge and Meaning-Making*. London: Routledge, 2018. https://doi.org/10.4324/9781315170008

Leipold, Sina, Peter H. Feindt, Georg Winkel, and Reiner Keller. "Discourse Analysis of Environmental Policy Revisited: Traditions, Trends, Perspectives." *Journal of Environmental Policy & Planning* 21, no. 5 (September 3, 2019): 445–463. https://doi.org/10.1080/1523908X.2019.1660462

Lippert, Ingmar, and Siddharth Sareen. "Alleviation of Energy Poverty Through Transitions to Low-Carbon Energy Infrastructure." *Energy Research & Social Science* 100 (2023): 103087. https://doi.org/10.1016/j.erss.2023.103087

Marvasti, Amir. *Qualitative Research in Sociology*. SAGE Publications, 2004. https://doi.org/10.4135/9781849209700

Moore, Richard. "Definitions of Fuel Poverty: Implications for Policy." *Energy Policy*, Special Section: Fuel Poverty Comes of Age: Commemorating 21 Years of Research and Policy, 49 (October 1, 2012): 19–26. https://doi.org/10.1016/j.enpol.2012.01.057

NECP, Austria. "Integrated National Energy and Climate Plan for Austria 2021-2030," 2019. https://ec.europa.eu/energy/sites/default/files/documents/at_final_necp_main_en.pdf

NECP, Belgium. "Belgian Integrated National Energy and Climate Plan 2021-2030 (Section A)," 2019. https://ec.europa.eu/energy/sites/default/files/documents/be_final_necp_parta_en.pdf

NECP, Czech Republic. "National Energy and Climate Plan of the Czech Republic," 2019. https://energy.ec.europa.eu/system/files/2020-03/cs_final_necp_main_en_0.pdf

NECP, Italy. "Italy's Integrated National Energy and Climate Plan," 2019. https://ec.europa.eu/energy/sites/default/files/documents/it_final_necp_main_en.pdf

NECP, Lithuania. "National Energy and Climate Action Plan of The Republic of Lithuania For 2021-2030," 2020. https://energy.ec.europa.eu/system/files/2020-06/lt_final_necp_main_en_0.pdf

NECP, Poland. "The National Energy and Climate Plan for 2021-2030 (Part 1-3) Objectives and Targets, and Policies and Measures," 2019. https://energy.ec.europa.eu/system/files/2020-08/pl_final_necp_part_1_3_en_0.pdf

Ntaintasis, E., S. Mirasgedis, and C. Tourkolias. "Comparing Different Methodological Approaches for Measuring Energy Poverty: Evidence from a Survey in the Region of Attika, Greece." *Energy Policy* 125 (February 1, 2019): 160–169. https://doi.org/10.1016/j.enpol.2018.10.048

Nussbaumer, Patrick, Morgan Bazilian, and Vijay Modi. "Measuring Energy Poverty: Focusing on What Matters." *Renewable and Sustainable Energy Reviews* 16, no. 1 (January 1, 2012): 231–243. https://doi.org/10.1016/j.rser.2011.07.150

Okushima, Shinichiro. "Gauging Energy Poverty: A Multidimensional Approach." *Energy* 137 (October 15, 2017): 1159–1166. https://doi.org/10.1016/j.energy.2017.05.137

Pachauri, Shonali, and Daniel Spreng. "Measuring and Monitoring Energy Poverty." *Energy Policy*, Clean Cooking Fuels and Technologies in Developing Economies, 39, no. 12 (December 1, 2011): 7497–7504. https://doi.org/10.1016/j.enpol.2011.07.008

Papada, Lefkothea, and Dimitris Kaliampakos. "Measuring Energy Poverty in Greece." *Energy Policy* 94 (July 1, 2016): 157–165. https://doi.org/10.1016/j.enpol.2016.04.004

Primc, Kaja, Renata Slabe-Erker, and Boris Majcen. "Constructing Energy Poverty Profiles for an Effective Energy Policy." *Energy Policy* 128 (May 1, 2019): 727–734. https://doi.org/10.1016/j.enpol.2019.01.059

Prior, Lindsay. *Using Documents in Social Research*. London: SAGE Publications, 2003.

Rosenthal, Gabriele. *Interpretive Social Research: An Introduction.* Göttingen: Göttingen University Press, 2018. https://doi.org/10.17875/gup2018-1103

Sareen, Siddharth, Harriet Thomson, Sergio Tirado Herrero, João Pedro Gouveia, Ingmar Lippert, and Aleksandra Lis. "European Energy Poverty Metrics: Scales, Prospects and Limits." *Global Transitions* 2 (January 1, 2020): 26–36. https://doi.org/10.1016/j.glt.2020.01.003

Scarpellini, Sabina, Pilar Rivera-Torres, Inés Suárez-Perales, and Alfonso Aranda-Usón. "Analysis of Energy Poverty Intensity from the Perspective of the Regional Administration: Empirical Evidence from Households in Southern Europe." *Energy Policy* 86 (November 1, 2015): 729–738. https://doi.org/10.1016/j.enpol.2015.08.009

Sovacool, Benjamin K., Christopher Cooper, Morgan Bazilian, Katie Johnson, David Zoppo, Shannon Clarke, Jay Eidsness, Meredith Crafton, Thiyagarajan Velumail, and Hilal A. Raza. "What Moves and Works: Broadening the Consideration of Energy Poverty." *Energy Policy* 42 (March 1, 2012): 715–719. https://doi.org/10.1016/j.enpol.2011.12.007

Stojilovska, Ana, Rachel Guyet, Katherine Mahoney, João Pedro Gouveia, Raúl Castaño-Rosa, Lidija Živčič, Ricardo Barbosa, and Tomislav Tkalec. "Energy Poverty and Emerging Debates: Beyond the Traditional Triangle of Energy Poverty Drivers." *Energy Policy* 169 (October 1, 2022): 113181. https://doi.org/10.1016/j.enpol.2022.113181

Thomson, Harriet, and Carolyn Snell. "Quantifying the Prevalence of Fuel Poverty Across the European Union." *Energy Policy*, Special Section: Transition Pathways to a Low Carbon Economy, 52 (January 1, 2013): 563–572. https://doi.org/10.1016/j.enpol.2012.10.009

20 What of Recognition Justice? An Empirical Analysis of the Role of Recognition Justice in Social Housing Smart City Projects in the Global North

Luca Lamonaca and Susana Batel

20.1 Introduction

Net-zero policies that foster smart city projects are increasingly being promoted across the EU and the rest of the Global North. These projects involve high-efficiency buildings and appliances, the digitalisation of neighbourhoods, and a shift to demand-oriented energy systems that include behavioural change and citizen empowerment (Levenda, 2019). However, there are concerns about the social aspects of these projects that could affect democratic processes and accountability (Hearn et al., 2021), including psychosocial and symbolic elements such as empowerment and epistemic dimensions (Fricker, 2007).

These concerns have often been investigated in the literature by adopting an energy justice (EJ) perspective to explore the social elements of energy transitions. EJ is predominantly, but not exclusively, considered from a three-tenet perspective: recognition, distributive, and procedural justice (Jenkins et al., 2021). However, as pointed out already (Levenda et al., 2021; Simcock et al., 2021), recognition justice is often considered absent or assumed in empirical investigations in the Global North; that is, it is not analysed based on individuals' and communities' lived experiences of recognition (in)justice. It has been used as a conceptual analytical framework (Simcock et al., 2021), but, as far as we know, not to empirically investigate vulnerable populations included in smart city projects. So, the chapter will focus on the tenet of recognition justice, taking into account the lived experience of energy vulnerability (Bouzarovski & Petrova, 2015). We adopt the notion of energy vulnerability as it challenges the narrow conception of energy poverty that is increasingly used in policymaking, which is overly focused on income, efficiency, and energy prices (Longhurst & Hargreaves, 2019). Instead, energy vulnerability considers issues of access, affordability, flexibility, energy efficiency, needs, and practices in domestic energy services (Bouzarovski & Petrova, 2015). As such, investigating the justice aspects of the lived experience of energy vulnerability can capture psychosocial disadvantages, specifically the agentic role of citizens. This role has been deemed essential for these smart city projects' success and social acceptance (Sovacool et al., 2021; Wittenberg & Fleury-Bahi, 2016).

Indeed, research on energy vulnerability has greatly benefited from adopting a lived experience framework (Gillard et al., 2017; Longhurst & Hargreaves, 2019; Middlemiss & Gillard, 2015). Middlemiss and Gillard (2015) identified the following attributes of vulnerabilities: power/empowerment, multidimensionality, capacity for action, challenge, and integrity (Middlemiss & Gillard, 2015; Spiers, 2000). In these circumstances, smart energy projects, once funding is secured, will likely address the main concerns regarding access, affordability, and energy efficiency, thereby addressing energy vulnerability's most material drivers (Middlemiss & Gillard, 2015), but normally leave behind the dimensions that a recognition justice lens would foreground: the role of citizens' agency, associated with aspects of needs, practices and

DOI: 10.4324/9781003408536-24

flexibility. These are the aspects that we aim to empirically investigate in this chapter, therefore contributing to a more explicit and substantive integration of recognition justice in energy vulnerability research.

20.2 Energy vulnerabilities and recognition justice

Recognition justice is one of the tenets of the so-called "triumvirate" of energy justice (EJ), along with distributional and procedural tenets (Jenkins et al., 2021). Distributional justice focuses on the equitable distribution of energy resources and services, including costs, risks, benefits, and energy poverty concerns (Jenkins et al., 2021; Sovacool et al., 2017). The procedural tenet of EJ focuses on how decisions and decision-making processes about energy resources and services are realised, ensuring that decisions are made by all (Jenkins et al., 2021; Sovacool et al., 2017). Procedural justice intersects with recognitional justice. The latter concerns the appreciation and representation of stakeholder and citizen groups in energy systems (Simcock et al., 2021). They should be granted full and equal social, civic, and political rights, and their identities and values should be recognised and included in every step of the process (Walker, 2009). Often, the EJ triumvirate serves to better analyse the principles of justice such as availability and affordability, due process, transparency and accountability, sustainability, intragenerational and intergenerational equity, responsibility, resistance, and intersectionality (Sovacool et al., 2017). The recognition tenet thus accounts for those less tangible, symbolic, and affective dimensions, including of these justice principles, of people's relations with the eco-social-systems where they live (Jenkins et al., 2021; Martin et al., 2016; van Uffelen, 2022).

In this analysis, the framework of recognition justice also assumes the centrality of the cultural status order, defined as the "institutionalised patterns of cultural value for their effects on the relative standing of social actors" (Fraser, 2000, p. 9) that prevents some individuals and groups from fully and fairly participating "as peers in social life" (Fraser, 2000; van Uffelen, 2022). This conceptualisation considers the power relationships embedded in social relationships and associated psychosocial impacts: agency, interpersonal treatment (Ruano-Chamorro et al., 2022), trust, and place attachment. Agency is concerned with the criteria of control, voice, and capabilities. Directly influencing decisions is known as decision control (Tyler, 2015). Voice is the capacity to communicate one's interests, needs, desires, and priorities and offer details that may indirectly affect decisions (Ruano-Chamorro et al., 2021). Capabilities refer to participants' access to the resources and capital they need to exercise agency, such as time, information, and human and material resources (George & Reed, 2017; Ruano-Chamorro et al., 2021). Interpersonal treatment concerns respect, politeness (Ruano-Chamorro et al., 2021), and dignity (Grossmann & Trubina, 2021). When treated with respect and politeness, people feel treated with dignity, which includes listening to them and caring about their needs and contributions (Tyler, 2015). Interpersonal relations create a "thick" type of trust (Swain & Tait, 2007), essential to engaging with proponents and empowering citizens (Talvitie, 2012). Place attachment can also be critical in recognition justice and its relationship with social acceptance of energy infrastructures (Batel & Devine-Wright, 2020; Manzo & Perkins, 2006). In the remainder of this chapter, we will, based on the data collected in Bolzano with social housing tenants involved in a smart city initiative, empirically explore the relationship between their lived experiences of recognition (in)justice – operationalised here through the principles of agency, interpersonal treatment and trust, and place attachment (Figure 20.1) – and their acceptance of these projects. It is timely to further conceptualise and empirically explore recognition justice because dismissing this tenet of EJ might result in the failure of projects and efforts towards greening the energy system in an inclusive and just way (Wittenberg & Fleury-Bahi, 2016).

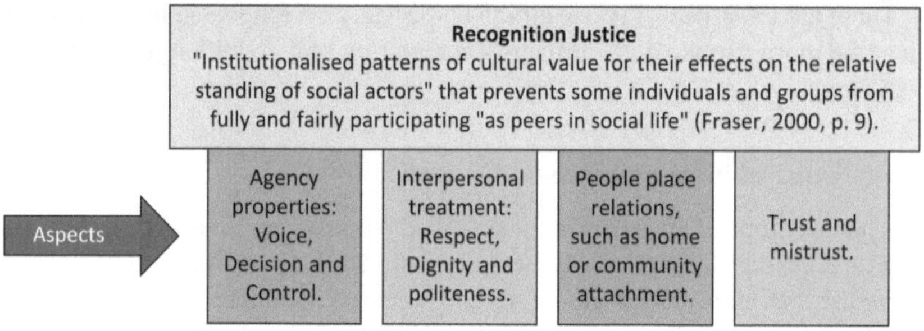

Figure 20.1 Conceptual framework.

Source: Adapted and modified from Ruano-Chamorro et al. (2021).

20.3 Case study

The case study is located in five different social housing complexes in two districts of Bolzano, North Italy, during the Sinfonia project (Caballero & Della Valle, 2021; Sinfonia smart cities, 2022). Bolzano is a midsized city and the capital of Alto Adige/South Tyrol (ASTAT, 2022). The share of social housing represents around 14% (ASTAT, 2022) against the national average of around 5% (Pittini, 2015). There are two social housing providers in Bolzano, one owned by the municipality and one by the province.

Sinfonia is a five-year pilot project that has been funded via the EU to achieve clean and affordable energy (Caballero & Della Valle, 2021; Sinfonia smart cities, 2022). The partnership is composed of the City Council, two social housing providers, owned by local polity: city council and the province, the local agency for Energy South Tyrol, two local research centres, and a local energy company. The declared objectives of the project were to reduce energy demand via retrofitting, behavioural change, and citizen engagement. Five social housing complexes were renovated, comprising around 300 flats. The project incorporated a deep renovation optimised for heating and cooling, smart electricity distribution, and PV panel installation. Retrofitting included home devices for Mechanical and Controlled Ventilation (MCV) and smart metering. Residents could voluntarily opt for a home tablet that monitors energy consumption and air quality to provide tips for optimal home comfort. MCV and tablets should require tenant interaction and a behavioural change in ventilation practices.

20.4 Methodology

We conducted 17 semi-structured interviews with households (mean length = 37 minutes), triangulated with 7 expert interviews (mean length = 40 minutes) (Creswell, 2009; Flyvbjerg, 2001), ethnographic observations, and official information from institutional sources. The experts were either involved in the Sinfonia project or similar projects in social housing districts (Creswell, 2009; Flyvbjerg, 2001). They provide further insights into tenants' lived experiences (Flyvbjerg, 2001; Martin et al., 2016; van Uffelen, 2022).

Fieldwork was carried out in November 2021 and January–February 2022. Procedures for data collection for the interviews were approved by the University Institute of Lisbon ethics committee (48/2021). Participants were provided with oral and written informed consent. The interviews were in Italian and were transcribed for coding and analysis. The authors translated quotations from Italian into English.

Interviews followed semi-structured guidelines in terms of topics. Questions revolved around the project itself, its design and implementation process, the new technologies installed, the relationship with, trust in, and ideas of the proponents, environmental and energy practices and place attachment. The participants were free to elaborate, and themes were constructed jointly between the researcher and the participants (Carter & Little, 2007; Patterson et al., 2012). Interviews were conducted adopting the saturation criterion, which is the criterion to stop sampling when no additional data is emerging from the interviews to develop themes (Glaser & Strauss, 1967; Saunders et al., 2018). The interviews with experts followed the same guidelines and reflections on preliminary insights from participants and tenants. Interviews were analysed on NVivo via thematic analysis (Braun & Clarke, 2006).

20.5 Analyses – lack of recognition of citizens' agency

Tenants mainly expressed satisfaction regarding the comfort of the renovated flats, whilst financial benefits were not always reported. Nevertheless, our research reveals how the agentic role of tenants was misrecognised in multiple ways. In turn, this misrecognition seems to affect citizen empowerment and lead to mistrust or, most frequently, indifference, a lack of appropriation, or confusion about the use of the new technologies, their goals, and the entire project. Feelings of anger, detachment, or disappointment sometimes emerged for different reasons: feeling disrespected because the proponents did not ask if they agreed with the renovation and due to long construction works, inconsistency in treatment, and a lack of clear accountability. Based on the interviews' analyses, this misrecognition seems to be mainly composed of four dimensions – influence of top-down governance mindset, epistemic dimension, people-place, and people-people relations – as described in detail below.

20.5.1 Influence of the top-down approach on misrecognition

The Sinfonia project was designed in a top-down fashion embedded in EU multilevel governance (Marks, 1996) and included citizen engagement activities solely for information, training, and environmental awareness. EU multilevel governance and regulations do not easily allow a bottom-up design to establish an open action plan or substantial modifications during the project. Whilst often in the literature these elements are merely considered part of decision-making processes in procedural injustice (Ruano-Chamorro et al., 2021), they are also a misrecognition of the agentic role of citizens, assuming the relevance of their selfhood (McAdams, 2013). Furthermore, the project was embedded in the existing dynamic of the SHP's mindset and its penalisation system, which has been increasingly adopted recently. These dynamics better show the impact of top-down governance on citizens' agency, as illustrated in the extracts below:

Extract 1. Participant 3 (expert – former Social Housing Provider Officer).

So, the ventilation [Mechanic and Controlled Ventilation – MCV] *is a problem that can be resolved. The remote control should not be given to tenants, but something centralised should be implemented using an external computer to control the MCV devices remotely. Do not give them* [the tenants] *any responsibility.*

Extract 1 shows the opinion of a former SHP officer: he clearly expresses that tenants' agency and control are problematic. It denotes a lack of trust towards them and reveals a strong centralistic and top-down mindset that the tenants are likely to perceive, possibly contributing to trust erosion. The dyadic relationship between agency and trust is well known, and the trust deficit is usually higher towards vulnerable or discriminated against sections of populations (Domenicucci, 2017).

Besides, this mindset materialised in some buildings where the top-down control of MCV was adopted or some measures were adopted to limit tenants' control. A degree of automation and control might be appropriate for the most vulnerable or people with other priorities. However, other residents seemed not to appreciate the lack of control over devices because it limits customisations and tenants' capacity for action (Middlemiss & Gillard, 2015). In these cases, it created feelings of confusion, frustration, or complete indifference towards some new devices. These processes affected trust towards new domestic energy infrastructures, as shown in extract 2:

Extract 2. Participant 15 (tenant).

> Yes, *the mechanical ventilation, but I don't use that much because .... It is annoying. I told them* [SHPs], *'Why don't you leave a remote control? So that I can control it myself. So now I can't control the MC that much* [the different features or turn on/off from a distance], *so it is always either on or off. You can turn it off/on just from the device itself, so ...* [he continues to talk about the impracticality, especially if someone adds the disturbance of the noise during the night, which results in always having it off].

The mindset is compounded and exacerbated by the penalisation system used by the SHPs, which further accentuates the already unbalanced power relationships between the social housing proponents and the tenants. Extract 3 shows how the penalisation systems greatly impact citizens' interactions with external entities and individuals.

Extract 3. Participant 3 (expert).

> *We always had to accompany* them [designers and workers that went to do door-to-door training for the smart energy tablet] ... *because if they go with the surveyor of the city council/SHP, they acquire a certain authority, and the tenants were a little scared that we would tell the provider that tenants refused to do something. As they are city council houses, it is easy to get them, but it is also easy to send the tenants away. So, as a consequence, they showed ... I would not say dread, but almost ....*

The system seemed to create fear and distrust towards the SHPs, making it more difficult to engage with them voluntarily without a sense of perceived obligation, often based on fear of losing one's home. As such, professionals working on citizen engagement had few opportunities for engagement, information, and empowerment. This system has traits of the so-called policing welfare (Headworth, 2021). There are well-known consequences of stigmatising those already vulnerable, exacerbating disparities of class, ethnicity, health, and so forth (Headworth, 2021). These processes produce mistrust towards the authorities and mixed emotions of fear, anger, and feeling degraded (Headworth, 2021). Some tenants expressed their concerns and manifested feelings of anger and impotence, typically towards bureaucracy and public authorities (Lapassade, 1977). The circumstances of being trapped in the bureaucracy have been thoroughly investigated, revealing an experience of injustice that leads to a sense of anger and detachment, affecting agency and self-esteem (van Uffelen, 2022). This is illustrated in the extract below:

Extract 4. Participant. 13 (tenant).

This year, the SHP called me [and said] *you did not pay the monthly rent last year' I responded that I did pay. They answered, "no, but that was for the previous house etc. etc., and you have to pay that one as well, 'look you must pay 101 Euros still left from the other time, and then you are done" And then you pay and shut up because they are always right. And I responded that I needed to check .... another bill of 45 euros arrived, and I was told 'no, you have to pay also this, but then you are done with the previous apartment ....*

The experience of penalisation or fear of penalisation in social housing combined with the lack of recognition in the implementation of the Sinfonia project seems to affect dimensions of self-esteem and respect (van Uffelen, 2022) and hinder agency, particularly in the aspect of "voice" (Ruano-Chamorro et al., 2021). These processes negatively affect the citizens' representations towards the SHPs, considered the only proponents. In turn, these representations seem to affect trust and engagement with new energy infrastructures and practices.

20.5.2 Misrecognising the epistemic dimension of energy justice

The lack of recognition of tenants as epistemic agents was also pointed out because adopting a standard and top-down design normally excludes this. In other words, tenants' knowledge about what worked and made sense and what did not in the context of their needs and lived experiences before and after the smart city-related interventions was dismissed and not included in the decisions taken at all. There is a perception of so-called deflated credibility (Fricker, 2007), which the literature shows might be associated with elitism or perceiving citizens as passive (Batel & Devine-Wright, 2020). The extracts below illustrate this issue:

Extract 5. Participant 16 (tenant).

I say to him [worker/architect], *'what if one day I want to change the furniture? What is going to happen? There is a hole behind this piece of furniture now; what should we do*?

Extract 6. Participant 13b (tenant).

It is not that the windows are in a high position; it is the different sizes. Before there were three windows, you know ... Now, they made a small window, like 20 cm and a huge window ... Each time my mother must open a window ... [he goes through the complicated procedure of opening the windows]

Participant 13a: "*They seem like hospital windows ....*"

The top-down design sometimes led to dissatisfaction, as shown by extract 5. The participant did not refer to any energy infrastructures or practices but to technical interventions that affected the aesthetics and control over their home. The sense of ownership of some tenants was affected; in turn, some tenants might be less engaged with the new energy infrastructures and practices that are considered alien to them or not specifically requested. More dramatically, neglect of the epistemic dimension and specific vulnerabilities lead to ageism and ableism (extract 6). This is aligned with the digitalisation of the SHPs that is common in public administration (Battisti, 2020), such as the increased use of email for communication, which appears to be particularly challenging for the elderly (Servon, 2002).

Misrecognising the epistemic dimensions and vulnerabilities of residents also materialised in the maintenance procedures of the new energy devices, namely the MCV and the tablet, which appear to be challenging for many tenants. These circumstances are especially true for those unfamiliar with the construction and manual labour industries and disengaged or detached from the Sinfonia participation process and final design.

20.5.3. Role of people-place relations

People-place relations have been proven to be greatly affected by EJ, given that if community members are not involved in the project's decision-making processes, important symbolic and affective aspects of their relations with their home and community will be neglected (Batel & Küpers, 2022). Impacts on people-place relations often emerge as a misrecognition of psychosocial needs in terms of social and place identities and meanings (Walker, 2009). Every interviewee presented a strong sense of attachment, either to the home, the neighbourhood, or community, even if often accompanied by a sense of somewhat not feeling at home due to a lack of agency, as explored above. This is illustrated through the extracts below:

Extract 7. Participant 12 (tenant).

> *it has always been like this* [SHP decided something and informed the residents]; ***I have been in municipal houses for 40 years.*** *They have already decided. Why should I go there?* [to workshop and meetings].

Extract 7 shows how person-house relations and past interactions with who owns the house can be an aspect that causes inertia and disempowerment.

This mixed feeling seems less present in long-term residents. Moreover, in former times, the social housing providers conceded a higher degree of informality and capacity for action to the tenants in their homes, with a higher share of interpersonal contacts and "common sense" agreements. These tenants usually exhibit a heightened sense of attachment to the home, neighbourhood, or community, showing more agency and confidence in their environment. However, they seem only to become active and engaged once their trust is acquired, when they feel there will be an improvement in their lives and that they could be part of it (see extract 8).

Extract 8. Participant 1 (expert).

> *So, there are those who have just arrived and are curious, that think that if they use things properly, they can better exploit the technology and context. Contrarily, some people who have lived there for a long time have higher satisfaction standards, and they say, 'I have always lived here', but when they understand that there is scope to have an improvement in their home, they become more active themselves.*

If they do not have that trust, their mindset seems to be more detached or dismissive, as extract 7 shows. The extract emphasises the "usual" verticality and top-down approach of the SHP governance and the representation of the tenants as passive, a perception well spread in energy projects (Lennon et al., 2020).

20.5.4. Role of social relationships

People-people relations emerge as crucial aspects that affect citizens' agentic function, affecting trust in multiple ways. The main proponents, namely the local city council, social housing providers, and local research centres, almost entirely neglect this dimension, including the critical role of relationships and attention to interpersonal treatment.

Most prominently, there is a neglect of the historical interactions with the SHP organisations and their officers (see extract 8) intertwined with the people-place relations.

The misrecognition of social dimensions reveals a lack of consideration for a holistic perspective on the residents' lived experiences. Indeed, lived experiences in Sinfonia are inseparable from the broader lived experience of the relationships with the SHPs. Past interactions greatly affect trust towards the proponents, specifically the SHPs. SHPs should be aware of the trust relationship and past interactions and held accountable for being public authorities. Low trust and, most dramatically, mistrust affect the citizen's agentic role (Talvitie, 2012). Issues in people-people relations are illustrated in the extracts below:

Extract 9. Participant (expert) responding to the question: would it have been possible to change anything in the project?

No, and additionally, the meetings were always poorly attended. When the director [of the SHP], *who is a difficult person, was there, they* [the tenants] *would ask things like, As you changed the windows, would it be possible to change the door? She would firmly answer "No", then get angry ....*

Extract 10 Participant 1 (expert).

I mean, the relations between tenants are just the result of individual and spontaneous efforts. There is no proper official structure that helps this. So, no, I don't know how the processes or SHP can be improved, but surely introducing a more "social" element inside the IPES [one of the two SHPs] *would help.*

Extract 9 shows the essential role of interpersonal treatment, specifically "politeness" (Ruano-Chamorro et al., 2021). Some interactions are perceived as distant or sometimes disrespectful. It appears to lead to a silencing attitude and a lack of mediation and procedures for feedback or complaints. It emerges as a violation of the principle of self-respect (van Uffelen, 2022). An expert who worked with SHPs on various occasions (extract 10) highlighted that there is no formal structure for mediation and relations in the SHPs. It reveals how the SHPs tend to be interested in the physical buildings rather than the socio-psychological aspects of the tenants. Furthermore, extracts 8 and 10 show how even other encounters with SHPs count and that smart city projects should not be considered outside the historical dimension (Küpers & Batel, 2023); that is, as embedded in particular contexts, with a given past and history that needs to be taken into account for these projects to be efficient and sustainable.

20.6. Conclusions and discussion

This exploratory study aimed to fill a gap in illustrating the role of recognition injustice empirically in smart city projects in the social housing context in the Global North, adopting a lived experience lens.

The recognition justice framework allowed us to identify an overreaching theme in tenants' lived experiences of the Sinfonia smart city project, the misrecognition of their agentic role, and disentangle its dimensions. Additionally, we exposed how the specific project, Sinfonia, is not happening in a vacuum but is incorporated into pre-existing dynamics and representations. Tenants cannot clearly distinguish the project from the wider relationship and interactions with their social landlords, similar to what has been found in other energy poverty case studies (Grossmann et al., 2021).

Tenants' agency misrecognition appeared to be associated with a lowered sense of ownership of the new energy devices and changes in their own homes, causing issues of disappropriation and disinterest towards these devices and mistrust towards the social landlords. We also exposed how the misrecognition and its dimensions created feelings of fear, anger, and frustration in the tenants, and, consequently, of, detachment and disengagement. As already shown in the literature (Longhurst & Hargreaves, 2019), the above-mentioned negative emotions, as relational practices, are not simply consequences of energy vulnerability but contribute to it. Indeed, the examined misrecognition emerges as a potent barrier to behavioural change and empowering environmental awareness, trust-building processes, and citizen engagement. It seems that failing these symbolic dimensions overrides cognitive factors, such as visibility of energy consumption (Hargreaves et al., 2013; Ingeborgrud & Ryghaug, 2019), or rational arguments on financial benefits from associated behavioural changes. We illustrated that the misrecognition of the agentic role is predominantly composed of four dimensions. Firstly, the influence on trust towards the SHPs following top-down governance and mindset and the current "policing welfare" (Headworth, 2021) of social housing. Secondly, we illustrated the role of people and place relations, which could be expected considering existing research (Batel & Devine-Wright, 2020). Thirdly, we also exposed the role of people-people ties, which seems under-researched but seems promising for future research. The fourth dimension is the epistemic dimension: tenants are not just misrecognised in their agentic role but also as epistemic agents that can contribute valuable knowledge to the discussion.

Misrecognising agency and its dimensions of injustice is then tied to the intersection between procedural and recognition justice (Jenkins et al., 2021; Ruano-Chamorro et al., 2021). As such, this chapter shows that addressing misrecognition is key to empowering citizens in the Global North and enabling a more active, autonomous, and caring energy citizenship. Whilst it is of primary importance to invest public resources in housing and energy efficiency, symbolic and psychosocial aspects also play a huge role in inclusive and democratic accountability and providing clean energy to all in a just way. In this regard, this chapter aimed to show the critical role of citizens' agency and its dimensions by examining the citizens' lived experience of recognition justice in a social housing smart city initiative.

References

ASTAT (2022). *Istituto provinciale di statistica – Provincia autonoma Bolzano – Alto Adige website (ASTAT)*. https://astat.provinz.bz.it/it/default.asp

Batel, S., & Devine-Wright, P. (2020). Using NIMBY Rhetoric as a Political Resource to Negotiate Responses to Local Energy Infrastructure: a Power Line Case Study. *Local Environment*, 25(5), 338–350.

Batel, S., & Küpers, S. (2022). Politicising Hydroelectric Power Plants in Portugal: Spatio-Temporal Injustices and Psychosocial Impacts of Renewable Energy Colonialism in the Global North. *Globalisations*, 1–20. https://doi.org/10.1080/14747731.2022.2070110

Battisti, D. (2020). The Digital Transformation of Italy's Public Sector: Government Cannot Be Left Behind. *JeDEM – eJournal of eDemocracy and Open Government*, 12(1), 25–39. https://doi.org/10.29379/jedem.v12i1.591

Bouzarovski, S., & Petrova, S. (2015). A Global Perspective on Domestic Energy Deprivation: Overcoming the Energy Poverty–Fuel Poverty Binary. *Energy Research & Social Science*, 10, 31–40. https://doi.org/10.1016/j.erss.2015.06.007

Braun, V., & Clarke, V. (2006). Using Thematic Analysis in Psychology. *Qualitative Research in Psychology*, 3(2), 77–101. https://doi.org/10.1191/1478088706qp063oa

Caballero, N., & Della Valle, N. (2021). Tackling Energy Poverty Through Behavioral Change: A Pilot Study on Social Comparison Interventions in Social Housing Districts. *Frontiers in Sustainable Cities*, 2, 601095. https://doi.org/10.3389/frsc.2020.601095

Carter, S., & Little, M. (2007). Justifying Knowledge, Justifying Method, Taking Action: Epistemologies, Methodologies, and Methods in Qualitative Research. *Qualitative Health Research*, 17(10), 1316–1328.

Creswell, J. W. (2009). *Research Design: Qualitative, Quantitative, and Mixed Methods Approaches* (3rd ed.). Thousand Oaks, CA: Sage Publications.

Domenicucci, J. (2017). Trust, Agency and Discrimination. *Rivista di estetica*, 64, 83–102. https://doi.org/10.4000/estetica.2067

Flyvbjerg, B. (2001). *Making Social Science Matter: Why Social Inquiry Fails and How It Can Succeed Again* (S. Sampson, Trad.; 1a ed.). Cambridge: Cambridge University Press. https://doi.org/10.1017/CBO9780511810503

Fraser, N. (2000). Rethinking Recognition. *New Left Rev*, 3, 107–120.

Fricker, M. (2007). *Epistemic Injustice: Power and the Ethics of Knowing*. Oxford: Oxford Press.

George, C., & Reed, M. G. (2017). Revealing Inadvertent Elitism in Stakeholder Models of Environmental Governance: Assessing Procedural Justice in Sustainability Organisations. *Journal of Environmental Planning and Management*, 60, 158–177. https://doi.org/10.1080/09640568.2016.1146576

Gillard, R., Snell, C., & Bevan, M. (2017). Advancing an Energy Justice Perspective of Fuel Poverty: Household Vulnerability and Domestic Retrofit Policy in the United Kingdom. *Energy Research & Social Science*, 29, 53–61. https://doi.org/10.1016/j.erss.2017.05.012

Glaser, B., & Strauss, A. L. (1967). *The Discovery of Grounded Theory: Strategies for Qualitative Research*. Chicago, IL: Aldine.

Grossmann, K., Jiglau, G., Dubois, U., Sinea, A., Martín-Consuegra, F., Dereniowska, M., Franke, R., Guyet, R., Horta, A., Katman, F., Papamikrouli, L., Castaño-Rosa, R., Sandmann, L., Stojilovska, A., & Varo, A. (2021). The Critical Role of Trust in Experiencing and Coping with Energy Poverty: Evidence from Across Europe. *Energy Research & Social Science*, *76*, 102064 https://doi.org/10.1016/j.erss.2021.102064

Grossmann, K., & Trubina, E. (2021). How the Concept of Dignity Is Relevant to the Study of Energy Poverty and Energy Justice. *Frontiers in Sustainable Cities*, 3. https://doi.org/10.3389/frsc.2021.644231

Hargreaves, T., Nye, M., & Burgess, J. (2013). Keeping Energy Visible? Exploring How Householders Interact with Feedback from Smart Energy Monitors in the Longer Term. *Energy Policy*, 52, 126–134. https://doi.org/10.1016/j.enpol.2012.03.027

Headworth, S. (2021). *Policing Welfare: Punitive Adversarialism in Public Assistance*. Chicago: The University of Chicago Press. https://doi.org/10.7208/chicago/9780226779539.001.0001

Hearn, A. X., Sohre, A., & Burger, P. (2021). Innovative but Unjust? Analysing the Opportunities and Justice Issues within Positive Energy Districts in Europe. *Energy Research & Social Science*, 78, 102127. https://doi.org/10.1016/j.erss.2021.102127

Ingeborgrud, L., & Ryghaug, M. (2019). The Role of Practical, Cognitive and Symbolic Factors in the Successful Implementation of Battery Electric Vehicles in Norway. *Transportation Research Part A: Policy and Practice*, 130, 507–516.

Jenkins, K. E. H., Sovacool, B. K., Mouter, N., Hacking, N., Burns, M.-K., & McCauley, D. (2021). The Methodologies, Geographies, and Technologies of Energy Justice: A Systematic and Comprehensive Review. *Environmental Research Letters*, 16(4), 043009. https://doi.org/10.1088/1748-9326/abd78c

Küpers, S., & Batel, S. (2023). Time, History and Meaning-Making in Research on People's Relations with Renewable Energy Technologies (RETs)–A Conceptual Proposal. *Energy Policy*, 173, 113358.

Lapassade, G. (1977). *Grupos, organizações e instituições*. Rio de Janeiro: Francisco Alves, 1977.

Lennon, B., Dunphy, N., Gaffney, C., Revez, A., Mullally, G., & O'Connor, P. (2020). Citizen or Consumer? Reconsidering Energy Citizenship. *Journal of Environmental Policy & Planning*, 22(2), 184–197.

Levenda, A. M. (2019). Thinking Critically About Smart City Experimentation: Entrepreneurialism and Responsibilisation in Urban Living Labs, *Local Environment*, 24(7), 565–579. https://doi.org/10.1080/13549839.2019.1598957

Levenda, A.M., Behrsin, I., & Disano, F. (2021). Renewable Energy for Whom? A Global Systematic Review of the Environmental Justice Implications of Renewable Energy Technologies. *Energy Research & Social Science*, 71, 101837.

Longhurst, N., & Hargreaves, T. (2019). Emotions and Fuel Poverty: The Lived Experience of Social Housing Tenants in the United Kingdom. *Energy Research & Social Science*, 56, 101207. https://doi.org/10.1016/j.erss.2019.05.017

Marks, G. (1996). An Actor-Centred Approach to Multi-Level Governance. *Regional & Federal Studies*, 6(2), 20–38.

Manzo, L. C., & Perkins, D. D. (2006). Finding Common Ground: The Importance of Place Attachment to Community Participation and Planning. *Journal of Planning Literature*, 20(4), 335–350. https://doi.org/10.1177/0885412205286160

Martin, A., Coolsaet, B., Corbera, E., Dawson, N. M., Fraser, J. A., Lehmann, I., & Rodriguez, I. (2016). Justice and Conservation: The Need to Incorporate Recognition. *Biological Conservation*, 197, 254–261. https://doi.org/10.1016/j.biocon.2016.03.021

McAdams, D. P. (2013). The Psychological Self as Actor, Agent, and Author. *Perspectives on Psychological Science*, 8(3), 272–295. https://doi.org/10.1177/1745691612464657

Middlemiss, L., & Gillard, R. (2015). Fuel Poverty from the Bottom-up: Characterising Household Energy Vulnerability Through the Lived Experience of the Fuel Poor. *Energy Research & Social Science*, 6, 146–154. https://doi.org/10.1016/j.erss.2015.02.001

Patterson, M. L., Markey, M. A., & Somers, J. M. (2012). Multiple Paths to Just Ends: Using Narrative Interviews and Timelines to Explore Health Equity and Homelessness. *International Journal of Qualitative Methods*, 11(2), 132–151. https://doi.org/10.1177/160940691201100202

Pittini, A. (2015). State of Housing in the EU 2015. A Housing Europe review. https://www.housingeurope.eu/resource-468/the-state-of-housing-in-the-eu-2015

Ruano-Chamorro, C., Gurney, G. G., & Cinner, J. E. (2021). Advancing Procedural Justice in Conservation. *Conservation Letters*, 15(3). https://doi.org/10.1111/conl.12861

Saunders, B., Sim, J., Kingstone, T., Baker, S., Waterfield, J., Bartlam, B., Burroughs, H., & Jinks, C. (2018). Saturation in Qualitative Research: Exploring Its Conceptualization and Operationalization. *Quality & Quantity*, 52, 1893–1907.

Servon, L. J. (2002). *Bridging the Digital Divide: Technology, Community, and Public Policy*. New York: Blackwell Publishing.

Simcock, N., Frankowski, J., & Bouzarovski, S. (2021). Rendered Invisible: Institutional Misrecognition and the Reproduction of Energy Poverty. *Geoforum*, 124, 1–9. https://doi.org/10.1016/j.geoforum.2021.05.005

Sinfonia Smart cities – Home. Available at: http://www.sinfonia-smartcities.eu/ (Accessed: November 10, 2022).

Sovacool, B. K., Burkec, M., Bakerb, L., Kumar Kotikalapudib, C., & Wloka, H. (2017). New Frontiers and Conceptual Frameworks for Energy justice. *Energy Policy*, 105, 677–91.

Sovacool, B. K., Martiskainen, M., & Furszyfer Del Rio, D. D. (2021). Knowledge, Energy Sustainability, and Vulnerability in the Demographics of Smart Home Technology Diffusion. *Energy Policy*, 153, 112196. https://doi.org/10.1016/j.enpol.2021.112196

Spiers, J. (2000). New Perspectives on Vulnerability Using Emic and Etic Approaches. *Journal of Advanced Nursing*, 31(3), 715–721. https://doi.org/10.1046/j.1365-2648.2000.01328

Swain, C., & Tait, M. (2007). The Crisis of Trust and Planning. *Planning Theory and Practice*, 8(2), 229–247.

Talvitie, A. (2012). The Problem of Trust in Planning. *Planning Theory*. https://doi.org/10.1177/1473095211430494

Tyler, T. R. (2015). Social Justice. In M. Mikulincer, P. R. Shaver, J. F. Dovidio, & J. A. Simpson (Eds.), *APA Handbook of Personality and Social Psychology; Vol. 2. Group Processes* (pp. 95–122). New York: American Psychological Association. https://doi.org/10.1037/14342-004

van Uffelen, N. (2022). Revisiting Recognition in Energy Justice. *Energy Research & Social Science*, 92, 102764. https://doi.org/10.1016/j.erss.2022.102764

Walker, G. (2009). Beyond Distribution and Proximity: Exploring the Multiple Spatialities of Environmental Justice. *Antipode*, 41(2009), 614–636.

Wittenberg, I., & Fleury-Bahi, G. (2016). Application of Sustainable Habitat: What Is the Appropriation and Utilisation of Equipment After Energy-Saving Renovations in Social Housing?: Application of Sustainable Habitat, *Journal of Community & Applied Social Psychology*, 26(5), 409–420. https://doi.org/10.1002/casp.2271

21 A Burning Desire

Trying to Achieve SDG 7 and Improving Access to Cleaner Cooking Fuels in Rural Ghana

Dickson Boateng, Julian Bloomer, and John Morrissey

21.1 Introduction

21.1.1 Energy and environment

As the global population increases, so does the demand for energy (UNDP, 2018). Energy consumption is, however, a major contributor to greenhouse gas emissions (UNDP, 2021; World Resources Institute, 2020). Hence, as the global population rises, an upsurge in GHG emissions from energy consumption is expected. This is the basis of the global call for cleaner energy as a catholicon to promote environmental sustainability (GPOBA, 2016; UNDP, 2021). Following this, energy, which was not explicitly considered in the Millennium Development Goals (MDGs), has been explicitly stated in the Sustainable Development Goals (SDG 7) (Shyu, 2014).

This notwithstanding, approximately one third of the world's population – about 2.8 billion people, still cooks without clean fuels and is thus excluded from the associated benefits. To achieve universal access to clean cooking fuels and efficient technologies by 2030, 130 million people will have to be transitioned from unclean fuels annually and a majority of these people live in Asia and Africa (International Energy Agency, 2022; UNDP, 2021). In Africa, 970 million people still lacked access to clean cooking fuels in 2021, and it is estimated that there should be a remarkable improvement in clean energy access across sub-Saharan Africa in the global drive to achieve SDG 7 (International Energy Agency, 2022). An increase in clean energy uptake in these regions is anticipated to be a major step towards reducing not only GHG emissions but also improving food security and promoting public health (Bisaga & Campbell, 2022; UNDP, 2021).

21.1.2 Focus on the energy situation of Ghana

In general, the adoption and investment in renewable energy in Ghana have been very slow (Bawakyilleno, 2017). While Ghana has a favourable landscape and is geographically well positioned to deploy wind and solar energy (Lin & Ankrah, 2019; Sun et al., 2020), there is a disconnect between the country's SDG7 aims of achieving clean cooking fuels and electricity-based technologies[1] (Newell & Daley, 2022). Using a political economy approach, Newell and Daley (2022) highlight intervention points for accelerating electricity-based cooking, specifically the need to realign governance and improve policy responses to tackle the issue, and they note: "this should be focussed less on individual behaviours and more on wider social and cultural settings and the gendered and other power dynamics that structure them" (Newell & Daley, 2022, p. 8). Cooking with electricity is more likely to be successful in urban areas due to the proximity of

DOI: 10.4324/9781003408536-25

communities to electricity supply, but will be more challenging in rural areas, given the high energy requirements for cooking in comparison to other domestic uses. As of 2017, less than 1% of Ghanaian households used electricity as their main source of cooking fuel (Ghana Statistical Service, 2019).

In the specific context of cooking fuels in Ghana, similar to many countries in the Global South, the cooking fuel mixes in the country tend to be woodfuels (including firewood and charcoal), Liquefied Petroleum Gas (LPG), kerosene, and harvested crop residues (Bawakyillenuo et al., 2021). Although access to LPG has been increasing in Ghana, woodfuels are the traditional source of energy and still remain the dominant source of cooking energy (Bawakyillenuo et al., 2021; Ghana Statistical Service, 2019). Woodfuel (firewood [33.3%] and charcoal [34.1%]) accounts for more than 60% of the country's total energy use (Energy Commission of Ghana, 2019; Ghana Statistical Service, 2019). Woodfuels are the only energy source where cooking appliances and most production equipment are locally sourced in Ghana. It thus contributes significantly to the Ghanaian economy and, as such, supports the livelihoods of many Ghanaians (Brobbey et al., 2021).

Firewood is the principal woodfuel used in rural areas (63%), while charcoal (44.3%) and LPG (34.8%) are predominantly used in urban areas (Ghana Statistical Service, 2019; Karimu et al., 2016; Wiedinmyer et al., 2017). Even so, the production of charcoal takes place in rural areas and is a source of income for many rural dwellers due to their proximity to forested areas (Anang et al., 2011; Bawakyillenuo et al., 2021; Brobbey et al., 2015). The majority of rural women are engaged in the charcoal business and play roles such as collection and burning of firewood, and handpicking and packing of charcoal (Bawakyillenuo et al., 2021). This notwithstanding, consumption of woodfuels is a major contributor to environmental degradation as well as poor health and food insecurity (McCray, 2018; Omer, 2017; Omiti & Laibuni, 2015; Sulaiman et al., 2017). In Ghana, biomass fuels (including woodfuels) account for 16,600 deaths annually and are a major contributor to ambient air pollution (Ghana Statistical Service, 2014a, 2019; Inkoom & Crentsil, 2015).

To reduce reliance on woodfuels, the Government of Ghana and other stakeholders have been rolling out programmes to target the use of clean cooking fuels in rural communities (Asante et al., 2018; IMANI, 2017). For example, the Rural LPG Promotion Program (RLP) aimed to improve LPG access in rural areas from 3% (in 2012) to 15% (by 2016). The RLP was introduced to specifically target rural communities with the use of LPG for cooking. Launched in 2013, the RLP sought to contribute to the nation's goal of expanding LPG coverage to about 50% of the population by 2020 (ENERGIA, 2015; GhanaWeb, 2016). Under the RLP, beneficiaries were freely provided with 6-kilogramme cylinders, single-burner stoves, hoses, and regulators. The RLP was coordinated by the Ministry of Energy. Beneficiaries were selected based on their ability to make an initial payment of GHC 22.00 (USD 5 then) to fill the cylinder. This selection was done by the District Assembly Officers and focal persons in the beneficiary communities.

However, the RLP failed to achieve its stated goal as beneficiaries failed to exclusively use or sustain the use of LPG (Adjei-Mantey & Takeuchi, 2021; Asante et al., 2018; Carrión et al., 2020; Energy Commission, 2016). Thus, many rural dwellers failed to shift to LPG and clung to the use of woodfuels (Makonese et al., 2018; Njenga et al., 2019; Wiedinmyer et al., 2017). This poses a challenge towards achieving SDG 2 (food security), SDG 3 (good health and well-being) and particularly, SDG 7 which aims to ensure universal access to affordable, reliable, and modern energy services. Currently, all interventions to promote LPG use in Ghana (both rural and urban) have been revised and consolidated in the LPG for Development (LPG4D) Programme which aims to achieve 50% access to LPG by 2030 (Ahunu, 2015; Ministry of Energy 2022). It remains to be seen how successful the LPG4D Programme will be in shifting rural fuel use to LPG.

Using primary data, this chapter interrogates why rural dwellers resort to woodfuels in particular and suggests ways of promoting LPG use in rural Ghana. It builds on and expands on research to-date in the area of just transitions, particularly in the social sciences (Sovacool, 2014a, 2014b), and provides a valuable Global South perspective on low-carbon transition challenges. This chapter is therefore, vital for policy formulation and implementation geared towards SDG 7.

21.2 Methodology

21.2.1 The research project and case study

This chapter is part of a wider study on political ecology, low-carbon transition, and energy justice in Ghana[2]. Ghana is considered a less economically developed country, with an estimated 31 million people spread across its 16 regions, including six new regions that were formed following a referendum in 2018 (The World Bank, 2019; UNDP, 2020). Compared with its neighbours, Ghana has a relatively large labour force, and about 67.9% of the working-age population is engaged in employment. Since 2017, the country has experienced a decline in GDP, which has significantly hampered efforts towards poverty alleviation (The World Bank, 2021). As such, the proportion of the country's population (13.3%) that is living below the international poverty line of $1.90 a day is expected to increase, especially in the aftermath of the COVID-19 pandemic and the Russian-Ukrainian war in early 2022 (Bukari et al., 2021; The World Bank, 2019; UNDP, 2020). Poverty in the country is primarily a rural phenomenon, with rural poverty levels being about four times higher than urban poverty rates (Cooke et al., 2016).

The chapter is based on a case study of the Kwahu Afram Plains North and South Districts in the Eastern Region of Ghana. The Districts are noted for agriculture, primarily fishing and farming, due to the rich soils and proximity to the Afram River. The Districts are predominantly rural with a few scattered small urban communities (Ghana Statistical Service, 2014b, 2014c). Fieldwork was conducted in five (5) rural communities: 3 in Kwahu Afram Plains South (Kwasi Addae, Asimpanyin, and Mmradan) and 2 in Kwahu Afram Plains North Districts (Abotanso 1 and Adukrom). The data were collected between September and December 2022. This chapter draws on empirical findings from field observations and qualitative interviews with key stakeholders in the Districts (Figure 21.1).

21.2.2 Sampling and participants

The research involved 45 key informant interviews with key community leaders, District Assembly officials, and improved/cleaner cookstove providers. Key community leaders consisted of chiefs, elders, and local government officials in the communities (Table 21.1).

These participants are mainly household heads who have lived and worked in the communities for years and have in-depth knowledge of the energy situation in the communities. Interviewees were recruited using a non-probabilistic sampling technique – purposive, snowballing, and convenience sampling (Mweshi & Sakyi, 2020).

21.2.3 Data collection

Interviews were conducted in the houses and offices of interviewees using a structured interview format. This helped to keep the interviews in line with the subject matter, with discussion directed along key thematic lines while also allowing the opportunity for unexpected new information to emerge (Brinkmann, 2014; Husband, 2020). Interviews were audio-recorded with

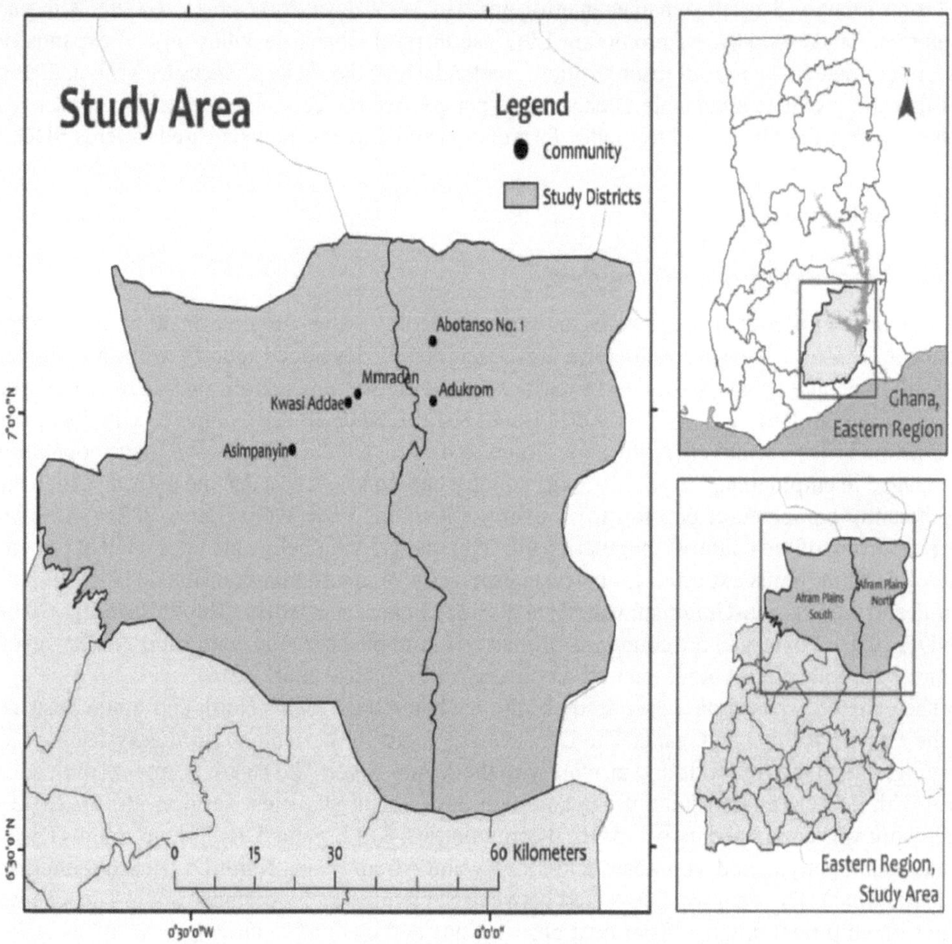

Figure 21.1 A map of Ghana showing the region, districts, and communities where the study was conducted.

Source: Authors' construct (2023) based on ESRI shapefiles using ArcGIS 10.6.

the consent of participants. Field notes were also taken to capture some relevant information during interviews. Since the majority of the interviewees were engaged in agriculture (farming and/or fishing), interviews were conducted early in the morning and late in the evening when interviewees had returned from the fields. Interviews were mainly conducted in Twi[3]. Each interview lasted approximately 45 minutes. Responses were written in English for validity and comprehension. All respondents were over 18 years old.

21.2.4 Ethical/Practical considerations

Since the study involved humans, ethical issues needed to be addressed prior to data collection. This was done in consultation with the Mary Immaculate College Research Ethics Committee (MIREC) with reference number "A21-056". Informed Consent (written or oral) was sought from all participants before engaging them in the study. Participants were informed about the

Table 21.1 Demographic characteristics of interviewees

Variables	*Category*	*Frequency*
Gender	Male	28
	Female	17
Age	31–40 years	12
	41–50 years	23
	51–60 years	8
	>61 years	2
Status	District Assembly officials	5
	Local Government official	3
	Chiefs	8
	Elders	27
	Improved cookstove providers	2
Size of household	<3	4
	3–5	15
	6–8	26
Education	Never schooled	11
	Basic	20
	Secondary	6
	Tertiary	8

Source: Authors' own construct (2023).

purpose of the study, and their voluntary involvement, and assured of confidentiality and anonymity. Audio records were put in a password protected folder on a secured laptop to prevent unauthorised access. Notes written during interviews were also compiled and stored securely. In the analysis of the data gathered, identifiers were used to ensure anonymity.

21.2.5 Data analysis

Audio-recorded interviews were transcribed and translated. After transcription, the analysis started with the reading of all the transcribed data and complementing it with the field notes by the first author. Every 45 minutes of recorded audio took approximately 7 hours to fully process and transcribe. A reflexive thematic analysis was conducted to identify emerging themes from the interviews. Reflexive thematic analysis was adopted because it is flexible, data-driven, and allows room to create new themes (Braun & Clarke, 2021). Data were inductively analysed by firstly, semantic coding – reading data and grouping relatable words or sentences. The codes were subsequently grouped into themes in relation to access to clean cooking fuels (Braun & Clarke, 2021). The codes and themes were discussed with the second and third authors, reviewed, and fine-tuned to reflect the views of respondents while ensuring anonymity. Excerpts of the responses by a few interviewees were used to support the findings using pseudonyms – common local names to ensure anonymity.

21.3 Results and discussion

21.3.1 The cooking fuel mix

There is comparatively limited access to cooking fuels in rural Ghana compared to urban Ghana (Bawakyillenuo et al., 2021). The study showed that firewood, charcoal, and LPG were the cooking energy sources for rural dwellers. However, woodfuels (firewood and charcoal) were predominant (Wiafe & Kwakwa, 2013; Wiedinmyer et al., 2017). This situation is not specific to rural Ghana as many developing countries find themselves in a similar situation (see Heltberg, 2005; Sola et al., 2019). The majority of rural dwellers interviewed used a combination of firewood and charcoal – the two dominant sources of cooking fuel in rural Ghana (Karakara & Osabuohien, 2021). Charcoal was especially important during rainy periods, when rural dwellers were unable to go to the forest to fetch firewood. Abena (a female), a community leader and a household head, stated:

> I heavily rely on firewood for cooking my food. While it is available for most times, I am unable to walk to the forest to get them during rainy periods. So, I have charcoal as an alternative.
>
> (Abena, October 2022)

Similar to Kenya and Uganda, collection of firewood and cooking in households in rural Ghana was done by females (wives and/or girl children) (Bamwesigye et al., 2020; Waswa et al., 2020). A lot of time was spent by females in collecting firewood as well as cooking traditional food, which took longer time to prepare. Households preferred to cook single meals instead of composite meals, and cooking was done in traditional kitchens, which are less ventilated. In the house, females cooked twice a day (mornings [before work] and evenings [after work]) while a few (specifically, wives) cooked on the farms in the afternoon. Each household cooked its own meals, and members of a household ate together. Household meals and cooking utensils were served and washed by females (especially, the girl child). These roles prevent females from investing time in other activities (e.g., civic engagements, employment, and education) and expose them to the health hazards associated with the combustion of fuels (Adjei-Mantey & Takeuchi, 2021; Bede-Ojimadu & Orisakwe, 2020; International Energy Agency, 2022). Yet, the knowledge of women about the effects of their exposure to smoke from firewood was limited to "irritation of the eyes". This notwithstanding, studies show that the effects can extend to respiratory diseases and affect the health of unborn children (Adjei-Mantey & Takeuchi, 2021; Bede-Ojimadu & Orisakwe, 2020; Hussein et al., 2020; Weber et al., 2020). Some men, on the other hand, despite cooking at times and frequenting the kitchen, trivialised the effects that exposure to smoke from woodfuels had on both women and men (Figure 21.2).

21.3.2 Why woodfuels?

A range of factors determined the type of cooking fuels (hereafter referred to as woodfuels) used by rural dwellers. The culture of a group is known to have an impact on their adoption of technologies (Panigrahi et al., 2018; Usman & Said, 2014). Rural communities have historically been known for the use of woodfuels. Likewise, the kind of fuels rural dwellers were exposed to at an early stage of life (custom) determined the type of cooking fuel they used presently (i.e., woodfuel). Some rural dwellers preferred to use woodfuel because of the longer time it took to prepare staple foods. Even in urban households where LPG is highly used, some households use charcoal when cooking certain stable foods to reduce LPG consumption (Wiedinmyer et al., 2017). The special taste of food cooked with woodfuels also made it favourable to some rural dwellers. Just as wood-fired ovens are regarded as superior for baking Neapolitan pizzas and wooden barrels add flavour to the liquids they contain (Ciarmiello & Morrone, 2016; Cisar-Erlach, 2019), so do some rural dwellers appreciate the value of woodfuels for cooking flavoured staple local meals. Consequently, many rural dwellers have a sense of cultural attachment to wood and have cultivated the habit of using woodfuels even when other alternatives are

Figure 21.2 A pile of chopped firewood; the dominant source of cooking fuel in rural Ghana.
Source: First author (2022).

available. Adwoa (female), a 50-year old indigene of the Kwahu Afram Plains South District, stated:

> I have since my childhood been using woodfuel for cooking. I know about other alternatives but I do not think I am ever going to try them. I find it comfortable to use woodfuel.
> (Adwoa, November 2022)

Cost was also a factor that rural dwellers used to decide the type of cooking fuel to use. Regardless of the significant time burden ssociated with accessing woodfuels, these fuels were nevertheless favoured by poor rural dwellers, not least as these groups found it difficult to afford the price of LPGs (Abdul-Wakeel Karakara & Dasmani, 2019; Karimu, 2015). Although acknowledging the time-benefits of using LPG, rural dwellers preferred to use woodfuels because their usage came at a much lower comparative economic price. The initial capital needed to get LPG – purchasing a cylinder, stove, and the LPG gas – prevented most rural dwellers from resorting to this method, despite widespread acknowledgement of the positives associated with it.

Availability was another factor rural dwellers used in deciding the type of fuel to use for cooking (Adjei-Mantey & Takeuchi, 2023; Karimu, 2015). Rural communities were close to forested areas, making woodfuels accessible to dwellers at any point in time. In the communities involved in the study, there was no gas station, and dwellers who preferred to use LPG had to travel long distances to fill their cylinders. To escape this inconvenience and the initial capital outlay involved in acquiring LPG, rural dwellers resorted to woodfuels.

> I can afford LPG but there are no gas stations nearby to fill the cylinder once it is emptied. I would have to travel to the city to get LPG which for me, is not economical.
>
> (Kwame [male household head], November 2022)

Even when available, explosions associated with LPG in urban areas posed risks to its adoption in rural Ghana as people were fearful of adopting it in their homes (Adjei-Mantey & Takeuchi,2023). Notable among such incidents was an explosion in Ghana's capital, Accra, in 2017, where seven lives were lost and 130 people sustained injuries (The Guardian, 2017). News of this event and of several other comparable accidents in urban communities made some rural dwellers concerned about potential safety issues from the use of LPG. Hence, some rural dwellers were risk-averse and committed to not using LPG even when it was made available (Figure 21.3). A District Assembly Officer, in an interview, stated that:

> Meeting the energy needs for cooking is not problematic for rural dwellers as they have an affordable, accessible, comfortable and risk-free source in woodfuels.
>
> (Akwasi, December 2022)

21.3.3 Charcoal contribution to rural livelihoods

Charcoal production was a major source of income for rural dwellers (Ablo et al., 2022; Asare et al., 2022; Nketiah & Asante, 2018). Despite farming and fishing being their major occupations,

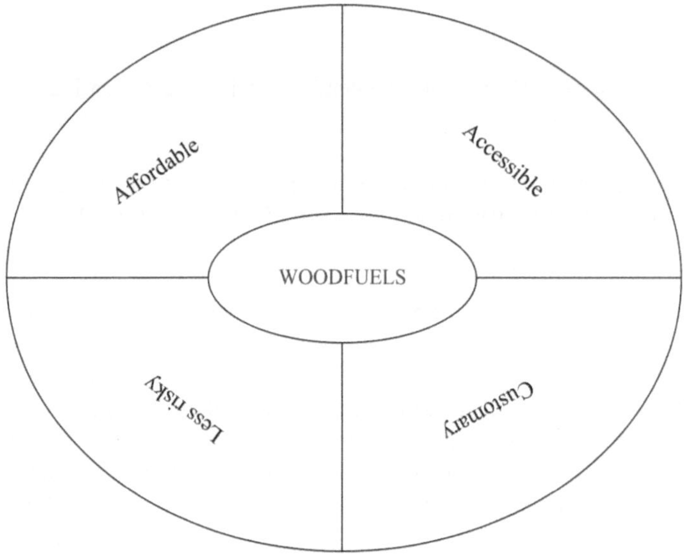

Figure 21.3 Why woodfuels are predominantly used by rural dwellers to meet their cooking needs.

Source: Authors' own construct (2023).

some rural dwellers were involved in the production of charcoal as a coping mechanism to deal with economic shocks, particularly those caused by climate change (Brobbey et al., 2019). Although heavily involved in the production of charcoal, these rural dwellers used significantly less charcoal compared to urban dwellers (Ghana Statistical Service, 2019). A lot of the charcoal produced in the rural communities is transported to nearby cities, such as Koforidua and Accra, to be sold. Charcoal production thereby creates employment opportunities for not only people in rural communities but also urban dwellers who are engaged in the commodity chain of charcoal production and consumption (Jones et al., 2016; Figure 21.4).

While of insufficient scale to meaningfully contribute to the eradication of poverty in rural communities, charcoal production remains an important poverty alleviation strategy (Kiruki et al., 2020). Charcoal production and sale provided additional income for rural dwellers, particularly women, who performed roles such as fetching and burning firewood, packing charcoal into bags, and trading charcoal. Not only women benefitted from the activity but also men who were involved in the loading and transportation of charcoal to the cities. Woodfuels (firewood and charcoal), aside from being critical to energy needs, support the livelihoods of rural dwellers (Ablo et al., 2022; Nketiah & Asante, 2018).

As a major driver of forest deforestation in sub-Saharan Africa, charcoal production causes biodiversity loss, reduces greenhouse gas sinks, and contributes to soil erosion (Bennett, 2017;

Figure 21.4 Bags of charcoal ready to be transported to the City for Sale.

Source: First author (2022).

Kumar et al., 2022; Sedano et al., 2016). Rural dwellers who engaged in the felling of trees for charcoal did not practise reforestation despite being reliant on trees for income and cooking fuel. Failure to replace cut trees can negatively impact the environment and further degenerate the living conditions of rural dwellers. In rural Kenya, for instance, households responded to the scarcity of woodfuels by cooking food with low nutritional value (since it took less time) and skipping meals (Mendum & Njenga, 2018; Waswa et al., 2020) even though there were available alternatives such as LPG.

21.3.4 Towards attaining SDG 7: implications for energy policy

As evident from the case of the rural adoption of LPG in Ghana, many rural dwellers are unable to transition to LPG, even when freely provided with LPG from the start. Also, LPG subsidy programmes tend to benefit people in urban communities more than rural dwellers, who find it difficult to afford even the subsidised cost of LPG technology (Greve & Lay, 2023). Thus, subsidy programmes do not exclusively support LPG uptake in rural communities. It is consequently recommended that other options, such as the promotion of improved cookstoves, could be used as a temporary pathway en route to a full transition to LPG in rural Ghana (Figure 21.5).

Improved cookstoves are innovations that aim to address the clean cooking energy needs of the poor, who cannot afford LPG and so are excluded from its benefits (Gill-Wiehl et al., 2021; Lambe et al., 2020). These technologies are locally produced and readily available for purchase in all parts of Ghana. In the rural communities involved in the study, all interviewees admitted to knowing and seeing the cookstoves. These cookstoves could be used as an alternative where

Figure 21.5 An example of an improved cookstove deployed in rural Ghana.

Source: Credit – Clean Cookstove Ghana Limited, 2023.

LPG uptake and/or use have been unsatisfactory. Currently, the distribution of improved cookstoves is championed by the Ministry of Energy and private distributors. However, coverage is very low as distribution is unable to reach many rural communities because of financial reasons. In an interview with a private distributor, he stated:

> We (the institution) are doing our best to save the environment by distributing improved cookstoves. But, we are limited in our efforts because of finances. With support from the government and other stakeholders, we can reach many rural communities.
>
> (Afua [an improved cookstove distributor] December, 2022)

Improved cookstoves are cheaper than LPG. The cost incurred in adopting LPG (i.e., to purchase a cylinder, stove, and gas) can be used to purchase two or more units of improved cookstoves. If the cost of improved cookstoves is problematic for rural dwellers, then it will be extremely difficult for them to shift to and sustain the use of LPG in their homes, given its higher cost. An improved cookstove that is relatively cheaper can provide a pathway to switching to LPG. This notwithstanding, improved cookstoves are more likely to be adopted by the few rich rural dwellers at the expense of the poor majority if they are not made affordable, as evidenced in Kenya and Ethiopia (Kedir et al., 2019; Waswa et al., 2020). Since rural dwellers have developed the habit of accessing free fuels for cooking, any substantial cost element attached to clean fuels is not likely to be accepted.

Improved cookstoves allow for more efficient use of woodfuels (firewood and charcoal). As an interim "stepping-stone" technology, improved cookstoves might help to avoid an abrupt disconnect from the traditional practices of many rural dwellers, closely associated with the use of woodfuels, as described. Since improved cookstoves use fewer woodfuels, the emissions from the fuels will be significantly reduced (Gill-Wiehl et al., 2021; Lambe et al., 2020). Rural dwellers are more comfortable with the use of woodfuel which stems from their historic and traditional use of resources in their home areas. Consequently, a rapid transition to LPG will probably be resisted, and, hence, not yield the intended results. A shift from woodfuels to improved cookstoves, on the other hand, indicates a steady change in the customs of rural dwellers, which will likely increase uptake and sustain the adoption of the technology. In an interview with a male household head who has been using improved cookstoves for the past few years, he stated:

> As a subscriber of improved cookstoves, I have observed that the amount of woodfuel I use for cooking reduces with an upgrade in the technology. I am always keen to purchase newer models.
>
> (Kofi, October 2022)

Being at the centre of charcoal production activities, rural dwellers will be negatively affected too by a rapid country-wide transition to LPG. A transition process that fails to address the livelihoods of these dwellers is likely to be unsuccessful, as many have such a critical economic stake in the production and sale of charcoal. Accordingly, rural dwellers who engage in charcoal production will continue to pollute the environment if that is their only means of survival. A male community leader and household head stated:

> I am aware of the environmental consequences of my decision to burn trees for charcoal. I do not have any other option since my household (7 in total) largely rely on charcoal production as a livelihood source.
>
> (Kwabena, December 2022)

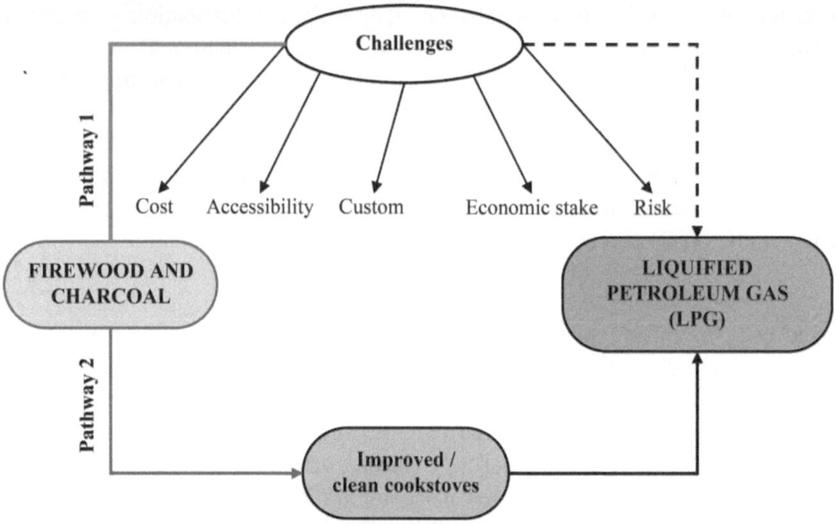

Figure 21.6 A proposed pathway to transitioning to LPG via improved cookstove.
Source: Authors' own construct (2023).

Unlike a direct transition to LPG, which collapses charcoal production activities, an improved cookstove stopgap allows the activities to continue – albeit at a slower pace. A move to improved cookstoves implies that rural dwellers engaged in the production of charcoal will still be in business and, therefore, be earning an income from charcoal production to support their livelihoods (Figure 21.6).

As illustrated in Figure 3.2, a direct transition to cleaner cooking fuels (such as LPG) is fraught with many challenges (for example, cost, accessibility, impacts on local custom, and disruption of income-generating activities), as already discussed. These challenges make it challenging to attain the ultimate goal of transitioning to the use of LPG via Pathway 1. Whereas similar challenges are encountered via Pathway 2, these challenges are relatively benign and can be coped with by rural dwellers since improved cookstoves are cheaper, more accessible, risk-free, and use traditional fuels, enabling a sustained and friendly transition to LPG.

21.4 Conclusion

The benefits of global clean cooking are considerable: "reducing premature deaths by about 500000 a year by 2030, drastically cutting time spent gathering fuel and cooking, and allowing millions of women to pursue education, employment and civic involvement" (International Energy Agency, 2022: 16). While a rapid shift to rolling out clean cooking technologies is demanded by the pace of action required to adapt to global climate change, a swift transition to LPG in rural sub-Saharan Africa is likely to backfire as rural dwellers prefer fuels that are more readily available, cheaper, and perceived as safer to use.

The "one-size-fits-all" approach to achieving SDG 7, particularly with regard to clean cooking fuels, is likely not attainable in rural sub-Saharan Africa, as evident in the case of Ghana. Rural dwellers lean towards the continued use of woodfuels (firewood and charcoal) mainly because they are readily accessible, affordable, customary, and risk-free. In addition, the livelihoods of some rural dwellers are tied to the trade of woodfuels and so there is resistance to any form of transition that will render them poorer in spite of the potential positive environmental impacts. Therefore, stakeholders who aim to support the drive to attain SDG 7 in sub-Saharan Africa should contextualise their efforts.

Countries should, depending on the specific situation, have a holistic approach that ensures enhanced access to cleaner cooking fuels. In Ghana, for instance, it is recommended that household-level energy efficiency be enhanced by improved cookstoves. These cookstoves should be made affordable to rural dwellers as the prices can push many to regress to woodfuels. Additionally, educational and training programmes on sustainable charcoal production and alternative income-generating activities should be run for people whose livelihoods are dependent on charcoal production. People should also be sensitised so that they become aware of the benefits of using LPG.

A "friendly" approach to transition to cleaner cooking fuels (LPG) that considers disparities and inequalities within societies (e.g., educational levels, income levels, customs, and traditions) an improved/clean cooking fuel pathway is encouraged (Jasanoff, 2018). A transition to LPG in rural sub-Saharan Africa requires huge sacrifices and investment (International Energy Agency, 2022). Indeed, "it will be hard and the road will be muddy and rough", a Ghanaian music group (Osibisa) famously said, "but we will get there".

Notes

1 E-cooking here refers to cooking with electricity, and the goal of this transition is to "increase the electrification of cooking using mainly solar-powered batteries and microgrids to power … stoves and hobs for cooking" (Newell & Daley, 2022).
2 The wider study is the first author's PhD research project.
3 The first author is Ghanaian and fluent in the Twi language.

References

Abdul-Wakeel Karakara, A., & Dasmani, I. (2019). An Econometric Analysis of Domestic Fuel Consumption in Ghana: Implications for Poverty Reduction. *Cogent Social Sciences*, 5(1), 1697499.

Ablo, A.D., Osei, L., Jarawura, F.X., & Yaro, J.A. (2022). Sacrificing the savannah woodlands for Energy and Livelihoods? Charcoal Production in Ghana. *Energy for Sustainable Development*, 70, 549–559.

Adjei-Mantey, K., & Takeuchi, K. (2021). The Effect of in Utero Exposure to Household Air Pollution on Child Health: Evidence from Ghana. *Health Policy OPEN*, 2, 100029.

Adjei-Mantey, K., & Takeuchi, K. (2023). Risk Aversion and Cleaner Cooking Fuel Choice: an Empirical Study in Ghana. *Environment and Development Economics*, 28(2), 130–148.

Ahunu, L. (2015). LPG Promotion Program. Africa Centre for Energy Policy. Retrieved 03/06/2019, from: http://acep.africa/wp-content/uploads/2019/11/THELPGPROMOTIONPROGRAMME1.pdf

Anang, B.T., Akuriba, M.A., & Alerigesane, A.A. (2011). Charcoal Production in Gushegu District, Northern Region, Ghana: Lessons for Sustainable Forest Management. *International Journal of Environmental Sciences*, 1(7), 1944–1953.

Asante, K.P., Afari-Asiedu, S., Abdulai, M.A., Dalaba, M.A., Carrión, D., Dickinson, K.L., Abeka, A.N., Sarpong, K., & Jack, D.W. (2018). Ghana's Rural Liquefied Petroleum Gas Program Scale up: A Case Study. *Energy for Sustainable Development*, 46, 94–102.

Asare, F., Owusu, F.W., & Gazo, R. (2022). Sustainable Charcoal Production Drive in Rural Communities in Ghana, West Africa. *Energy for Sustainable Development*, 68, 364–372.

Bamwesigye, D., Kupec, P., kuimo, G., vlis, J., Asamoah, O., Darkwah, S.A., & Hlaváčková, P. (2020). Charcoal and Wood Biomass Utilization in Uganda: the Socioeconomic and Environmental Dynamics and Implications. *Sustainability*, 12(20), 8337.

Bawakyilleno, S. (2017). The Political Economy of Renewable Energy Investment in Ghana. *IDS Bulletin*, 48, 141–163.

Bawakyillenuo, S., Crentsil, A.O., Innocent, K., Danquah, S., & Boakye, E. (2021). The Landscape of Energy for Cooking in Ghana: A Review, Institute of Statistical, Social, and Economic Research, University of Ghana, 2021. *Modern Energy Cooking Services*, 1–56.

Bede-Ojimadu, O., & Orisakwe, O.E. (2020). Exposure to Wood Smoke and Associated Health Effects in Sub-Saharan Africa: a Systematic Review. *Annals of Global Health*, 86(1), 32.

Bennett, L. (2017). Deforestation and climate change. *A publication of climate institute, 1400*. Retrieved 07/03/2023, from: https://climate.org/wp-content/uploads/2017/04/deforestation-final_r1.pdf

Bisaga, I., & Campbell, K. (2022). Clean and modern energy for cooking: A path to food security and sustainable development, World Food Programme & Modern Energy Cooking Services, Position paper. Retrieved 02/12/2022, from: https://docs.wfp.org/api/documents/WFP-0000140194/download/?_ga=2.228545055.329401895.1670759387-1790221718.1670759387

Braun, V., & Clarke, V. (2021). One Size Fits All? What Counts as Quality Practice in (reflexive) Thematic Analysis? *Qualitative Research in Psychology*, 18(3), 328–352.

Brinkmann, S. (2014). Unstructured and Semi-Structured Interviewing. *The Oxford Handbook of Qualitative Research*, 2, 277–299.

Brobbey, L.K., Asante, J., Sampong, K.E., Kumeh, E.M., & Nketiah, K.S. (2015). Securing Rights to Wood Resources for Charcoal Production in Ghana. *IIED Small and Medium Forest Enterprise. Tropenbos International Ghana*. Retrieved 24/03/2021, from: https://www.iied.org/sites/default/files/pdfs/migrate/G04060.pdf?

Brobbey, L.K., Hansen, C.P., & Kyereh, B. (2021). The Dynamics of Property and Other Mechanisms of Access: The Case of Charcoal Production and Trade in Ghana. *Land Use Policy*, 101, 105152.

Brobbey, L.K., Hansen, C.P., Kyereh, B., & Pouliot, M. (2019). The Economic Importance of Charcoal to Rural Livelihoods: Evidence from a Key Charcoal-Producing Area in Ghana. *Forest Policy and Economics*, 101, 19–31.

Bukari, D., Kemausuor, F., Quansah, D.A., & Adaramola, M.S. (2021). Towards Accelerating the Deployment of Decentralised Renewable Energy Mini-Grids in Ghana: Review and Analysis of Barriers. *Renewable and Sustainable Energy Reviews*, 135, 110408.

Carrión, D., Prah, R., Gould, C.F., Agbokey, F., Mujtaba, M., Pillarisetti, A., Tumasi, M., Agyei, O., Chillrud, S., Tawiah, T. and Jack, D. (2020). Using Longitudinal Survey and Sensor Data to Understand the Social and Ecological Determinants of Clean Fuels Use and Discontinuance in Rural Ghana. *Environmental Research Communications*, 2(9), 095003.

Ciarmiello, M., & Morrone, B. (2016). Why Not Using Electric Ovens for Neapolitan Pizzas? A Thermal Analysis of a High Temperature Electric Pizza Oven. *Energy Procedia*, 101, 1010–1017.

Cisar-Erlach, A. (2019). *The Flavor of Wood: In Search of the Wild Taste of Trees from Smoke and Sap to Root and Bark*. Abrams, New York.

Cooke, E., Hague, S., & McKay, A. (2016). *The Ghana Poverty and Inequality Report: Using the 6th Ghana Living Standards Survey*. University of Sussex. pp. 1–43.

ENERGIA (2015). LPG: Increasing the energy options benefitting women worldwide. Retrieved 02/02/2022, from: https://energia.org/assets/2015/10/ENERGIA-FINAL-lr.pdf

Energy Commission of Ghana (2019). National Energy Statistics (2009–2018). Retrieved 30/01/2021, from: http://energycom.gov.gh/files/ENERGY_STATISTICS_2019_Updated.pdf

Energy Commission (2016). 2016 energy (Supply and Demand) outlook for Ghana. Retrieved 03/06/2019, from: http://www.energycom.gov.gh/files/Energy%20Commission%20-%202016Energy%20Outlook%20for%20Ghana_final.pdf

Ghana Statistical Service (2014a). Ghana Demographic and Health Survey 2014. Rockville, Maryland, USA: GSS, GHS, and ICF International. Retrieved 10/04/2020, from: https://dhsprogram.com/pubs/pdf/fr307/fr307.pdf

Ghana Statistical Service (2014b). 2010 Population and Housing Census. District Analytical Report, Kwahu Afram Plains North District. Retrieved 25/08/2020, from: https://www2.statsghana.gov.gh/docfiles/2010_District_Report/Eastern/KWAHU%20AFRAM%20PLAINS%20NORTH.pdf

Ghana Statistical Service (2014c). 2010 Population and Housing Census. District Analytical Report, Kwahu Afram Plains South District. Retrieved 25/08/2020, from: https://www2.statsghana.gov.gh/docfiles/2010_District_Report/Eastern/KWAHU%20AFRAM%20PLAINS%20SOUTH.pdf

Ghana Statistical Service (2019). Ghana Living Standards Survey Round 7 (GLSS 7). Main Report. Ghana Statistical Service. Accra-Ghana. Retrieved 03/02/2021, from: https://www.statsghana.gov.gh/gssmain/fileUpload/pressrelease/GLSS7%20MAIN%20REPORT_FINAL.pdf

GhanaWeb(2016).Nomorelongqueuesforgas–PetroleumMinister.Retrieved21/01/2019,from:https://www.ghanaweb.com/GhanaHomePage/business/No-more-long-queues-for-gas-Petroleum-Minister-477848

Gill-Wiehl, A., Ray, I., & Kammen, D. (2021). Is Clean Cooking Affordable? A Review. *Renewable and Sustainable Energy Reviews*, 151, 111537.

Global Partnership on Active-Based Aid (GPOBA) (2016). Improving Rural Energy Access through solar home systems in Ghana. Retrieved 02/10/2019, from: https://www.gpoba.org/sites/gpoba/files/LL12_GhanaSHS.pdf

Greve, H., & Lay, J. (2023). "Stepping Down the Ladder": The Impacts of Fossil Fuel Subsidy Removal in a Developing Country. *Journal of the Association of Environmental and Resource Economists*, 10(1), 121–158.

Heltberg, R. (2005). Factors Determining Household Fuel Choice in Guatemala. *Environment and Development Economics*, 10(3), 337–361.

Husband, G. (2020). Ethical Data Collection and Recognizing the Impact of Semi-Structured Interviews on Research Respondents. *Education Sciences*, 10(8), 206.

Hussein, H., Shamsipour, M., Yunesian, M., Hasanvand, M.S., & Fotouhi, A. (2020). Association of Adverse Birth Outcomes With Exposure to Fuel Type Use: A Prospective Cohort Study in the Northern Region of Ghana. *Heliyon*, 6(6), 04169.

IMANI (2017). IMANI: Ideas for making government's gas cylinder exchange programme work in Ghana. Retrieved 02/02/2022, from: https://www.modernghana.com/news/794866/imani-ideas-for-making-governments-gas-cylinder-exchange-p.html

Inkoom, D.K., & Crentsil, A.O. (2015). Estimation of indoor air pollution and health impacts due to biomass burning in rural northern Ghana. In Case studies for developing globally responsible engineers, GDEE (eds.), Global Dimension in Engineering Education, Barcelona 2015. Retrieved 03/10/2022, from: https://upcommons.upc.edu/bitstream/handle/2117/89157/CS_14.pdf

International Energy Agency (2022). Africa Energy Outlook 2022. Retrieved 07/03/2023, from: https://iea.blob.core.windows.net/assets/27f568cc-1f9e-4c5b-9b09-b18a55fc850b/AfricaEnergyOutlook2022.pdf

Jasanoff, S. (2018). Just Transitions: A Humble Approach to Global Energy Futures. *Energy Research & Social Science*, 35, 11–14.

Jones, D., Ryan, C.M., & Fisher, J. (2016). Charcoal as a Diversification Strategy: The Flexible Role of Charcoal Production in the Livelihoods of Smallholders in Central Mozambique. *Energy for Sustainable Development*, 32, 14–21.

Karakara, A.A., & Osabuohien, E.S. (2021). Clean Versus Dirty Energy: Empirical Evidence from Fuel Adoption and Usage by Households in Ghana. *African Journal of Science, Technology, Innovation and Development*, 13(7), 785–795.

Karimu, A. (2015). Cooking Fuel Preferences Among Ghanaian Households: an Empirical Analysis. *Energy for Sustainable Development*, 27, 10–17.

Karimu, A., Mensah, J.T., & Adu, G. (2016). Who Adopts LPG as the Main Cooking Fuel and Why? Empirical Evidence on Ghana Based on National Survey. *World Development*, 85, 43–57.

Kedir, M.F., Bekele, T., & Feleke, S. (2019). Problems of Mirt, and Potentials of Improved Gonzie and Traditional Open Cook Stoves in Biomass Consumption and End Use Emission in Rural Wooden Houses of Southern Ethiopia. *Scientific African*, 3, 00064.

Kiruki, H.M., van der Zanden, E.H., Kariuki, P., & Verburg, P.H. (2020). The Contribution of Charcoal Production to Rural Livelihoods in a Semi-Arid Area in Kenya. *Environment, Development and Sustainability*, 22, 6931–6960.

Kumar, R., Kumar, A., & Saikia, P. (2022). Deforestation and Forests Degradation Impacts on the Environment. In *Environmental Degradation: Challenges and Strategies for Mitigation*. Cham: Springer International Publishing. Switzerland, pp. 19–46.

Lambe, F., Ran, Y., Kwamboka, E., Holmlid, S., Lycke, K., Ringström, S., Annebäck, J., Ghosh, E., O'Conner, M., & Bailis, R. (2020). Opening the Black Pot: a Service Design-Driven Approach to Understanding the Use of Cleaner Cookstoves in Peri-Urban Kenya. *Energy Research & Social Science*, 70, 101754.

Lin, B., & Ankrah, I. (2019). Renewable Energy (electricity) Development in Ghana: Observations, Concerns, Substitution Possibilities, and Implications for the Economy. *Journal of Cleaner Production*, 233, 1396–1409.
Makonese, T., Ifegbesan, A.P., & Rampedi, I.T. (2018). Household Cooking Fuel Use Patterns and Determinants Across Southern Africa: Evidence from the Demographic and Health Survey Data. *Energy & Environment*, 29(1), 29–48.
McCray, N. (2018). Exploratory Analysis of Influencing Variables on Deforestation in Ghana. Intermediate Quantitative Methods Final Project submitted to Clark University.
Mendum, R., & Njenga, M. (2018). Integrating Wood Fuels into Agriculture and Food Security Agendas and Research in Sub-Saharan Africa. *Facets*, 3(1), 1–11.
Ministry of Energy (2022). Medium Term Expenditure Framework (MTEF) For 2022 – 2025; Ministry of Energy Programme Based Budget Estimates For 2022. Retrieved 03/01/2023, from: https://mofep.gov.gh/sites/default/files/pbb-estimates/2022/2022-PBB-MoEn.pdf
Mweshi, G.K., & Sakyi, K. (2020). Application of Sampling Methods for the Research Design. *Archives of Business Research*, 8(11), 180–193.
Newell, P., & Daley, F. (2022). Cooking up an Electric Revolution: The Political Economy of e-Cooking. *Energy Research & Social Science*, 91, 1–9.
Njenga, M., Gitau, J.K., Iiyama, M., Jamnadassa, R., Mahmoud, Y., & Karanja, N. (2019). Innovative Biomass Cooking Approaches for Sub-Saharan Africa. *African Journal of Food, Agriculture, Nutrition and Development*, 19(1), 14066–14087.
Nketiah, K.S., & Asante, J. (2018). Estimating national charcoal production in Ghana. *Ghana: Tropenbos*. Retrieved 07/02/2023, from: https://ifro.ku.dk/english/research/projects/ax/publications/Estimating-national-charcoal-production-in-Ghana_report-2018.pdf
Omer, A.M. (2017). Sustainable Development and Environmentally Friendly Energy Systems. *International Journal of Physical Sciences and Engineering*, 1(1), 1–39.
Omiti, J.M., & Laibuni, N.M. (2015). Feeding Africa's Growing Population. *AFRICA at a Fork in the Road*, 25, 435.
Panigrahi, R., Srivastava, P.R., & Sharma, D. (2018). Online Learning: Adoption, Continuance, and Learning Outcome—A Review of Literature. *International Journal of Information Management*, 43, 1–14.
Sedano, F., Silva, J.A., Machoco, R., Meque, C.H., Sitoe, A., Ribeiro, N., Anderson, K., Ombe, Z.A., Baule, S.H., & Tucker, C.J. (2016). The Impact of Charcoal Production on Forest Degradation: a Case Study in Tete, Mozambique. *Environmental Research Letters*, 11(9), 094020.
Shyu, C-W. (2014). Ensuring Access to Electricity and Minimum Basic Energy Needs as a Goal for the Post-MDG Development Agenda after 2015. *Energy for Sustainable Development*, 19, 29–38.
Sola, P., Schure, J., Eba'a Atyi, R., Gumbo, D., Okeyo, I., & Awono, A. (2019). Woodfuel Policies and Practices in Selected Countries in Sub-Saharan Africa—A Critical Review. *Bois et Forêts des Tropiques*, 340, 5–19.
Sovacool, B.K. (2014a). What Are We Doing Here? Analyzing Fifteen Years of Energy Scholarship and Proposing a Social Science Research Agenda. *Energy Research & Social Science*, 1, 1–29.
Sovacool, B.K. (2014b). Diversity: Energy Studies Need Social Science. *Nature News*, 511(7511), 529. Retrieved 21/04/2021, from: http://web.mit.edu/12.000/www/m2018/pdfs/energy_social%20science%20.pdf
Sulaiman, C., Abdul-Rahim, A.S., Chin, L., & Mohd-Shahwahid, H.O. (2017). Wood Fuel Consumption and Mortality Rates in Sub-Saharan Africa: Evidence from a Dynamic Panel Study. *Chemosphere*, 177, 224–231.
Sun, H., Khan, A.R., Bashir, A., Alemzero, D.A., Abbas, Q., & Abudu, H. (2020). Energy Insecurity, Pollution Mitigation, and Renewable Energy Integration: Prospective of Wind Energy in Ghana. *Environmental Science and Pollution Research*, 27, 38259–38275.
The Guardian (2017). Ghana vows improved safety to stop gas fires, blasts. Retrieved 17/09/2019, from: https://guardian.ng/news/ghana-vows-improved-safety-to-stop-gas-fires-blasts/
The World Bank (2019). Poverty and Equity Brief, Sub-Saharan Africa, Ghana. Retrieved 18/05/2021, from: https://databank.worldbank.org/data/download/poverty/33EF03BB-9722-4AE2-ABC7-AA2972D68AFE/Archives-2019/Global_POVEQ_GHA.pdf

The World Bank (2021). GDP Growth (annual %) Ghana. Retrieved 27/07/2021, from: https://data.worldbank.org/indicator/NY.GDP.MKTP.KD.ZG?locations=GH

United Nations Development Programme (UNDP) (2018). Sustainable Development Goals. Retrieved 21/11/2019, from: http://www.undp.org/content/undp/en/home/sustainable-development-goals/goal-7-affordable-and-clean-energy.html

United Nations Development Programme (UNDP) (2020). Resilience in a Pandemic UNDP Ghana 2020 Annual Report. Retrieved 13/11/2021, from: https://annualreport.undp.org.gh/gh2020/

United Nations Development Programme (UNDP) (2021). Goal 7: Affordable and clean energy. Retrieved 18/01/2021, from: https://www.undp.org/content/undp/en/home/sustainable-development-goals/goal-7-affordable-and-clean-energy.html

Usman, N., & Said, I. (2014). Key Factors That Affects Adoption of Technology in the Nigerian Construction Firms: A Theoretical Framework. *International Journal of Accounting, Business and Management*, 2, 26–38.

Waswa, F., Mcharo, M., & Mworia, M. (2020). Declining Wood Fuel and Implications for Household Cooking and Diets in Tigania Sub-County Kenya. *Scientific African*, 8, 00417.

Weber, E., Adu-Bonsaffoh, K., Vermeulen, R., Klipstein-Grobusch, K., Grobbee, D.E., Browne, J.L., & Downward, G.S. (2020). Household Fuel Use and Adverse Pregnancy Outcomes in a Ghanaian Cohort Study. *Reproductive Health*, 17, 1–8.

Wiafe, E.D., & Kwakwa, P.A. (2013). Fuel-Wood Usage Assessment Among Rural Households in Ghana. *Spanish Journal of Rural Development*, 4(1), 41–48.

Wiedinmyer, C., Dickinson, K., Piedrahita, R., Kanyomse, E., Coffey, E., Hannigan, M., Alirigia, R., & Oduro, A. (2017). Rural–Urban Differences in Cooking Practices and Exposures in Northern Ghana. *Environmental Research Letters*, 12(6), 065009.

World Resources Institute (2020). 4 Charts Explain Greenhouse Gas Emissions by Countries and Sectors. Retrieved 01/02/2020, from: https://www.wri.org/blog/2020/02/greenhouse-gas-emissions-by-country-sector

Conclusion

22 Towards a Better Understanding of Energy Poverty

Niall P. Dunphy, Breffní Lennon, and Paola Velasco-Herrejón

22.1 Defining energy poverty

As we have seen in previous chapters, energy poverty can be succinctly described as the inability to access the levels of clean energy required for essential basic household energy services[1] such as heating, lighting, cooking, etc., which can constrain people's ability to live the life they value most. While linked, energy poverty is considered by many to be distinct from income poverty – Palmer et al. (2008) for instance, observe that not all those who are energy poor are necessarily income poor and that not all those under income poverty thresholds are in energy poverty. Rather than simply being about income level, energy poverty is a complex and dynamic condition resulting from the interaction of multiple factors for any given household, including energy availability and prices, income, building efficiency, appliance efficiency, specific household energy needs, and householder capabilities. The influence of these different factors varies from household to household and is strongly motivated by social context and economic conditions.

Energy poverty can manifest itself in households unable, for reasons of access and/or affordability, to source clean energy for necessities such as heat, light, cooling, cooking, and appliance use, or having to use an excessive portion of their disposable income to provide these essentials (Thomson et al., 2016). The condition results in cold and uncomfortable homes as well as indoor air pollution that has an important impact on well-being and quality of life standards globally. This phenomenon can lead to significant deteriorations in physical and mental health and premature deaths, in addition to more restricted lifestyles and social exclusion.

As mentioned earlier in the book, for the first time in decades, the number of households without access to affordable energy is increasing (Velasco-Herrejón et al., Chapter 1, this volume). Developing more effective responses to this social challenge necessitates a deeper appreciation of energy poverty and the different ways in which it manifests. While there has been some arguing for the importance of appreciating the lived experience of the energy poor (see, e.g., Middlemiss & Gillard, 2015; Middlemiss et al., 2018), much of the literature on energy poverty has tended to be quantitative in nature. A lot of the discourse has tended to focus on questions such as, what constitutes energy poverty? How can a household be deemed energy poor? How can the prevalence of energy poverty be quantified? Defining and measuring the phenomenon is, of course, vitally important for the design and evaluation of effective alleviation programmes, but on their own such numerically grounded perspectives are not sufficient to really understand energy poverty.

This book project emerged from two engaging conference panels[2] that attempted to extend understanding of the lived experiences of household energy poverty in both developed and developing country contexts. The book builds on the dialogue that commenced from these panels

DOI: 10.4324/9781003408536-27

and gives voice to diverse perspectives from across the Global South and Global North. In doing so, it offers insights into the reality of energy poverty and the experiences of the energy poor. The aim of this concluding chapter is to reflect on the perspectives forwarded and consider what they might mean for action on energy poverty.

22.2 Identifying energy poverty

Work within the EnergyMeasures project identified a gap between the macro- and meso-level analysis of energy poverty and the identification of individual energy poor households (See Dunphy et al., Chapter 2, this volume). There is a substantial discourse on defining and measuring the rate of energy poverty, but a dearth of literature on how to actually identify and engage households is in need of support. Identifying energy poverty and engaging with the energy poor is not an easy task; it is a sensitive private condition that is confined to the home (Thomson et al., 2017) and one with a significant amount of stigma and shame attached to it (Longhurst & Hargreaves, 2019). Indeed, the term 'energy poverty' itself while perhaps useful in encouraging policy-makers to take action, can be quite counterproductive when reaching out to those who may be suffering from it. In many socio-cultural contexts, poverty is quite simply a taboo issue.

There are three principal approaches to measuring energy poverty. The first of these are methods based on energy expenditure thresholds. Expenditure-based metrics include a high share of income expended on energy (e.g., twice the median expenditure) and insufficient energy spending among those who self-restrain so-called hidden energy poverty (e.g., one-half the median expenditure on energy)[3]. The second approach is the so-called 'consensual' method, involving self-reporting of living conditions (e.g., inability to heat or cool) and financial distress (e.g., arrears on energy bills). Such self-reporting approaches are often criticised for their basis in residents' perceptions and supposed subjectivity, but as Dunphy (2020) argues, this is in fact 'exactly the point' of such metrics, reporting on the lived experience as they do. Consensual methods do not necessarily mean over-reporting; for example, Scott et al. (2008) observe that the self-reported measure of energy deprivation in Ireland was less than half the official energy poverty rate. A third approach is needs-assessment modelling, where the emphasis is not on actual expenditure incurred but rather on the expenditure that would be required to achieve acceptable levels of warmth[4]. This approach, although rather resource-intensive, provides excellent granularity of data.

Notwithstanding the possible advantages of other approaches, energy poverty metrics based on expenditure have tended to predominate, with relative thresholds used as proxies for energy deprivation. This has been explained by some on the basis of the assumed objectivity of such data and its perceived superiority compared to consensual data (Sareen et al., 2020). The reliance on (certain) expenditure indicators is not without contestation, including in this volume. See, for instance, Ibáñez Martín et al. (Chapter 3) for a discussion of the limitations of such measures; Barrella and Romero's (Chapter 4) critique of the quantitative methods used in Europe; and Mejía-Montero and Soriano-Hernández's (Chapter 5) call for greater a methodological diversity in studies of energy poverty. In subsequent chapters, Gayoso Heredia et al. (Chapter 6) and Antadze (Chapter 7) discuss how they used qualitative methods (such as participatory action research techniques and video ethnography) in their explorations of energy poverty.

22.3 Appreciating energy poverty

Energy poverty is fundamentally a human condition. The various definitions of energy poverty speak of people being unable to access or afford sufficient energy to meet their basic service needs. Yet, much of the discourse around the phenomenon has focused on the 'energy' part of the description and to a large extent disregarded the equally (if not more) important component, 'people's needs'. There is rightly an increasing acknowledgement of the value of understanding the lived experience of those who find themselves in energy poverty. For instance, Middlemiss et al. (2018) argue that the integration of knowledge from lived experience into policy and practice design simply gives better outcomes. Longhurst and Hargreaves (2019) concur, arguing that ignoring evidence from lived experience results in energy poverty policies that focus on narrow technical framings and neglect the important human-centric factors that shape everyday energy use.

Lived experience is a term that is increasingly used in the policy domain. While its meaning is almost intuitive, McIntosh and Wright (2019, 459–462) suggest that it is often used as the '*basis of knowledge or form of evidence, which is usually invoked without exploration or clarification about what the term itself might mean or imply*'[5]. Drawing from a range of literature[6] relating to experiential knowledge and its meaning, they suggest that lived experience can perhaps be understood as being '*rooted in the everyday and routine*'. They argue that lived experience is not necessarily unique, intimating that the intersubjectivity of daily life can be reflected in some forms of lived experience, which might be '*usefully understood as involving clusters of commonality*'. In this way, the researcher can find '*patterns and typical forms of behaviour and concerns*', which are useful in understanding social phenomena.

There are significant gendered differentials in patterns of energy use and energy-related practices, and women typically have a very different relationship with energy than men (Dunphy et al., 2017). These differences, combined with wider societal gender social norms and related societal structure, mean that women face a higher risk of energy poverty, are impacted more severely by it, and find it harder to escape from. Gendered experiences of energy poverty are explored by Jodoin and Mang-Benza (Chapter 8) in sub-Saharan Africa and by Stojilovska and Feenstra in Austria and North Macedonia (Chapter 15). Although studying in the Global South and Global North, respectively, both chapters support the need for women's voices to be heard on energy poverty and for their (specific) needs to be better reflected in policy and practice.

People suffering from deprivation of any sort, including energy poverty, find ways to make their condition less unbearable and cope with difficulties. These so-called coping strategies are typically not purposive-rational actions but generally reactive ways of responding to the issue at hand (Brunner et al., 2012). Araya et al. (Chapter 14), speaking of the Southern Cone region of South America, note that many coping mechanisms span generations and are perceived as normal skills to overcome the cold. However, Chard and Walker (2016) note that many coping mechanisms mean that households settle for adapting rather than recognising that their situation should be improved. Berger's (Chapter 13) observation that heat tolerance is used as a coping strategy in India irrespective of its health risks is illustrative of the potential issues arising from coping mechanisms. Stojilovska et al. (2021) furthermore observe that while some strategies can be empowering experiences for householders, supporting their participation in the energy transition, others can act to reduce household resilience and keep them locked into energy poverty. Robert (Chapter 12) made an interesting point noting that not everybody has the agency to deploy effective coping strategies, illustrating how energy poverty impacts groups differentially.

It is important to understand the specifics of energy poverty in a particular setting. Cultural norms, socio-economic environments and the socio-political context combine to influence the occurrence of energy poverty and its impact on the citizenry. David and Kod'ousková (Chapter 9), for instance, analysed the reactions of vulnerable households transferred to a 'supplier of last instance'. They noted that many householders did not have the skill set to navigate offers from potential suppliers or apply for benefits, which left them poorly equipped to respond to difficult market conditions. Setyawati (Chapter 10) drawing on the lived experiences of street vendors in Indonesia, showed that even among vulnerable energy users, there is a hierarchy of needs that impacts levels of resilience and empowerment. Mohlakoana and Wolpe (Chapter 11) reflecting on energy poverty in South Africa, comment that economic policies since democracy that might have been expected to improve matters actually slowed or even prevented a redistributive transformation of the energy system.

22.4 Addressing energy poverty

As we have discussed throughout this volume, energy poverty is a multi-faceted phenomenon that can hypothetically at least be addressed through multiple approaches. Infrastructure can be constructed, high energy costs can be subsidised through social tariffs (see, e.g., Giuliano et al., 2020), low incomes can be supplemented, for example, social welfare fuel allowances (see, e.g., Pillaia et al., 2022); poor quality buildings can be improved through grants and green loans (see, e.g., Bertoldi et al., 2021), capabilities can be enhanced, etc. All of these approaches can be deployed individually or in combination and can be realised through an appropriate and coherent mix of policy and practice initiatives.

Arguably, the most cost-effective and sustainable solution to energy poverty is reducing energy demand, primarily through building renovation and upgrading appliances. While acknowledging there is a social need for other types of support (if only as a bridging measure), governmental expenditure on subsidising energy costs (e.g., social electricity tariff in Belgium, free basic water and electricity services in South Africa) and supplementing income (e.g., fuel allowance in Ireland) remains very high, not least because that's what voters want. In keeping with an energy demand-oriented approach, Dervishi et al. (Chapter 18) present an assessment of the prevalence of energy poverty in housing stock in Tirana and highlight the vulnerability of households while also providing potential passive retrofitting measures for its resolution.

Noaman El Sherbini and Lippert (Chapter 19) conduct discourse analysis on EU member states' National Energy and Climate Plans (NECPs). They consider whether the NECPs demonstrate ambiguity, complexity, and/or multidimensionality in the way that energy poverty is expressed in terms of language and framing. They suggest changing the nature of the energy poverty diagnosis from measuring and defining to one that aims to identify the 'vulnerable consumer', reasons for their vulnerability, and measures to tackle specific types of vulnerability[7]. These types of on-the-ground supports are very well received in energy-vulnerable communities. Our experience in the EnergyMeasures Project demonstrates that building trust and developing relationships are central to engaging with a community on energy; working through a trusted gatekeeper is a great way of kick-starting initial engagement. Kenner et al. (Chapter 16) describe neighbourhood energy centres in Philadelphia, USA, which assist energy-poor households through a support programme involving budget counselling, service referrals, energy conservation workshops, and assistance with applications for energy aid programmes. Such centres provide an important network of embedded, flexible, and responsive care actors, alongside the existing stratified and ossified government support structures.

Innovation is a key part of tackling energy poverty, but it does not have to involve high technology. Boateng et al. (Chapter 21), for instance, explored the use of wood fuels in Ghana, examined possible uses of cleaner alternatives for cooking, and considered how these would affect rural livelihoods. Taking account of technological availability, costs, and socio-cultural aspects, they propose the use of improved cookstoves as a first step in the pathway to introducing LPG. Cornelis et al. (Chapter 17) discuss the experiences of two electrification projects in Africa.

The first is a top-down innovation in Burkina Faso that aims to prevent power outages through an optimisation solution based on smart metres and the Internet of things. It will prevent local overloads through smart demand management, sharing the power the grid can handle among consumers according to their needs and means – leading to more sustainable consumption practices based on resource sharing. The second is a bottom-up innovation. In Northern Madagascar, in the absence of national grid links, the deployment of a local nanogrid[8]-based system coupled with an innovative 'energy as a service' business model has disrupted the traditional consumption paradigm where consumers purchase services provided by the electricity services rather than the electricity itself.

Feelings of fairness or justice are key to the acceptability of decisions by public authorities and others with power. A key element of this justice (albeit one that is often side-lined) is recognition. Recognition justice calls for acknowledgement of differences while achieving social equity in procedures and outcomes. This requires appreciating social differences and can be concerned with re-valorising unjustly devaluated identities[9] (Fraser, 2007). Schlosberg (2004) notes that it may present itself not only as a failure to recognise but also misrecognise. Cultural and institutional processes of disrespect that devalue some people more than others are at the centre of misrecognition (Velasco-Herrejón et al., 2022). Lamonaca and Batel (Chapter 20) empirically examine the role of recognition justice in social housing within smart city projects in the Global North (Bolzano, Italy). They highlight the role of symbolic and psychosocial aspects in promoting inclusive and democratic accountability, and the importance of addressing misrecognition as a means of empowering citizens to enable a more active and autonomous energy citizenship, is also noted.

22.5 Conclusion

Appreciating the experiential knowledge of energy poverty is vital for any attempt to alleviate it. To this end, this volume aimed to expand collective understandings of energy poverty and deepen recognition of the phenomenon by engaging with the lived experiences of energy-poor households across different contexts. Drawing contributors from the Global South and the Global North broadened the analysis of the condition through the supply and demand sides, as well as through outcomes such as energy services, capabilities, and (in)justices. This provided a wide analysis of the drivers of energy deprivation at the household level and the risks and vulnerabilities people face as a result, facilitating a useful dialogue on the nature of energy poverty and one that we hope speaks to diverse audiences.

Acknowledgement

This work has been conducted in the context of the EnergyMeasures Project funded by the European Union's Horizon 2020 Programme for Research and Innovation under Grant Agreement No. 894759. Further information about the project can be found at https://doi.org/10.3030/894759.

Notes

1 This description draws from Brenda Boardman's (2010) characterisation of energy poverty from over a decade ago, albeit with her emphasis on affordability expanded to 'inability to access' (which inherently includes affordability).

2 Development Studies Association Conference 2022, Panels 12a & 12b on *'Understanding the lived experiences of energy poverty in the Global North and South'*.

3 The two expenditure-based metrics outlined above relate to absolute spend, a related approach would be as Hills (2011) suggests using a combination of expenditure over the median level and residual income below the official poverty line.
4 Such as the Irish Government's 2016 review of residential energy (SEAI, 2018, p. 37).
5 Although they do acknowledge an assumed meaning of living through phenomena, e.g., in qualitative longitudinal research.
6 Ultimately, focusing on three areas of work, which offered the most insights, namely: phenomenology, feminist theory and ethnography.
7 In this respect they concur with the approach forwarded by Dunphy et al. (Chapter 2, this volume).
8 This 'bottom-up approach' will involve interconnection of several nanogrids within a village-wide balancing microgrid.
9 Relating to identities such as gender, sexuality, life stage, race, and ethnicity.

References

Bertoldi, P., Economidou, M., Palermo, V., Boza-Kiss, B., & Todeschi, V. (2021). How to Finance Energy Renovation of Residential Buildings: Review of Current and Emerging Financing Instruments in the EU. *Wiley Interdisciplinary Reviews: Energy and Environment*, 10(1). https://doi.org/10.1002/wene.384

Boardman, B. (2010). *Fixing Fuel Poverty: Challenges and Solutions*. London: Earthscan.

Brunner, K. M., Spitzer, M., & Christanell, A. (2012). Experiencing Fuel Poverty. Coping Strategies of Low-Income Households in Vienna/Austria. *Energy Policy*, 49, 53–59. https://doi.org/10.1016/j.enpol.2011.11.076

Chard, R., & Walker, G. (2016). Living with Fuel Poverty in Older Age: Coping Strategies and Their Problematic Implications. *Energy Research and Social Science*, 18, 62–70. https://doi.org/10.1016/j.erss.2016.03.004

Dunphy, N. P., Revez, A., Gaffney, C., Lennon, B., Ramis Aguilo, A., Morrissey, J. E., & Axon, S. (2017). *Intersectional Analysis of Energy Practices*. A research report of the ENTRUST H2020 Project. https://doi.org/10.5281/zenodo.3479295

Dunphy, N. P. (2020). Review of Methods of Identifying Energy Poor Households. A Research Report of the EnergyMeasures H2020 Project. https://doi.org/10.5281/zenodo.6867434

Fraser, N. (2007). Feminist Politics in the Age of Recognition: A Two-Dimensional Approach to Gender Justice. *Studies in Social Justice*, 1(1), 23–35. https://doi.org/10.26522/ssj.v1i1.979

Giuliano, F., Lugo, M. A., Masut, A., & Puig, J. (2020). Distributional Effects of Reducing Energy Subsidies: Evidence from Recent Policy Reform in Argentina. *Energy Economics*, 92, 104980. https://doi.org/10.1016/j.eneco.2020.104980

Hills, J. (2011). *Fuel Poverty: The Problem and Its Measurement. Interim Report of the Fuel Poverty Review*. London: London School of Economics and Political Science. Available at: https://bit.ly/42CePwU

Longhurst, N., & Hargreaves, T. (2019). Emotions and Fuel Poverty: The Lived Experience of Social Housing Tenants in the United Kingdom. *Energy Research* and *Social Science*, 56, 101207. https://doi.org/10.1016/j.erss.2019.05.017

McIntosh, I., & Wright, S. (2019). Exploring What the Notion of "lived Experience" Offers for Social Policy Analysis. *Journal of Social Policy*, 48(3), 449. https://doi.org/10.1017/S0047279418000570

Middlemiss, L., & Gillard, R. (2015). Fuel Poverty from the Bottom-up: Characterising Household Energy Vulnerability Through the Lived Experience of the Fuel Poor. *Energy Research* and *Social Science*, 6, 146–154. https://doi.org/10.1016/j.erss.2015.02.001

Middlemiss, L., Gillard, R., Pellicer, V., & Straver, K. (2018). Plugging the Gap Between Energy Policy and the Lived Experience of Energy Poverty: Five Principles for a Multidisciplinary Approach. In *Advancing Energy Policy* (pp. 15–29). Cham: Palgrave Macmillan. https://doi.org/10.1007/978-3-319-99097-2_2

Palmer, G., Macinnes, T., & Kenqay, P. (2008). *Cold and Poor: An Analysis of the Link between Fuel Poverty and Low Income*. London: New Policy Institute.

Pillaia, A., Tovar Reaños, M. A., & Curtis, J. (2022). *Fuel Poverty in Ireland: An Analysis of Trends and Profiles*, ESRI Working Paper No. 729. Dublin: Economic and Social Research Institute. http://hdl.handle.net/10419/265908

Sareen, S., Thomson, H., Tirado Herrero, S., Gouveia, J. P., Lippert, I., & Lis, A. (2020). European Energy Poverty Metrics: Scales, Prospects and Limits. *Global Transitions*, 2, 26–36. https://doi.org/10.1016/j.glt.2020.01.003

Schlosberg, D. (2004). Reconceiving Environmental Justice: Global Movements and Political Theories. *Environmental Politics*, 13(3), 517–540. https://doi.org/10.1080/0964401042000229025

Scott, S., Lyons, S., Keane, C., McCarthy, D., & Tol, R., 2008. *Fuel Poverty in Ireland: Extent, Affected Groups and Policy Issues*. ERSI Working Paper No. 262. Dublin: Economic and Social Research Institute. https://bit.ly/3MqHgs2

SEAI. (2018). *Energy in the Residential Sector – 2018 Report*. Dublin: Sustainable Energy Authority of Ireland.

Stojilovska, A., Yoon, H., & Robert, C. (2021). Out of the margins, into the light: Exploring energy poverty and household coping strategies in Austria, North Macedonia, France, and Spain. *Energy Research and Social Science*, 82, 102279. https://doi.org/10.1016/j.erss.2021.102279

Thomson, H., Bouzarovski, S., & Snell, C. (2017). Rethinking the Measurement of Energy Poverty in Europe: A Critical Analysis of Indicators and Data, *Indoor and Built Environment,* 26(7), 879–901. https://doi.org/10.1177/1420326X17699260

Thomson, H., Snell, C., & Liddell, C. (2016). Fuel Poverty in the European Union: A Concept in Need of Definition? *People; Place and Policy*, 10(1), 5–24. https://doi.org/10.3351/ppp.0010.0001.0002

Velasco-Herrejón, P., Dunphy, N.P., Lennon, B., & Quinlivan, L. (2022). *Citizen perceptions of energy justice*. A research report of EnergyPolities an SEAI RD&D funded project. https://doi.org/10.5281/zenodo.7352699

Index

Note: Page references in italics denote figures, in bold tables and with "n" endnotes.